# Flexibility in Electric Power Distribution Networks

# Flexibility in Electric Power Distribution Networks

**Edited by**

*Hassan Haes Alhelou*
*Ehsan Heydarian-Forushani*
*Pierluigi Siano*

CRC Press
Taylor & Francis Group
Boca Raton London New York

CRC Press is an imprint of the
Taylor & Francis Group, an **informa** business

First edition published 2021
by CRC Press
6000 Broken Sound Parkway NW, Suite 300, Boca Raton, FL 33487-2742
and by CRC Press

2 Park Square, Milton Park, Abingdon, Oxon, OX14 4RN

CRC Press is an imprint of Taylor & Francis Group, LLC

ISBN: 978-0-367-64141-2 (hbk)
ISBN: 978-0-367-64144-3 (pbk)
ISBN: 978-1-003-12232-6 (ebk)

Typeset in Times LT Std
by KnowledgeWorks Global Ltd.

# Contents

# Acknowledgment

The work of Hassan Haes Alhelou was supported in part by the Science Foundation Ireland (SFI) through the SFI Strategic Partnership Programme under Grant SFI/15/SPP/E3125, and in part by the UCD Energy Institute. The opinions, findings and conclusions or recommendations expressed in this material are those of the authors and do not necessarily reflect the views of the Science Foundation Ireland.

# Preface

In an effort to achieve sustainability, ensure supply security, and attain environmental goals, a global trend has been created currently, whereby electricity generation portfolios are moving toward renewable energy sources (RESs), particularly at distribution level. High penetration of RESs imposes several techno-economic challenges to distribution system operators (DSOs) due to their variability in power generation, and hence, increases the need for additional operational flexibility. Operational flexibility aims at securely covering the possible variations at the minimum cost using emerging flexible alternatives or designing novel local market mechanisms to incentivize flexibility providers. In such a situation, the DSOs can use the potential of flexible options such as energy storages (ESs), demand response (DR), plug-in electric vehicles (PEVs) or on-site fast run generators. However, each of the mentioned flexible resources has its own specific characteristics and requirements that should be taken into account, and this raises complexity. Optimal network reconfiguration schemes are the other solution for increasing power system flexibility at distribution level. To sum up, there is a great research gap related to renewable-based distribution network planning from flexibility point of view. Therefore, this book aims to discuss the additional flexibility needs introduced by RESs and describe general approaches to analyze the need for and provision of additional flexibility in future distribution networks at both planning and operational time frames.

# About the Editor

**Dr. Hassan Haes Alhelou** is with the School of Electrical and Electronic Engineering, University College Dublin (UCD), Dublin, Ireland. He received a B.Sc. degree (ranked first) from Tishreen University, Lattakia, Syria, in 2011 and a M.Sc. degree from Isfahan University of Technology (IUT), Isfahan, in 2016; all in Electrical Power Engineering, power systems (with honors). In 2016, he also started his Ph.D. at the Isfahan University of Technology, Isfahan, Iran. His name is included in the 2018 & 2019 Publons's list of top 1% best reviewers and researchers in the field of engineering. He was the recipient of the Outstanding Reviewer Award from Energy Conversion and Management Journal in 2016, ISA Transactions Journal in 2018, Applied Energy Journal in 2019, and many other awards. He was the recipient of the best young researcher in the Arab Student Forum Creative among 61 researchers from 16 countries at Alexandria University, Egypt, 2011. Dr. Alhelou has published more than 100 research papers in high-quality peer-reviewed journals and international conferences. He has also performed more than 600 reviews for high prestigious journals, including *IEEE Transactions on Power Systems, IEEE Transactions on Smart Grid, IEEE Transactions on Industrial Informatics, IEEE Transactions on Industrial Electronics, Energy Conversion and Management, Applied Energy,* and *International Journal of Electrical Power & Energy Systems.* He has participated in more than 15 industrial projects worldwide. His major research interests are power systems, power system dynamics, power system operation and control, dynamic state estimation, frequency control, smart grids, micro-grids, demand response, load shedding, and power system protection.

**Dr. Ehsan Heydarian-Forushani** is with Qom University of Technology (QUT), Qom, Iran & Esfahan Electricity Power Distribution Company (EEPDC), Isfahan 81737-51387, Iran. He received M.Sc. and Ph.D. degrees in electrical engineering from Tarbiat Modares University, Tehran, Iran, and Isfahan University of Technology, Isfahan, Iran, in 2013 and 2017, respectively. He was a visiting researcher in the University of Salerno, Salerno, Italy, in 2016–2017. In 2019, he also received the first postdoctorate from Isfahan University of Technology, Isfahan. His research interests include power system flexibility, renewables integration, demand response, smart grids, and electricity market.

**Professor Pierluigi Siano** received M.Sc. degree in electronic engineering and Ph.D. degree in information and electrical engineering from the University of Salerno, Salerno, Italy, in 2001 and 2006, respectively. He is a professor and scientific director of the Smart Grids and Smart Cities Laboratory in the Department of Management & Innovation Systems, University of Salerno. His research activities are centered on demand response, integration of distributed energy resources in smart grids, and planning and management of power systems. He has co-authored more than 450 papers, including more than 250 international journal papers that received more than 8400 citations with an H-index equal to 46. He has been the chair of the IES TC on Smart Grids. He is an editor for the Power & Energy Society Section of *IEEE Access, IEEE Transactions on Industrial Informatics, IEEE Transactions on Industrial Electronics, Open Journal of the IEEE IES and of IET Renewable Power Generation*. He received the award of 2019 and 2020 Highly Cited Researcher by ISI Web of Science Group.

# Contributors

**Chapter 1:**   **Social and Economic Factors in Demand-Side Flexibility**

Z. Kaheh[1], H. Arasteh[2,*], P. Siano[3]
[1]Department of Industrial Engineering, Tarbiat Modares University, Tehran, Iran.
[2]Power System Planning and Operation Group, Niroo Research Institute, Tehran, Iran.
[3]Department of Management & Innovation Systems, University of Salerno, Fisciano (SA). Italy.
*Corresponding author: H. Arasteh

**Chapter 2:**   **Role of Smart Homes and Smart Communities in Flexibility Provision**

Hosna Khajeh[1,*], Ran Zheng[1], Hooman Firoozi[1], Hannu Laaksonen[1], Miadreza Shafie-khah[1]
[1]Flexible Energy Resources, School of Technology and Innovations, University of Vaasa, Vaasa, Finland
*Corresponding author: Hosna Khajeh

**Chapter 3:**   **A System-of-Systems Planning Platform for Enabling Flexibility Provision at Distribution Level**

H. Arasteh[1,*], S. Bahramara[2], Z. Kaheh[3], S. M. Hashemi[4], V. Vahidinasab[5], P. Siano[6], M. S. Sepasian[7]
[1]Power System Planning and Operation Group, Niroo Research Institute, Tehran, Iran.
[2]Department of Electrical Engineering, Sanandaj Branch, Islamic Azad University, Sanandaj, Iran.
[3]Department of Industrial Engineering, Tarbiat Modares University, Tehran, Iran.
[4]Power System Planning and Operation Group, Niroo Research Institute, Tehran, Iran.
[5]Department of Electrical Engineering, Abbaspour School of Engineering, Shahid Beheshti University, Tehran, Iran, and School of Engineering, Newcastle University, Newcastle upon Tyne, United Kingdom.
[6]Department of Management & Innovation Systems, University of Salerno, Fisciano (SA), Italy.
[7]Department of Electrical Engineering, Abbaspour School of Engineering, Shahid Beheshti University, Tehran, Iran.
*Corresponding author: H. Arasteh

**Chapter 9:**　**Decongestion of Active Distribution Grids via D-PMUs-based Reactive Power Control and Electric Vehicle Chargers**

Gabriel Mejia-Ruiz[1], Mario Arrieta Paternina[2,*], Juan M Ramirez[3], Juan Ramon Rodriguez Rodriguez[2], Romel Angel Cardenas Javier[4], Alejandro Zamora Mendez[5]
[1]Graduated Program in Engineering-Doctorate at National Autonomous University of Mexico, Mexico City, Mexico.
[2]Department of Electrical Engineering, at National Autonomous University of Mexico, Mexico City, Mexico.
[3]Department of Electrical Engineering at CINVESTAV, Guadalajara, Mexico.
[4]Graduated Program in Electrical Engineering-Master at National Autonomous University of Mexico, Mexico City, Mexico.
[5]Faculty of Electrical Engineering, at the Michoacan University of Saint Nicholas of Hidalgo, Morelia, Michoacan, Mexico.
*Corresponding author: Mario Arrieta Paternina

**Chapter 10:**　**Blockchain-based Decentralized, Flexible, and Transparent Energy Market**

Mohd Adil Sheikh[1,*], Vaishali Kamuni[2], Mohit Fulpagare[2], Udaykumar Suryawanshi[2], Sushama Wagh[3], Navdeep M. Singh[4]
[1]Ph.D. Research Scholar, CDRC, EED, VJTI Mumbai, India.
[2]M.Tech Research Scholar, CDRC, EED, VJTI Mumbai, India.
[3]Asst. Professor, CDRC, EED, VJTI Mumbai, India.
[4]Adjunct Professor, CDRC, EED, VJTI Mumbai, India.
*Corresponding author: Mohd Adil Sheikh

**Chapter 11:**　**Integrated Operation and Planning Model of Renewable Energy Sources with Flexible devices in Active Distribution Networks**

Vijay Babu Pamshetti[1,*] and Shiv Pujan Singh[2]
[1]Department of Electrical and Electronics Engineering, B V Raju Institute of Technology, Narsapur, Telangana.
[2]Department of Electrical Engineering, Indian Institute of Technology (BHU), Varanasi.
*Corresponding author: Vijay Babu Pamshetti

Mahnaz Moradijoz*
Tarbiat Modares University, Tehran, Iran.
*Corresponding author: Mahnaz Moradi

Meisam Mahdavi[1,*], Pierluigi Siano[2], and
Hassan Haes Alhelou[3]
[1]Bioenergy Research Institute, São Paulo State University, Ilha Solteira, SP, Brazil.
[2]Department of Management & Innovation Systems, University of Salerno, Fisciano (SA), Italy.
[3]School of Electrical and Electronic Engineering, University College Dublin, Dublin 4, Ireland.
*Corresponding author: Meisam Mahdavi

Meisam Mahdavi[1,*], Pierluigi Siano[2], and
Hassan Haes Alhelou[3]
[1]Bioenergy Research Institute, São Paulo State University, Ilha Solteira, SP, Brazil.
[2]Department of Management & Innovation Systems, University of Salerno, Fisciano (SA), Italy.
[3]School of Electrical and Electronic Engineering, University College Dublin, Dublin 4, Ireland.
*Corresponding author: Meisam Mahdavi

# Introduction

This book discusses the additional flexibility needs introduced by Renewable Energy Sources (RESs) and describes general approaches to analyze the need for and provision of additional flexibility in future distribution networks at both planning and operational time frames. Each chapter begins with the fundamental structure of the problem required for a rudimentary understanding of the methods described.

**Chapter 1:** Demand-side flexibility refers to the voluntary swift alternations in the grid electricity demand, which are activated by some sort of incentives. This chapter focuses on the socio-economic aspects of consumer-side flexibility procurement. To this end, different types of demand-side flexibility, the associated drivers and actors, the plausible physical trading market, as well as the possible opportunities and obstacles of demand-side flexibility from a socio-economic perspective are examined. The main aim of this chapter is paving the way for further researches in assessing the effects of socio-economic drivers. To this end, we introduce the causal inference method as a lesser-known but powerful method for assessing the demand-side flexibility from socio-economic aspects.

**Chapter 2:** Distribution-network-located flexible energy resources such as smart homes and energy communities are potential resources that can contribute to enhancing the flexibility of power systems. In this regard, this chapter firstly discusses the available demand-side management methods and demand response programs used to incentivize demand-side resources. Then, the role of smart homes and communities in the provision of system-wide and local flexibilities are analyzed. In this way, smart homes and communities are proposed to reshape their consumption and production through their flexible energy resources, including battery energy storage systems, electric vehicles, or their controllable appliances. If they provide local flexibilities, they help the distribution system operator to control voltages and manage congestion of the distribution network. If they provide system-wide flexibilities, they assist the transmission system operator with regulating the frequency of the whole system. Finally, the flexible capacities of a small residential community are estimated as a case study.

**Chapter 3:** This chapter aims to study the distribution expansion planning (DEP) problem considering the flexibility providers. In this study, demand response providers (DRPs) are introduced as the flexibility service providers which will play an important role besides other independent entities to satisfy the increasing electrical demand. Moreover, there are different autonomous entities in distribution systems according to which their behaviors should be modeled. It means, several independent entities are playing their roles in the distribution systems with their objectives (that may conflict with the objectives of other players), while they can have interoperability with each other. Hence, an efficient modeling architecture is essential to optimally model such interactions. In this chapter, a system-of-systems (SoS) approach will be proposed as an efficient framework to model the behavior of these independent entities. The SoS is an approach for modeling the collaboration of different entities with distinct objective functions, which are able to cooperate. Independent systems under the SoS framework aim at achieving their own goals. The outcome of the SoS is an optimum solution that will determine the strategies of all the entities in an optimal manner.

**Chapter 4:** The electrical load prediction is necessary for distributed network energy management and finding opportunity for flexibility in shifting the operation of non-critical power intensive loads. The application of regression tools has shown to be promising for predicting electric load within distributed network as well as for flexibility analysis. The distributed network is a low-capacity network with low amount of data that need flexible operation and analysis. Random forest (RF) regressor, k-nearest neighbor (kNN) regressor, and linear regression are considered for analyzing electrical energy demand forecasting. The methodology used in this chapter is developed to deal with the problems of irregularities and randomness in the time series for urban and rural area. RF regressor yields good result on hourly time prediction in load forecasting. The kNN has shown precise prediction due to its capability to capture the nearest step in a time series based on the nearest neighbor principle. The presented vertical time approach uses seasonal data for training and inference, as opposed to continuous time approach that utilizes all data in a continuum from the start of the dataset until the time used for inference. The regression tools can handle the low amount of data, and the prediction accuracy matches with the other techniques.

**Chapter 5:** Due to high importance of flexibility in modern power systems, this chapter is focused on several effective strategies of flexibility in modern distribution systems. In this chapter, the impact of three new technologies (i.e., renewable energy sources, plug-in electric vehicles, and feeder reconfiguration process) on the operation and flexibility of electric distribution systems is investigated. The operation of the electric distribution system is formulated as a constraint optimization problem, which incorporates the overall cost of the system, including reliability cost, and operation cost. Also, the unscented transform is utilized to capture uncertainties associated with stochastic parameters of the system. Finally, the IEEE 69-bus test system is employed to show how much the proposed strategies can affect a practical case.

**Chapter 6:** In this chapter, an active distribution network (ADN) expansion planning framework is presented considering the flexibility provided by resource-driven ones, i.e., distributed energy resources (DERs). To this end, first, the investment environment governing DER deployment is determined. In bundled environment, in which distribution network operators (DNOs) are allowed to perform investment in DER deployment, DERs are modeled as operational and structural flexibility providers in ADN expansion planning problem. Then, the ADN expansion planning problem is modeled in an unbundled environment, where DNO and DER operator are two independent entities. In both environments, the conditional value at risk, a quite popular risk index used in financial research, is taken into account to measure the financial risk of the planning schemes.

**Chapter 7:** This chapter focuses on the distribution network emergency operation in the presence of flexible resources and the microgrid concept. The concept of flexibility at distribution level, the occurrence of emergency, and general approaches for enhancing the flexibility are reviewed in this chapter. The configuration of microgrids, including basic roles and strategies applied by microgrids for enhancing the flexibility of the distribution networks in single, networked, and multi-energy system frameworks, the flexibility of the communication system, and the decision-making framework in the presence of the microgrids are investigated in detail. Furthermore, the islanding strategies and demand response programs are also presented for an emergency operation as the local emergency resources. Various mobile and stationary power generation sources including mobile emergency generators, mobile energy storage systems, battery energy storage system, and plug-in electric vehicles, and their impacts on the distribution networks flexibility during emergency operation are evaluated. Besides, the emergency power dispatch and balancing, and the restoration strategies are presented to comprehensively investigate the operating conditions in the emergencies. Then, the IEC 61850-based energy management system during emergency is analyzed as a communication standard at the distribution level.

**Chapter 8:** This chapter studies harvesting flexibility from DR aggregation through a three-layer approach: the first layer deals with consumer/prosumer stage as the first step to attain the benefits of DR programs, and is to increase awareness level of the customers; the second layer deals with customers' demand control and flexibility management between aggregators and consumers; the third layer provides a competitive trading platform, i.e., local flexibility market (LFM) to facilitate flexibility trading between flexibility buyers (i.e., the distribution system operator (DSO)) and flexibility sellers. In the LFM, the market operator is assumed to have access to network parameters making market clearing solution technically feasible. To protect the privacy of network parameters, this chapter proposes an alternating direction method of multipliers-based market clearing strategy, in which market operator communicates with the DSO to clear the market such that the market clearing solution respects network operation constraints without revealing network parameters to the market operator. The case studies were conducted on the Roy Billinton Test System with electrical vehicles and heat pumps as flexibility resources. The results demonstrate that the proposed LFM framework is effective to perform day-ahead congestion management of distribution networks and is profitable to aggregators and end-users.

**Chapter 9:** This chapter presents an application to increase the operational flexibility of active distribution grids (DGs) by exploiting the potential of electric vehicles (EVs) chargers. EVs chargers are flexible resources available in modern DGs with the ability to decongest and increase the network capacity, using a grid-side controller to coordinate the reactive power injection. Therefore, a new reactive power flow management framework, including the individual injection capacity of EVs chargers along with the DG is proposed in this work. The proposed hierarchical control architecture may intelligently handle voltage regulation, even when measurements are made at different nodes than the point of common coupling (PCC) between the EVs chargers and the grid, by taking advantage of time-synchronized measurement devices remotely allocated. A typical linear quadratic Gaussian (LQG) control structure that employs a data-driven system identification technique is adopted to coordinately provide the injection of reactive power from EVs chargers to precisely and timely regulate the dynamic voltage response in DGs. The effectiveness and feasibility of the proposal are demonstrated by employing simulated scenarios on the IEEE 13 nodes test feeder, illustrating that the plan can compensate for voltage variations under highly unbalanced conditions in short time.

**Chapter 10:** To address the issue of demand-supply mismatch, various distributed generators (DG) are installed in the traditional power system. The coordination between DGs and the grid in view of energy flow and financial transaction exchange are coordinated with help of aggregators which act as an intermediary and lead to low transparency, high risk of data modification, and high operating cost. Hence, a decentralized mechanism is required which can overcome these issues and make the overall system transparent, secure, and reliable leading to flexibility in the operation of power system. In this chapter, the dependency of aggregator facilitating energy exchange between DGs and the traditional grid is eliminated by utilizing the blockchain framework. The electricity consumption data available from smart meters are stored in a tamper-proof manner in the blockchain. The smart contracts formulate program code for the financial aspect of energy trading with amalgamated penalties or rewards and the rules for matching the energy demand with generation. The different consensus mechanisms are used in blockchain for validating as well as achieving the desired flexibility between DGs and the grid. The chapter further explores the effectiveness of blockchain in providing flexibility and this is represented by considering different scenarios of energy and financial transaction exchange.

**Chapter 11:** To cope with the intermittency and uncertainty nature associated with renewable energy sources (RES), it is indispensable to introduce more additional operational flexibility into the system. Though operation of soft open point (SOP), battery energy storage systems (BESS), demand response (DR), and dynamic network reconfiguration (NR) has emerged as potential solution for increasing the operational flexibility in the system, an efficient operation and planning strategy is much needed for proper coordination of multiple aforementioned devices. In this context, a new two-stage coordinated optimization model has been developed for integrated operation and planning of RES, BESS, and SOP devices. The objective of the proposed methodology is to minimize the total investment and operating cost of RES, BESS, and SOP devices. Meanwhile, to address high-level uncertainties related to RES and load demands, a stochastic module has been adopted. The proposed framework has been implemented on IEEE 33-bus distribution system and solved by using proposed hybrid optimization solver.

**Chapter 12:** The contribution of distributed energy resources (DERs) in reliability enhancement generally depends on the island mode operation of these resources. Optimal placement of sectionalizing switches (SSs) enables active distribution networks (ADNs) to be operated more reliably by islanding the faulted part of the system in the form of flexible microgrids (FMGs). However, practical methods are needed to handle the complex problem of the SS placement in ADNs. This chapter presents a risk-based model for the optimal placement of the network-driven flexibility providers, i.e., SSs, in ADNs taking into account the limits of the resource-driven flexibility providers, i.e., DERs. In the placement model, SSs are considered as devices determining the boundaries of FMGs in ADNs.

**Chapter 13:** Distribution system reconfiguration (DSR) is a large-scale combinational optimization problem including decision variables, an objective function, and a set of constraints that can often contain nonlinearities. The feasible search space in DSR is typically large, non-convex, and hard to explore. Hence, determining good-quality solutions for the DSR problem is always a challenging task. In order to cope with this issue, distribution system researchers have dedicated their efforts to develop efficient methodologies to find the best possible solution for the DSR. In this regard, classical optimization methods, heuristics, and metaheuristics have played prominent roles in the DSR solution. Since the DSR problem was first proposed in 1975, classical optimization has been presented as an important tool in order to find good-quality solutions for this problem. Later, heuristic methods were adopted in the DSR as a solution strategy to avoid limitations presented by classical optimization. Finally, by improving heuristic performances in the DSR, metaheuristics were introduced.

**Chapter 14:** Distribution network is an important part of the power system that delivers the generated electricity to individual customers. Electric power distribution systems are operated radially because of better operation, control (voltage and power flow control), and protection (easier coordination of protective relays and lower short circuit currents). In distribution network reconfiguration, configuration of the system is changed by closing and opening tie and sectional switches, respectively. Power losses of distribution system are more than other parts of power systems, and these losses can affect the operational costs and power quality. One of common ways to minimize these losses is reconfiguration of distribution system. It is very important that the configuration of the system is remained as radial during reconfiguration. Since the DSR problem was proposed, classical methods such as mathematical programing have been presented as an important tool for solving the problem. Today, one of important tools for solving this large-scale combinatorial optimization problem is A Mathematical Programming Language (AMPL).

# 1 Social and Economic Factors in Demand-Side Flexibility

*Z. Kaheh¹, H. Arasteh², and P. Siano³*
¹Tarbiat Modares University, Tehran, Iran
²Niroo Research Institute, Tehran, Iran
³University of Salerno, Fisciano (SA), Italy

## CONTENTS

## 1.1 INTRODUCTION

For the sake of environmental issues, the deployment of renewable energy sources (RESs) such as wind and solar farms has been highly recommended in recent decades. Decarbonizing policies have changed the energy system structure and necessitate rapid integration of RESs in different sectors. High penetration of RESs has given rise to new challenges due to their intermittent generation patterns, as their progressing deployment potentially upsurges contingencies [1].

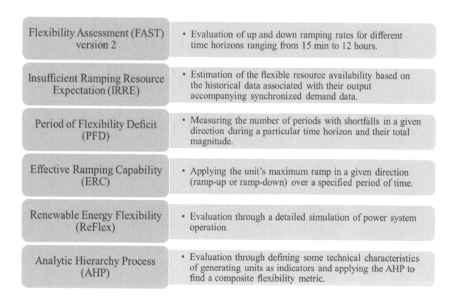

| Flexibility Assessment (FAST) version 2 | • Evaluation of up and down ramping rates for different time horizons ranging from 15 min to 12 hours. |
|---|---|
| Insufficient Ramping Resource Expectation (IRRE) | • Estimation of the flexible resource availability based on the historical data associated with their output accompanying synchronized demand data. |
| Period of Flexibility Deficit (PFD) | • Measuring the number of periods with shortfalls in a given direction during a particular time horizon and their total magnitude. |
| Effective Ramping Capability (ERC) | • Applying the unit's maximum ramp in a given direction (ramp-up or ramp-down) over a specified period of time. |
| Renewable Energy Flexibility (ReFlex) | • Evaluation through a detailed simulation of power system operation. |
| Analytic Hierarchy Process (AHP) | • Evaluation through defining some technical characteristics of generating units as indicators and applying the AHP to find a composite flexibility metric. |

**FIGURE 1.1**    Common metrics for flexibility availability evaluation [3–8].

A common method for estimating the ramp flexibility required by the system is based on the expected level of variability and the uncertainty associated with the forecasting errors [2]. However, there are several methods to evaluate the availability of flexible resources in power systems (Figure 1.1) [3].

To cope with the high penetration of RESs, it has been recognized that the conventional operation methods are inadequate for covering the net load variations [9]. Therefore, flexibility is introduced as one of the main concerns of future renewable-dominated power systems. The major resources of ramp flexibility are (1) conventional power plants, (2) distributed generation (DG), (3) demand-side resources, (4) energy storage systems, and (5) virtual flexibility as mechanisms and regulations for enhancing the system flexibility [3, 10]. Figure 1.2 illustrates the flexibility resources and uses.

Demand-side flexibility refers to the voluntary swift alternations in the grid electricity demand, which are activated by some sort of incentives. In this definition, only the voluntary consumer-side flexibility is considered, in which consumers actively decide to participate in supplying demand-side flexibility. Demand-side flexibility makes customers adapt their consumption regarding different price signals. It is established beyond doubt that demand-side flexibility is a complicated area. Because most consumers do not have a firm grasp of demand-side flexibility, the assessment of their capability for supplying demand-side flexibility from the socio-economic aspect is of paramount importance.

There is a wide array of qualitative and quantitative methods for analyzing the effect of social and economic factors on supplying demand-side flexibility. System dynamic and causal inference are the two powerful methods for such assessment. A common approach for the estimation of demand flexibility is conducting experiments

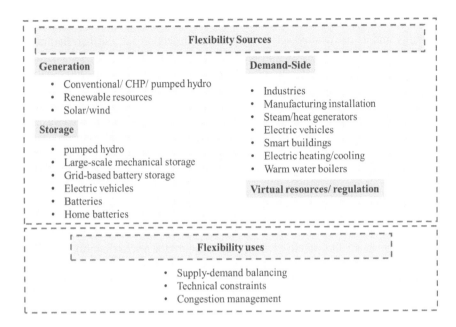

**FIGURE 1.2**  Flexibility sources and uses.

based on two groups that are identical in most aspects of the consumption except one factor. One group is the control group, and the other one is the experimental group (the demand respose tariff is tested on the experimental group). Since this approach is expensive and time-consuming, a more appropriate approach is based on the causal effects [11].

In causal inference, as a counterfactual prediction, data are used to predict certain features of the world as if the world had been different [12]. To be more specific, an action or intervention may have an effect, and the causes of these effects should be studied to understand how the intervention resulted in the effect [11]. Through causality inference, it is possible to estimate the flexibility of each customer and indicate to what extent the socio-economic factors influence consumers' flexibility.

This chapter focuses on the socio-economic aspects of consumer-side flexibility procurement. To this end, different types of demand-side flexibility, the associated drivers and actors, the plausible physical trading market, as well as the possible opportunities and obstacles of demand-side flexibility from a socio-economic perspective are examined. When it comes to the physical electricity trading, the whole of the day-ahead market, the intra-day market, and the balancing market should be considered; therefore, the measurements associated with these areas are also taken into account.

The main aim of this chapter is to pave the way for further researches on analyzing the effects of socio-economic drivers on demand-side flexibility. To this end, among different methods, we introduce the causal inference method as a lesser-known but powerful method for analyzing demand-side flexibility.

The rest of this chapter is organized as follows. An overview of demand-side flexibility and its drivers and actors is presented in section 1.2. Further, the trading demand-side flexibility in different markets is explained in section 1.3. The barriers to demand-side flexibility are described in section 1.4. Demand-side flexibility enablers from the socio-economic aspect in different areas ranging from customer level to network level, as well as taxing and subsidizing are explained in section 1.5. The available approaches including our preferred method, causal inference, are presented in section 1.6. In this section, among different available approaches, the causal inference approach is recommended to analyze the socio-economic effects on demand-side flexibility. Finally, section 1.7 presents the status quo and outlook.

## 1.2  DEMAND-SIDE FLEXIBILITY AND ITS DRIVERS AND ACTORS

Demand response (DR) is an important activity from the demand side, in which electricity customers are motivated to be responsive against the price/reliability signals and to modify their consumption patterns when they are called [13–15]. The demand side as a major resource of flexibility could participate in different markets, including day-ahead, intraday, and balancing markets [5]. Thence, avoiding deployment of demand-side flexibility is an essential reason for energy market disruptions [16, 17]. Therefore, DR can facilitate the deployment of distributed energy resources and increase the flexibility of the power system through very fast upward/downward changes in the demand.

Small-scale customers with different types of flexible resources such as storage, controllable and flexible appliances, and electric vehicles can be considered as demand-side flexibility providers. In this regard, the concept of "virtual batteries" has been presented [18]. In the absence of expensive batteries, to track zero-mean regulation signals or balancing reserves, the flexibility can be exploited from the power consumption of the most of electric loads such as heating, ventilation, and air conditioning systems [18].

Demand-side flexibility has been examined through different modeling; e.g. rebound effect (RE) refers to the changes in consumers' demand because of the previous and future price reactions, and is also related to the technical constraints of loads and consumers' preferences (Figure 1.3). In other words, RE refers to the increment/decrement of power consumption following an event of up- (down-) regulation [19].

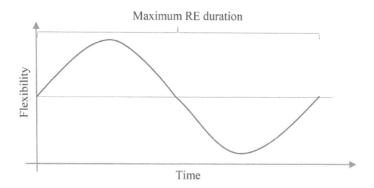

**FIGURE 1.3**   A simple scheme of the rebound effect.

RE is commonly studied for thermal loads or refrigerators [19], which need to recover their consumption immediately after a decrease in their consumption. However, according to ref. [20], RE is of relevance to study in demand-side flexibility and for shiftable loads.

With high penetration of RES in the electricity system, the power system confronts many profound difficulties in dealing with frequency regulation, power deficit risks, and local network problems. Demand-side flexibility is potentially able to rectify these problems. Although the initial challenges of flexibility supplying from the demand-side are mainly associated with the infrastructure issues, including developing the advanced metering, digital technologies, and internet of things (IoT) in smart grids [21, 22], consumers' training and prediction of their uncertain behavior are other problems.

From the socio-economic aspect, it is preferable that flexible electricity customers reduce their demand during shortage periods with high prices and increase it during low-price periods to avoid costly investments and losses. Besides, demand-side flexibility decreases price fluctuation and price spikes, and provides more efficient deployment of RESs.

Since flexible customers could impact the price formation in the day-ahead market, it is important that these customers should be exposed to the price volatilities [23]. In fact, the price fluctuations on the spot market determine to what extent the customer could save from flexibility bidding on the day-ahead market. However, the price flattening effect of energy storage technologies may mitigate the effect of bidding and associated revenue in these markets on interrelated markets.

The most critical challenges including power shortage risks, local network problems, and frequency regulations that can be rectified through demand-side flexibility are as follows:

- Power deficit risks: Power deficit refers to a situation, in which the demand for electricity is greater than the available production. This imbalance can be stabilized via demand-side flexibility. In high penetration of the RESs, a power deficit may even occur during a normal situation, not only during peak load. However, by providing correct price signals during scarcity events, demand-side flexibility will be helpful for balancing the system.
- Local network problems: A high penetration of RESs and electric vehicles will exert a significant effect on networks and capacity limitations. Thence, time-differentiated network tariffs have been introduced to reduce network losses and costs.
- Frequency regulation: Demand-side flexibility grants maintaining frequency in the electricity system via reducing/increasing electricity consumption in shortage/surplus situations.

The most important requirements for demand-side flexibility are as follows:

- Fair pricing and fair payment system encourage customers to participate in the demand-side flexibility programs.
- Technical potential including smart meters and novel technologies facilitate the demand-side flexibility programs.

- Clear information is a vital requirement for enabling the customer to participate in demand-side flexibility. Thence, network companies are required to clearly report network tariffs on customers' bills.
- Novel solutions should be presented to deal with the problems associated with demand-side flexibility. For instance, local peer-to-peer (P2P) flexibility trading is a novel solution to cover the uncertainties stemmed from environmental factors as well as the unforeseeable human behavior [24, 25].

Three kinds of driver for participating in demand-side flexibility are (1) economic drivers, (2) societal responsibility, and (3) technology drivers. It stands to reason that the economic driver of demand-side flexibility is the most important driver for many consumers. Industrial companies as demand-side flexibility providers are so price-sensitive that in response to high prices, they decrease their demand from the network by temporarily stopping the electricity-intensive production processes in those periods or by instigating their own electricity production. That being said, household consumers have other drivers such as manifesting their social responsibility and contributing to environment preservation via participation in flexibility providing programs.

Figure 1.4 illustrates the players participating in the development of demand-side flexibility [23, 26].

Aggregators are one of the main players in supplying demand-side flexibility. An aggregator can aggregate the demand-side flexibility from several customers into larger volumes and submit bids in associated markets. Therefore, aggregators can incentivize demand-side flexibility through solving some of the barriers such as minimum bid size and fees, and provide an effective market with low-entry barriers.

A general barrier, regardless of the aggregator business model, is the risk associated with the investment for the required equipment for balancing market. This may give a rise to enforcing consumers to involve in long-term contracts by the investor, which is not desirable in the long run. Besides, if an aggregator only has a few balance responsible parties (BPR) to select from, the BPR may dominate the

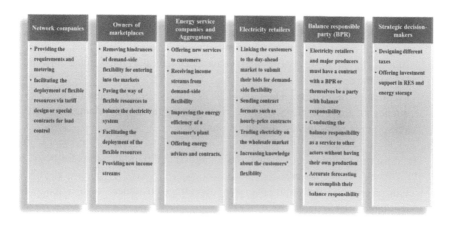

| Network companies | Owners of marketplaces | Energy service companies and Aggregators | Electricity retailers | Balance responsible party (BPR) | Strategic decision-makers |
|---|---|---|---|---|---|
| • Providing the requirements and metering <br> • facilitating the deployment of flexible resources via tariff design or special contracts for load control | • Removing hindrances of demand-side flexibility for entering into the markets <br> • Paving the way of flexible resources to balance the electricity system <br> • Facilitating the deployment of the flexible resources <br> • Providing new income streams | • Offering new services to customers <br> • Receiving income streams from demand-side flexibility <br> • Improving the energy efficiency of a customer's plant <br> • Offering energy advices and contracts. | • Linking the customers to the day-ahead market to submit their bids for demand-side flexibility <br> • Sending contract formats such as hourly-price contracts <br> • Trading electricity on the wholesale market <br> • Increasing knowledge about the customers' flexibility | • Electricity retailers and major producers must have a contract with a BPR or themselves be a party with balance responsibility <br> • Conducting the balance responsibility as a service to other actors without having their own production <br> • Accurate forecasting to accomplish their balance responsibility | • Designing different taxes <br> • Offering investment support in RES and energy storage |

**FIGURE 1.4** Role of different actors participating in demand-side flexibility development.

aggregator's business. As a consequence, an aggregator may be uninterested in the collaborations that entail revealing a lot of information to the BPR [23]. In the next section, demand-side flexibility is examined in different markets.

## 1.3  DEMAND-SIDE FLEXIBILITY IN DIFFERENT MARKETS

It is indispensable to exploit the require d ramp capability in electricity markets to cover fluctuations and uncertainties, and decrease the probability of ramp shortages and price spike occurrences. The most effective incentive for suppliers of demand-side flexibility is a fair payment system, which encourages customers to participate in the market.

Although intra-hourly and real-time markets mitigate ramp shortage, an ineffective day-ahead market threatens real-time markets by increasing contingencies [3]. The integrated flexibility market with interrelated markets could be encountered with several challenges. The fundamental features of effectual markets are the sustainability of transactions between partners, the existence of incentives and assurance, the market capability for the entrance of new suppliers, and the possibility of their winning the market share from incumbents [3].

Demand-side flexibility can impact all time frames from very short term (respond in sub-seconds) to long term (seasonal demand flexibility in the monthly time frame) [27].

### 1.3.1  DAY-AHEAD MARKET AND INTRADAY

Demand-side flexibility in the day-ahead market can be divided into proactive and reactive demand-side flexibility. Proactive demand-side flexibility has a direct effect on price formation, though reactive demand-side flexibility means that customers react once the prices on the day-ahead market have been announced [23]. Since proactive demand-side flexibility has a direct effect on price formation, providing the consumer with the right price signals via an hourly price contract is important to enable demand-side flexibility to bid into the day-ahead market [23, 26].

Proactive demand-side flexibility can be effective in price formation in the day-ahead market. For example, if the price rises above a certain level, consumers reduce the heat load; thence, the BPR can make bids for such situations. Proactive demand-side flexibility is useful for decreasing peak loads and price spikes. On the other hand, reactive demand-side flexibility, in which customers react once the prices on the day-ahead market have been announced, has a significant influence. If the BPR cannot forecast the customers' reactions, this leads to increased imbalance costs for BPR. Therefore, BPRs usually prefer the proactive demand-side flexibility because of the challenges around the reactive demand side.

It should be mentioned that the intraday market for flexibility trading is a correction market to the day-ahead market, and used primarily by BPR. In this market, actors including electricity consumers and aggregators have the opportunity to offer their flexibility and modify any earlier trading if the previous forecasts were not accurate.

Accurate forecasting and appropriate bidding formats for demand-side flexibility pave the way for electricity retailers to deploy the customers' flexibility in the day-ahead market. Besides, by informing the customers about their probable demand in real time, they will realize how different activities in the home have an influence on their usage. Nevertheless, the current markets are not effectively tailored for some actors who aim to offer their flexibility based on the flexibility characteristics of some technologies that enable load shifting [23].

### 1.3.2 BALANCING POWER MARKET

Bids in the real-time market consist of volume [MW], price [$/MWh], geographic location, and the speed of their full activation. The automatic reserves for correcting the minor frequency differences and manual reserves for correcting the larger frequency differences are the two ways of providing flexibility to the balancing market. To substitute regulation volume from the production side with that of the demand side, it must be guaranteed that demand-side flexibility is permanently accessible.

Besides, defining appropriate laws for competition (e.g., pricing method) on the automatic reserve market would facilitate the demand-side flexibility deployment. For instance, shifting from pay-as-bid to marginal pricing, which incentivizes aggregators to invest in administration and communication systems for control. Applying marginal pricing means that all activated upward regulation bids are priced the same as the most expensive activated bid. Demand-side bids are rarely activated because they are commonly priced higher than the production bids [23].

## 1.4 BARRIERS OF THE DEMAND-SIDE FLEXIBILITY

The demand-side flexibility is influenced by several factors. There are several entry barriers, which make consumers uninterested in participating in demand-side flexibility. Similar to the DR barriers, demand-side flexibility barriers could be divided into fundamental and secondary barriers. Fundamental barriers are economic, social (including behavioral and organizational), and technological barriers. On the other hand, secondary barriers relate to political regulatory aspects, design of markets, system feedbacks, and a general understanding of DR and demand-side flexibility [28]. In what follows, the socio-economic barriers which influence the demand-side flexibility are discussed (Figure 1.5).

### 1.4.1 BEHAVIORAL BARRIERS

Behavioral barriers refer to those unpredictable behaviors which deviate from preferable rational behaviors. Rational means profit-maximizing for companies; however, it means utility-maximizing for residential customers, which is determined based on the resulted comfort and revenue [23, 28].

- Lack of awareness: Despite electricity-intensive industries, rarely do residential consumers know about the concept of demand-side flexibility. Inordinate lack of knowledge about demand-side flexibility as well as the complexity of demand-side flexibility contracts prevents consumers from participating in the market [23].

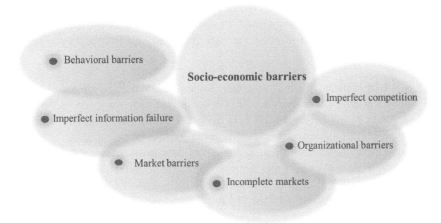

**FIGURE 1.5** Socio-economic barriers which impact the demand-side flexibility.

- Communication problems: Since the prompt response of DR providers to DR signals is of paramount importance, the inappropriate way of imparting the information (e.g., an inappropriate interface) is potentially an obstacle [28].
- Lack of trust: Trust in the electricity market and its actors play a vital role to encourage the customers to be flexible and enter into a demand-side flexibility contract. Lack of trustworthiness is an obstacle for DR and its flexibility.
- Transient ideas: Residential DR providers decide based on their comfort and convenience; however, the relationship between values and preferences is ambiguous to a great extent; and consumers' values may change in the course of time.
- Inertia barrier: Despite knowing its benefits, consumers usually are uninterested in changing their normal behavior. For instance, a DR provider may refrain from increasing their flexibility by installing more energy storage, whereas the investment will pay back rapidly.
- Bounded rationality: If the business case is complicated, the DR provider may prefer to simply take action based on the evidence of profitability in the recent past. Bounded rationality refers to non-optimal satisficing behavior in confronting with multiple complex alternatives [28].
- Customers' difficulties for juxtaposing offers: Clear information (e.g., price signals and tariffs) is a vital requirement for enabling a customer to participate in demand-side flexibility. Thence, network companies are required to clearly report network tariffs on customers' bills. However, customers may have difficulty comparing different network tariffs for which some startups have emerged to solve this problem.

## 1.4.2 IMPERFECT INFORMATION FAILURE

Distorted and ambiguous information produces imperfect information failure (IIF), in which one party acts on behalf of other parties, but does not precisely indicate their interests. Split incentives cannot be settled via a contract because they are not clearly or completely defined. Based on the principal-agent problem, in contracts between a consumer and an aggregator, this problem could occur if the aggregator cannot fully recognize user preferences. IIF appears for several reasons, including [28]:

- Lack of knowledge about potential demand-side flexibility providers and their variable and uncertain. The immense cost of information gathering and processing in smart grids may dissuade parties from gathering all the relevant information, and leads to adverse consequences.
- Immature markets, in which the historical data is not sufficient or there is no motivation to collect and distribute data. Small potential DR providers such as small commercial business which cannot predict the value of their flexible load or cannot find market access.

### 1.4.3   INCOMPLETE MARKETS AND IMPERFECT COMPETITION

Incomplete markets appear when property rights are vague or the benefits of an asset are non-excludable [28]. This happens when a curtailment in peak load turns to other actors' advantage without paying for it, called free-riding [29]. This issue may exert a mental pressure on customers who provide demand-side flexibility.

On the other hand, imperfect competition happens when a party (or parties) is able to exercise market power and impose prices above its marginal costs due to its huge market share.

### 1.4.4   MARKET BARRIERS

There are different types of market barriers, which can be considered as economic barriers to the demand-side flexibility, including:

- Lack of access to investment sources.
- Increased uncertainty.
- Hidden costs associated with participation in markets (especially for small parties).
- Uncertain value of flexibility (depending on the rate of RES penetration).

There are difficulties to gain access to hourly values. Many residential customers may not have regular access to their meters. Nonetheless, historical meter data per hour is indispensable for (1) estimating a customer's flexibility potential, (2) designing captivating offers, and (3) calculating costs and benefits of demand-side flexibility for each cluster of customers.

### 1.4.5   ORGANIZATIONAL BARRIERS

Organizational barriers consist of power and culture barriers. The level of power or effectiveness of a person, who is in charge of DR programs implementation is a determining factor in installing enabling technologies, instructing behavior change, and investing in increased flexibility. An ineffective in charge responsible for DR programs implementation, especially in organizations in which environmental matters are not generally regarded as important, gives rise to resistance to providing demand-side flexibility [28].

## 1.5  DEMAND-SIDE FLEXIBILITY ENABLERS FROM SOCIO-ECONOMIC ASPECT

### 1.5.1  CUSTOMER AREA

It is of paramount importance to edify customers about their external and internal information, to enable them to make a knowledgeable decision on using their demand-side flexibility potential. External information is supplied by other actors and includes information such as price signals. Internal information refers to the outcomes of the customers' actions such as load shifting [23]. In addition to financial drivers, customers may have other drivers; therefore, providing information about other types of incentives may also be of relevance. Large-scale industrial companies can usually generate their own strategies for optimal demand-side flexibility. However, residential customers should consider a third party to help them find an optimal strategy for demand-side flexibility. To this end, the network companies pave the way for customers by providing continuous free access to meter values and voltage values. The required data to facilitate juxtaposition of the various tariffs and contracts are [23]:

- The information about available network tariffs and hourly price contracts.
- The information about the effect of different tariffs on the overall electricity cost.
- Simulation tool and decision support systems to help customers compare different offered options. Data access for other actors (like companies aiming to develop applications to facilitate the contract comparing for customers) to facilitate the development of smart services.

### 1.5.2  ELECTRICITY NETWORK AREA

The network tariff can directly impact the behavior of customers by imparting different price signals. Network companies can pave the way for customers to be flexible by edifying customers about the design of the tariff and their effect on costs. The design of a network tariff differs based on different features, including parameter tuning for fixed and variable parts.

According to the analyses of the tariffs, it has been recognized that different tariffs provide different levels of encouragement for customers to provide demand-side flexibility and different benefits for the network company. High fixed costs in the tariff give rise to limited opportunities for flexibility of customers, though the variable component of the tariff in time-differentiated tariffs is vital for providing demand-side flexibility [23]. More information about taxing and subsidizing will be provided in the next subsection.

### 1.5.2.1 Taxing

Taxes can strengthen price signals to customers, and promote investing in equipment that facilitates demand-side flexibility. There is a wide array of taxes, which are introduced based on the size of the network charge, the retail contract, and the geographical location of consumers [23].

Fixed taxes is practical for those flexible consumers (such as industrial customers), who increase or decrease their demand temporarily during hours with high or low prices. For this group of customers, the fixed electricity tax distorts the price signal, as the electricity price no longer reflects society's real costs for production. Nevertheless, dynamic energy taxes affect the behavior of those flexible consumers who shift their load between hours by reacting to price variations over a 24-hour period. This type of customers only react to price variations between hours; thence, the dynamic energy tax strengthens the price variations on the retail market and lead to increased demand-side flexibility for this group [23]. Moreover, the energy tax also has an effect on the behavior of customers who can switch from electricity to other energy carriers such as district heating producers [23].

### 1.5.2.2 Support and Subsidy

Decision-makers can use a wide array of support systems, such as funding various technologies, to enable demand-side flexibility and encourage customers to be more flexible. Based on different analyses, they determine the benefits of establishing the market or investing for technologies associated with demand-side flexibility in different contexts such as finance and security. The market for trading demand-side flexibility requires investing an immense amount of resources in edifying customers about the advantages of demand-side flexibility. This investment cost can disperse among different companies, so that a lower entry cost imposes on actors. This lowered cost as well as the environmental effects justify this issue socio-economically [23].

That being said, (1) financial incentives and subsidy, (2) marketing of demand-side flexibility, and (3) edifying the customers about the benefits of adapting demand according to different signals for promoting demand-side flexibility may lead to the electricity customers becoming more price-sensitive. Therefore, a subsidy is more appropriate for early adopters, since without a subsidy, they have to pay a higher cost for equipment and information management to be flexible, which slows down the development. In addition, it is more likely that companies and consumers delay the investment until the cost of the equipment and information management drops down. Therefore, from socio-economic aspect, investment support for energy storage or control systems can be justified if the socio-economic benefits from increased price sensitivity are at least equal to the cost of the subsidy [23].

## 1.6 DEMAND-SIDE FLEXIBILITY ASSESSMENT

There is a wide array of qualitative and quantitative methods (e.g., system dynamic and causal inference) for analyzing the effect of social and economic factors on providing demand-side flexibility. Figure 1.6 illustrates a simple scheme of the evaluation framework for analyzing demand-side flexibility from the economical-behavioral-organizational aspect.

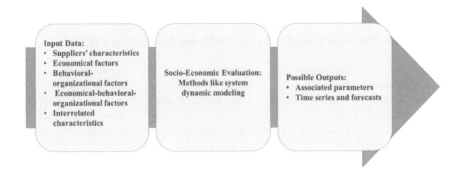

**FIGURE 1.6** A simple scheme of the evaluation framework for analyzing demand-side flexibility from the socio-economic aspect.

System dynamic modeling is commonly used to get a solid grasp of complex systems under a given set of conditions or scenarios to avoid probable disturbances. System dynamic modeling is an appropriate tool to study the dynamic behavior of the system, which is the result of an interconnected web of feedback loops. Feedback loops demonstrate the connection between different variables. To this end, two types of loops, including positive (reinforcing) and negative (balancing), should be developed. Positive feedback loops reinforce the feedback of information, and negative feedback loops are goal-seeking and tend to resist any change in the system [30, 31]. When it comes to the examination of socio-economic factors in demand-side flexibility, a system dynamics approach could be used to illuminate the interrelated complexities and dynamic interactions.

In addition, there are some other useful methods such as statistical methods and causal inference. In fact, since the demand-side flexibility is influenced by several factors, statistical methods such as regression and design of experiments (DOE) are also of relevance to consider for evaluation of these factors. Our proposed method for this problem will be presented in the following.

Here, we aim to introduce a lesser-known method for assessing demand-side flexibility (Causal inference). Causal inference is a powerful method for assessing the demand-side flexibility. The authors in ref. [11] have taken a similar approach to estimate the elasticity of each customer. Therefore, through causality inference between different DR tariffs and electricity consumption, it is possible to estimate the flexibility of each customer and indicate to what extent the tariffs offered by each DR program have an influence on the consumption of consumers [11].

A well-known approach for the estimation of the demand elasticity is conducting experiments based on two groups that are identical in most aspects of consumption; in such an experiment, one group is the control group, and the other one is the experimental group (the DR tariff is tested on the experimental group). Since this approach is expensive and time-consuming, a more appropriate approach is based on the causal inference [11]. To be more specific, an action or intervention may have an effect, and the causes of these effects should be studied to understand how the

**TABLE 1.1**

**Comparison between Causal Inference with Similar Concepts [12]**

| | Description | Prediction | Causal Inference |
|---|---|---|---|
| **Research Question** | What happened? | What will happen? | What will happen if …? |
| **Subject** | What was affected? | What will be affected? | What were the causes of these effects? (Why) |
| **Result** | Subject with X had Y | Subject with X are more likely to have Y. | If X changed, how would it change Y? |

intervention resulted in the effect, which is the aim of causal inference. The causality is compared with similar concepts in Tables 1.1 and 1.2.

In causal inference, as a counterfactual prediction, data are used to predict certain features of the world as if the world had been different [12]. Temporal precedence has been known as the most important feature of causation, which distinguishes causal from other types of associations [32]. Once a causal model is formed, it defines a joint probability distribution over the variables in the system, which reflect some features of the causal structure [32]. Reinforcement learning (RL) may produce decent causal inference in some simple settings, though RL is fruitless in complex causal settings. Therefore, causal inference and RL are not interchangeable [12].

A causal structure of a set of variables V is a directed acyclic graph (DAG) in which each node corresponds to a distinct element of V, and each link represents a direct functional relationship among the corresponding variables [32]. The common method of estimating causal effects in observational studies is to adjust for a set

**TABLE 1.2**

**Comparison between Causality with Similar Concepts [32]**

| | Probabilistic | Statistical | Causal |
|---|---|---|---|
| **Parameter** | Any quantity that is defined in terms of a joint probability function. | Any quantity that is defined in terms of a joint probability distribution of observed variables, making no assumption whatsoever regarding the existence or nonexistence of unobserved variables. | Any quantity that is defined in terms of a causal model and is not a statistical parameter. |
| **Assumption** | – | Any constraint on a joint distribution of observed variable | Causal assumptions may or may not have statistical implications. Often, though not always, causal assumptions can be disproved from experimental studies. |

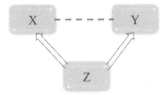

**FIGURE 1.7**   Causal graph. [11]

of confounder variables. Confounder variables are those variables, which have an impact on both the dependent and independent variables.

The research question may be deciding if two sets of variables in a DAG are equally valuable for adjustment. This question is important in many aspects. First, prior to taking any measurement, the researcher may aim to assess whether two candidate sets of variables, differing considerably in dimensionality, measurement error, cost, or sample variability are equally valuable in their bias-reduction potential. Second, assuming that the structure of the underlying DAG is only partially known, one may wish to assess whether a given structure is compatible with the available data [33].

In ref. [11], price (X) and consumption (Y) are statistically correlated with a common cause Z (Figure 1.7). Variable set Z includes calendar variables ($Z_1$: hour, $Z_2$: day, and $Z_3$: week).

It is necessary for confounder variable Z to satisfy the back-door criterion, which means it should block all non-causal paths (backdoor paths) from independent variables to the dependent variable and it should not include any descendants of the independent variable [32]. If Z satisfies this back-door criterion, then the effect of X on Y can be shown as Equation (1.1), which provides the distribution of Y.

$$P(Y \mid X = x) = \sum_{Z} P(Y \mid X, Z) P(Z) \tag{1.1}$$

Causal effect refers to the conditional expectation of Y given X, controlling for a sufficient sets of variables (Z) to make the effect recognizable.

$$E[Y \mid X = x] = \sum_{Y,Z} Y P(Y \mid X, Z) P(Z)$$

$$E[Y \mid X = x] = \sum_{Z} E[Y \mid X, Z] \tag{1.2}$$

The calendar variables ($Z_1$: hour, $Z_2$: day, and $Z_3$: week) are confounders that have effects on both X and Y. Therefore, the causal analysis model works by controlling the Z, according to Equation (1.3).

$$E[Y \mid X_j] = \sum_{Z} E[Y \mid X_j, (Z_1, Z_2, Z_3)] \tag{1.3}$$

Causal-inference methods can be divided into three categories: (1) Formal models of causation, especially potential-outcome models; (2) canonical inference, in which causality is a property of an association to be diagnosed by signs; and (3) methodologic modeling [34]. The interested reader can find more information about these methods in ref. [34].

Based on this introduction and the strengths of causal inference, this approach is highly recommended for the assessment of demand-side flexibility under the influence of socio-economic effects. Even though there is no direct research in this area in the literature, as we mentioned earlier, authors in ref. [11] have taken a similar approach to estimate the elasticity of each customer.

## 1.7  CONCLUSION AND FUTURE TRENDS

Demand-side flexibility makes it possible to deploy solar and wind power stronger than ever; which provides a great opportunity for investment in this sector. Besides, it reduces the price spikes, which are common in high penetration of RES. At the first stage, it stands to reason that conducting a cost-benefit analysis is required to guarantee its cost-effectiveness, and determine its possible unintended consequences. However, it is not straightforward to approximate the socio-economic costs and benefits of demand-side flexibility. Generally speaking, if demand-side flexibility rectifies the problems associated with inefficient resource deployment and local network problems, and these benefits outweigh its total costs, the benefits to society will be meaningful; but it is not what we pursue in a formal analysis.

As mentioned in this chapter, there is a wide array of qualitative and quantitative methods for analyzing the effect of social and economic factors on demand-side flexibility, including system dynamics, statistical methods such as regression and design of experiments (DOE), and causal inference. In causal inference, data are used to predict certain features of the world as if the world had been different. Therefore, it reduces the costs in comparison with experiments with a control group and an experimental group.

Regardless of our approach, considering the following points are of value in deploying the demand-side flexibility:

- Development of new markets or adjustment of existing markets is of paramount importance to enhance the effectiveness of markets for the trading of demand-side flexibility, and reduce search costs by linking the buyers and suppliers.
- Establishing an information portal from related public authorities and actors for demand-side flexibility is of relevance, which provides information on how various flexibility resources can participate in flexibility providing.
- The enablers for diminishing market barriers require market intervention to deal with the barriers associated with access to capital, uncertainty, hidden costs, or value generally. If market interventions result in a social benefit, they can be justified by governments (e.g., low-rate loans offered or guaranteed by the government, contracts for difference (CFD), and subsidies).

- Employing some technologies to conduct DR by making operational decisions on behalf of users is of significance due to their capability for presenting precise information.
- The probability of leaving programs associated with demand-side flexibility can be decreased via a precise selection of DR business models.
- It is indispensable that all customers, regardless of their size and the type of their contract, access to hourly metering without extra cost.
- Since the customers' preferences are ambiguous, time-varying, and probably not understood fully by the users themselves, introducing some metrics and measurements for user preferences and quantifying the acceptable comfort [35] would be beneficial, to reduce the impact of split incentives.
- The network companies and electricity retailers should have the authority to edify customers about effective and optimal behavior.
- Designing appropriate contracts are useful for conveying efficient behavior to demand-side flexibility providers.
- Since the main objective of firms is profit-maximizing, the manifestation of revenue gained by demand-side flexibility is of value.
- There are several ways to exert an effect on the values and preferences of users for enabling DR including (1) changing social norms by influencing perceptions of energy use and (2) changing laws and regulation to communicate those changed norms.
- Changing regulations may displease customers who feel that they are either not allowed to use or forced into using a new tariff. This risk could be minimized through finding the appropriate audience as well as removing the financial effect of a new tariff on individual customers [23].
- Progressive change toward dynamic real-time tariffs is indispensable to deal with the inertia barrier.
- Despite its benefits, the growing trust between DR providers and aggregators may threaten privacy. However, through data anonymization, consumers could be ensured that their data cannot be used for gaining their personal information [28].

Taking all aforementioned points into account, it is important that operators realize the dynamics of flexible loads and apply new models for analyzing socio-economic effects on the demand-side flexibility.

## REFERENCES

[1] M. R. Hesamzadeh, O. Galland, and D. Biggar, "Short-run economic dispatch with mathematical modelling of the adjustment cost," *International Journal of Electrical Power & Energy Systems*, vol. 58, pp. 9–18, 2014.

[2] E. Ela *et al.*, "Wholesale electricity market design with increasing levels of renewable generation: Incentivizing flexibility in system operations," *The Electricity Journal*, vol. 29, no. 4, pp. 51–60, 2016.

[3] Z. Kaheh, R. B. Kazemzadeh, and M. K. Sheikh-El-Eslami, "Flexible ramping services in power systems: Background, challenges, and procurement methods," *Iranian Journal of Science and Technology, Transactions of Electrical Engineering*, pp. 1–13, 2020.

[4] H. Chandler, "*Harnessing Variable Renewables: A Guide to the Balancing Challenge*," Paris, France: International Energy Agency, 2011.

[5] M. Milligan, B. Frew, E. Zhou, and D. J. Arent, "*Advancing System Flexibility for High Penetration Renewable Integration (Chinese Translation)*," Golden, CO (United States): National Renewable Energy Lab. (NREL), 2015.

[6] E. Lannoye *et al.*, "Integration of variable generation: Capacity value and evaluation of flexibility," in *IEEE PES General Meeting*, 2010, pp. 1–6: IEEE.

[7] J. Hargreaves, E. K. Hart, R. Jones, and A. Olson, "REFLEX: An adapted production simulation methodology for flexible capacity planning," *IEEE Transactions on Power Systems*, vol. 30, no. 3, pp. 1306–1315, 2015.

[8] V. Oree and S. Z. S. Hassen, "A composite metric for assessing flexibility available in conventional generators of power systems," *Applied Energy*, vol. 177, pp. 683–691, 2016.

[9] Z. Kaheh, R. B. Kazemzadeh, and M. K. Sheikh-El-Eslami, "Simultaneous consideration of the balancing market and day-ahead market in Stackelberg game for flexiramp procurement problem in the presence of the wind farms and a DR aggregator," *IET Generation, Transmission & Distribution*, vol. 13, no. 18, pp. 4099–4113, 2019.

[10] M. Alizadeh, M. P. Moghaddam, N. Amjady, P. Siano, and M. Sheikh-El-Eslami, "Flexibility in future power systems with high renewable penetration: A review," *Renewable and Sustainable Energy Reviews*, vol. 57, pp. 1186–1193, 2016.

[11] K. Ganesan, J. Tomé Saraiva, and R. J. Bessa, "On the use of causality inference in designing tariffs to implement more effective behavioral demand response programs," *Energies*, vol. 12, no. 14, p. 2666, 2019.

[12] M. A. Hernán, J. Hsu, and B. Healy, "A second chance to get causal inference right: a classification of data science tasks," *Chance*, vol. 32, no. 1, pp. 42–49, 2019.

[13] H. Arasteh, V. Vahidinasab, M. S. Sepasian, and J. Aghaei, "Stochastic system of systems architecture for adaptive expansion of smart distribution grids," *IEEE Transactions on Industrial Informatics*, vol. 15, no. 1, pp. 377–389, 2018.

[14] H. R. Arasteh, M. P. Moghaddam, M. K. Sheikh-El-Eslami, and A. Abdollahi, "Integrating commercial demand response resources with unit commitment," *International Journal of Electrical Power & Energy Systems*, vol. 51, pp. 153–161, 2013.

[15] M. Kia, M. Shafiekhani, H. Arasteh, S. Hashemi, M. Shafie-khah, and J. Catalão, "Short-term operation of microgrids with thermal and electrical loads under different uncertainties using information gap decision theory," *Energy*, vol. 208, p. 118418, 2020.

[16] F. Rahimi and A. Ipakchi, "Demand response as a market resource under the smart grid paradigm," *IEEE Transactions on smart grid*, vol. 1, no. 1, pp. 82–88, 2010.

[17] H. Wu, M. Shahidehpour, A. Alabdulwahab, and A. Abusorrah, "Thermal generation flexibility with ramping costs and hourly demand response in stochastic security-constrained scheduling of variable energy sources," *IEEE Transactions on Power Systems*, vol. 30, no. 6, pp. 2955–2964, 2015.

[18] P. Barooah, A. Buic, and S. Meyn, "Spectral decomposition of demand-side flexibility for reliable ancillary services in a smart grid," in *48th Hawaii International Conference on System Sciences*, 2015, pp. 2700–2709: IEEE.

[19] N. O'Connell, H. Madsen, P. Pinson, M. O'Malley, and T. Green, "Regulating power from supermarket refrigeration," in *IEEE PES Innovative Smart Grid Technologies, Europe*, 2014, pp. 1–6: IEEE.

[20] G. De Zotti, D. Guericke, S. A. Pourmousavi, J. M. Morales, H. Madsen, and N. K. Poulsen, "Analysis of rebound effect modelling for flexible electrical consumers," *IFAC-PapersOnLine*, vol. 52, no. 4, pp. 6–11, 2019.

[21] M. H. Amini, H. Arasteh, and P. Siano, "Sustainable smart cities through the lens of complex interdependent infrastructures: Panorama and state-of-the-art," in *Sustainable*

*Interdependent Networks II: From Smart Power Grids to Intelligent Transportation Networks, Studies in Systems, Decision and Control*, Chapter: 3, Publisher: Springer International Publishing,, 2019, pp. 45–68.

[22] H. Arasteh *et al.*, "IoT-based smart cities: A survey," in *IEEE 16th International Conference on Environment and Electrical Engineering (EEEIC)*, 2016, pp. 1–6: IEEE.

[23] K. Alvehag, "Measures to increase demand side flexibility in the Swedish electricity system," *Swedish Energy Markets Inspectorate*, vol. 10, p. 2017, 2017.

[24] D. Fischer, A. Harbrecht, A. Surmann, and R. McKenna, "Electric vehicles' impacts on residential electric local profiles–A stochastic modelling approach considering socio-economic, behavioural and spatial factors," *Applied Energy*, vol. 233, pp. 644–658, 2019.

[25] S. Zhou, F. Zou, Z. Wu, W. Gu, Q. Hong, and C. Booth, "A smart community energy management scheme considering user dominated demand side response and P2P trading," *International Journal of Electrical Power & Energy Systems*, vol. 114, p. 105378, 2020.

[26] S. S. Torbaghan *et al.*, "A market-based framework for demand side flexibility scheduling and dispatching," *Sustainable Energy, Grids and Networks*, vol. 14, pp. 47–61, 2018.

[27] D.-s. f. f. p. s. t. IRENA (2019), *International Renewable Energy Agency*, Abu Dhabi.

[28] N. Good, K. A. Ellis, and P. Mancarella, "Review and classification of barriers and enablers of demand response in the smart grid," *Renewable and Sustainable Energy Reviews*, vol. 72, pp. 57–72, 2017.

[29] D. T. Nguyen, M. Negnevitsky, and M. de Groot, "Market-based demand response scheduling in a deregulated environment," *IEEE Transactions on Smart Grid*, vol. 4, no. 4, pp. 1948–1956, 2013.

[30] N. S. Ebnouf, *"The System Dynamics of Electricity Demand-Supply Gap in Sudan: The Socio-Economic Impacts and Integrated Modelling Approach,"* Pan African University, Publisher: PAUWES, Master Thesis, 2019.

[31] H. M. Mousavian, G. H. Shakouri, A.-N. Mashayekhi, and A. Kazemi, "Does the short-term boost of renewable energies guarantee their stable long-term growth? Assessment of the dynamics of feed-in tariff policy," *Renewable Energy*, vol. 159, pp. 1252–1268, 2020.

[32] J. Pearl, *"Causality: Models, Reasoning and Inference Cambridge University Press,"* Cambridge, MA, USA, vol. 9, pp. 10–11, 2000.

[33] J. Pearl and A. Paz, "Confounding equivalence in causal inference," *Journal of Causal Inference*, vol. 2, no. 1, pp. 75–93, 2014.

[34] A. Gelman and X.-L. Meng, *"Applied Bayesian Modeling and Causal Inference from Incomplete-Data Perspectives."* Publisher: Wiley; 1st edition, John Wiley & Sons, Series: Wiley Series in Probability and Statistics, DOI:10.1002/0470090456, 2004.

[35] K. Aduda, T. Labeodan, W. Zeiler, G. Boxem, and Y. Zhao, "Demand side flexibility: Potentials and building performance implications," *Sustainable Cities and Society*, vol. 22, pp. 146–163, 2016.

# 2 Role of Smart Homes and Smart Communities in Flexibility Provision

*Hosna Khajeh[1], Ran Zheng[1], Hooman Firoozi[1], Hannu Laaksonen[1], and Miadreza Shafie-khah[1]*
[1]School of Technology and Innovations,
University of Vaasa, Vaasa, Finland

## CONTENTS

## 2.1   INTRODUCTION

The impacts of decentralization as well as the high penetration of renewable energy resources into power systems have created a necessity to change the operation and management of power systems in recent years. Although the effects of renewables at all power system levels can be seen, it is more likely to become an issue, especially in distribution networks, due to the increased integration of small-scale renewable generation units (e.g., domestic PV panels). In addition, the output of renewable energy sources (RESs) is naturally variable and intermittent in very short-term periods due to their dependency on environmental weather-related factors. Therefore, there might exist a constant imbalance between generation and demand, which results in the instability of power system. In order to prevent these instabilities, the electrical systems must become more flexible, meaning that the system must respond to the expected and unexpected fluctuations rapidly, consistently, and adequately. To do so, all the potential flexibilities in the electrical systems should be deployed. Thus, the flexible capacities have to be effectively managed and controlled in a smart way by the system operators, including distribution system operators (DSO) and transmission system operators (TSO) [1].

At the distribution system level, the flexible energy resources are mostly located in the demand side where the majority of consumers are generally in residential households/areas. These flexible resources (which can include battery energy storage systems, electric vehicles, flexible loads, demand response, etc.) are able to contribute to enhancing the grid's flexibility by means of different operation and management methods. It is noticeable that in distribution networks, the complexity of network and the high amount of end-users and flexible resources worsen the process of dealing with all the end-users individually. Therefore, in order to unlock a considerable amount of flexibility in distribution networks, the system operators need to manage these flexibilities in an aggregated manner in order to be able to utilize them during critical situations. The aggregated management is possible with advanced information and communication technology (ICT) such as smart management systems in homes or in energy communities with group of homes [2]. In this regard, some literatures proposed the contribution of demand-side resources in providing flexibility services. Some of them analyzed the contribution of these resources to help the DSO increase flexibility of distribution networks while others assessed the participation of demand-side resources in providing TSO-level ancillary services.

For instance, in terms of TSO-level flexibility provision of demand-side resources, the authors in ref. [3] assessed the deployment of grid-connected photovoltaic panels and a battery storage system in order to follow the signals sent by the TSO. These signals aim to regulate the frequency of the system. The participation of electric vehicle (EV) charging stations in TSO-level flexibility services was also analyzed by the authors in ref. [4]. The study tried to estimate the maximum TSO-level flexibility that can be procured from the charging cycles of electric vehicles. The authors in ref. [5] also proposed a novel method considering the potential of flexible demand so as to provide frequency regulation capacity.

In terms of DSO-level flexibility provision, for example, it was proposed in ref. [6] that a DSO assigns its operation-related responsibilities to some aggregators located in the distribution networks. The proposed aggregators deployed EVs, RESs, and responsive loads to operate their assigned local networks. In a similar study, the authors in ref. [7] introduced a model using the flexibility of the resources connected to distribution networks. In this study, household controllable appliances were rescheduled to provide the DSO with its needed flexibility. In ref. [8], DSOs utilized energy hubs of demand response in order to operate the distribution network in a more effective way. In addition, it was proposed in ref. [9] that end-users of the distribution network can provide voltage control services for the DSO.

To complement and review the research associated with the participation of demand-side resources in providing flexibility services, this chapter is focused on the role of smart homes and communities in providing demand-side flexibility to the grid. Several energy management methods and approaches for smart homes using home energy management system (HEMS) for flexibility provision have been previously proposed [10]. These approaches are mostly based on demand-side management by providing dynamic prices to the rational end-users equipped with HEMS and/or taking advantage of demand response programs (DRPs). DRPs are mostly deployed to incentivize customers to react to external signals. The chapter presents popular DRPs that can reshape the consumption pattern of smart homes and communities. In addition, the flexible resources of residential households and communities are discussed. In this way, smart homes are considered to be equipped with battery energy storages (BESs) and EVs. Furthermore, the constraints related to the physics of loads as well as storage-based resources (BES and EV) are introduced.

Finally, the chapter focuses on the services that smart homes and communities can provide to the system operators. In this regard, DSOs can buy local flexibility services from these resources while TSOs can also utilize distribution-network-connected smart homes and communities to provide system-wide services for their needs. Our proposed case study consists of a small community with two types of flexible energy resources, including batteries and plug-in electric vehicles (PEVs). The community tries to exploit upward and downward flexible capacities from these flexible energy resources. The upward and downward flexibilities as well as the optimal schedule of these resources are discussed in the chapter.

## 2.2   DEMAND-SIDE MANAGEMENT FOR FLEXIBILITY SERVICES

Demand-side management (DSM) is the measure or implementation which can be utilized to influence the customers at the demand side with the aim to change load patterns to optimize the energy consumption and improve energy efficiency. This can be realized by using incentives, monetary benefits or administrative actions [11]. DSM can achieve energy cost reduction, energy conservation, optimizing energy consumption, and help to maintain the power system more reliable. In addition, DSM postpones the investment in the reinforcement of the power system by shifting or reducing the demand. This will also assist the TSO with its balancing-related responsibilities. Previously, increased investments in generation were needed in order to fulfill the increasing demand [12]. Demand response (DR) can be applied to change the load consumption patterns in short term, which can be implemented by load shifting or demand reduction [13]. Compared with DR, DSM includes the response

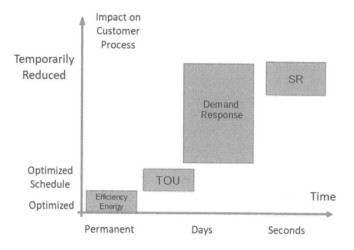

**FIGURE 2.1**   Classification of DSM programs.

of customers to change the load patterns in short term and long term. Hence, DR is a potent solution of DSM that can be implemented at some specific period [12]. Based on the effect on the customer, DSM can be classified into four types, as shown in Figure 2.1 [14]:

- Energy efficiency which is recognized as long-term energy conversation implementation that will be used to achieve energy-saving and demand reduction.
- Time-of-use (TOU).
- DR which can be implemented by the activities, including load shifting, peak clipping or valley filling or the combination of the mentioned activities.
- Spinning reserve (SR).

Depending on the different impacts on the end-users, DR programs consist of two types: price-based and incentive-based. In price- or time-based DR programs, the end-users will reshape the demand according to external signals such as price signals. This kind of programs can motivate customers indirectly to follow system changes. In incentive-based DR programs, the end-users can receive some incentives, cash benefits, payments, or preferential prices if they can shift or reduce the energy consumption according to the system needs. Market-based programs are based on specific electricity market places where the price structures are set up and the electricity is traded [12]. As presented in Figure 2.2, price-based DR programs consist of TOU rates, critical peak pricing (CPP), and real time pricing (RTP) programs. For the incentive-based programs (IBP), they can be further classified into two types, namely, market-based and classical programs. Classical programs consist of direct load control (DLC) programs and interruptible/curtailable programs, while market-based programs consist of emergency demand response programs (EDR), capacity market, demand bidding/buyback programs, and ancillary market services programs [15, 16].

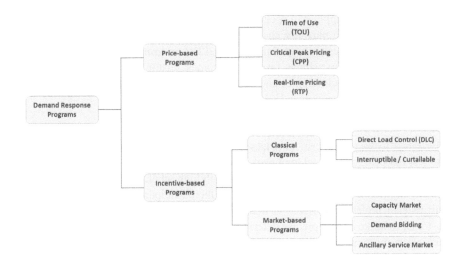

**FIGURE 2.2**   Classification of DR programs.

## 2.2.1   PRICE- OR TIME-BASED

Price-based programs depend on the dynamic pricing where the tariff of electricity is not flat, and the pricing rate changes in response to the real-time energy cost. In this kind of program, the high pricing at peak demand time or low pricing at the off-demand time are offered to flatten the curve of the consumption. Priced-based DR programs can be applied in the industrial sector where the energy costs are regarded as partial production cost. Rational customers equipped with ICT technologies can also react to these programs. Increased electricity costs result in more production costs. Hence, the manufacturer with larger energy consumption would like to attend price-based programs with the purpose of energy cost reduction. With the fast-growing deployment of information and communication in residential sectors, the application of DR programs becomes more available for smart homes and communities as well [17].

### 2.2.1.1   Time-of-Use Rates

TOU rates refer to the varying electricity prices designed for a different time. The prices vary over the day and are well-defined in advance [15]. The time period of a day can be split up into three periods, including peak, off-peak, and shoulder according to different load-demand levels [18]. Generally, the prices are specified based on the historical data in which the prices of high demand time are defined to be higher while the prices at off-peak periods are defined to have low rates. This is a core point for the grid utilities as it will encourage customers to decrease the load demand during peak periods and increase the consumption during off-peak periods. Thus, the end-users can gain the energy cost reduction by reducing the load demands. TOU rates are favorable for the grid operator due to the constant price for each period of the day compared with other price-based programs [18]. This is the basic DR program and can be applied in specific periods even in the whole year. The application of TOU rates can partially change the load demand patterns of the end-users, but actually, TOU rates are regarded as beyond DR programs due to their long-term application. In order to achieve favorable incomes, customers need to be equipped with advanced ICT technologies and smart controllers which can be used to control the operation of their resources [14].

### 2.2.1.2   Critical Peak Pricing

CPP refers to very high critical prices which are pre-set electricity prices added to TOU or flat prices at critical load peak times [19]. CPP-based DR programs are usually applied in some specific hours or days with high wholesale market prices or the periods in system contingencies. In practice, the participants for the CPP programs will receive preferential prices at non-critical peak demand times. This helps the customer to make a more precise determination to utilize the energy more efficiently. CPP offers more precise information on the energy cost for the end-users, especially at peak load times; a more precise determination is made by customers on energy consumption and then energy cost reduction can be achieved [8]. CPP may happen in extreme situations with very high loads in the summertime and wintertime, some festival days or holidays such as Christmas or Easter day.

### 2.2.1.3   Real-Time Pricing

RTP rates refer to, for example, the wholesale electricity prices which vary continuously according to the time during the day [20]. In RTP programs, the end-users will receive the bills based on the hourly day-ahead pricing that reflects the real-time energy bill cost. Price-based DR is a key demand response that will use price-based signals to control the energy consumption of customers.

### 2.2.1.4   Comparison of Different Priced-Based Demand Response Programs

In ref. [17], the merits and demerits of price-based DR programs have been addressed. The TOU program has two advantages. First, it is easy for the participants to understand the TOU pricing portfolios and make the plan for the daily energy consumption since the same price scheme is maintained by TOU pricing. Second, compared with the other two DR programs, the levels of participants attending the programs are stable. The disadvantage is that when the peak load is reduced by TOU pricing DR programs, the new peak may be created.

Regarding the CPP program, it consists of three merits. First, it is easy to understand and follow the CPP portfolio; second, CPP programs can help the operator to shift the peak load demand effectively; and third, the incentive payment for the participant can be estimated. However, this program is not without disadvantages. CPP programs cannot be applied on a daily basis because the system does not face severe situations every day. Hence, the CPP programs will not reduce the energy bill costs effectively.

For RTP programs, electricity prices vary continuously with the market prices. The dynamic prices can help participants reduce or shift their loads in order to maximize their profits or reduce their costs. But full utilization of RTP programs need the deployment of smart metering and automated control because it is difficult to control the load manually in response to time-differentiated pricing [21].

## 2.2.2 Incentive-Based

### 2.2.2.1 Direct Load Control

DLC means the grid operators or utilities can access the electrical equipment freely and control it remotely and directly on short notice. The remotely controlled equipment typically refers to house appliances such as thermostatically controllable ones, including water heaters and air conditioners. This kind of DR program has been implemented in small commercial customers or residential customers [22]. In this regard, the operator can switch the appliances on or off in the limited periods of each year or season. Generally, this type of program will be utilized on the remotely controlled devices such as water boilers and air conditioners by residential houses or some small commercial customers [16].

### 2.2.2.2 Interruptible/Curtailable Rates

Interruptible/curtailable rates program means the end-users agree on this program to gain some incentives, bill discount or cash payments in return by reducing or shifting their loads to pre-set values during the periods in which the systems are in contingencies or need flexibility. The customers who do not follow the agreement are penalized. Typically, there exists a minimum capacity for participants who want to contribute to this program. The load needs to be curtailed by the participants during a specified period after being notified [12].

### 2.2.2.3 Emergency Demand Response Programs

EDR programs are utilized to keep the reliability of the system within certain limits, but it is voluntary for customers to curtail their load. The incentives will be provided to the customers that reduce the load. Recently, many studies have been carried out on optimization of energy management of microgrid, but they do not consider demand response programs in operation. This will result in suboptimal operation and control of microgrid. The operation cost and peak load will decrease while applying EDR programs in the operation of microgrid. But the running cost of EDR programs will increase when the level of participation is over 50% due to incentive payment to the customers [23]. In ref. [17], the contribution of EDR programs has been investigated. They can help balance market power on the supply side and provide benefits for the end-users due to demand reduction. Meanwhile, the reliability of system will be fulfilled by load shedding at peak times.

### 2.2.2.4 Capacity Market Programs

The capacity market programs are formed so that the end-users are committed to reducing the load demand to the pre-defined values in system contingencies. The customers will receive penalties if they refuse to curtail when being notified by the utilities. This kind of program can be considered as insurance. In addition, the guaranteed payments will be offered to the participants [12].

### 2.2.2.5 Demand Bidding/Buyback Programs

Demand bidding/buyback programs refer to those programs in which the large end-users are encouraged to provide some specific load curtailment at bidding prices in the wholesale electricity market. When the bidding price is less than the wholesale electricity price, this bid would be accepted. The agreed specific load curtailment should be made when the customers offer the bid or they will be subjected to penalties. This kind of program is based on the day-ahead prediction of demand. The program is applied when the end-users are willing to reduce energy consumption at a specific preset price. One example of demand bidding programs is the application of programmable thermostat that can be used to control heating and air-conditioning systems. The setting of the thermostat can be changed based on the electricity pricing levels. This setting can also need to be modified with seasons [24].

### 2.2.2.6 Ancillary Services Market

Ancillary services market programs refer to those programs in which the end-users are allowed to offer their flexible capacities in upward and downward directions using their available flexible energy resources [25].

### 2.2.3 BENEFITS OF DEMAND RESPONSE PROGRAMS

As presented in Figure 2.3, the potential benefits from DR programs are summarized under three categories: (1) end-users, (2) system costs, and (3) system reliability [16].

End-users or customers involved in DR programs can reduce their energy bill costs when they lower the electricity consumption at peak times or offer flexibility services [26]. The end-users can also increase their energy saving by load shifting from peak times to off-peak times. In addition, the customers participating in incentive-based DR programs will get incentives. The incentive payments for the participants in DR programs, capacity market programs, demand bidding programs,

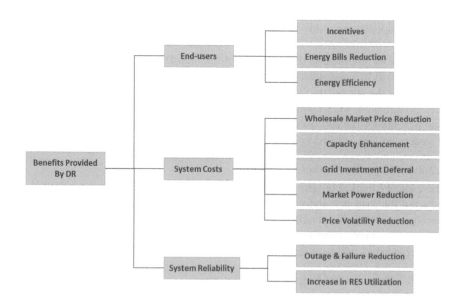

**FIGURE 2.3** Benefits provided by demand response programs.

and ancillary services market programs are dependent on their performance and the system's need.

Regarding the system costs-related benefits, the wholesale electricity prices may be reduced due to the reaction of demand-side resources during peak hours. In addition, the grid capacity can be increased using price-based DR programs, and thus, the capacity costs will be reduced or postponed. In other words, these programs will defer the investment for grid upgrade or enforcement. The participants in market-based DR and price-based programs can affect the electricity market by rescheduling the operation of their flexible energy resources. DR programs can also offer another benefit through price volatility reduction [27].

Participants can enhance the reliability of power system as well. If DR programs are well-defined, the customers will mitigate the outage risks, and at the same time, the customers can reduce the risk for the blackout. As a result, the system operators can have more resources to maintain the system in a reliable state [4].

## 2.3   DEMAND-SIDE FLEXIBLE ENERGY RESOURCES

The conventional types of electrical power and energy systems were based on the unidirectional energy supply starting from bulk generation units to the final consumers at the end. These energy systems are increasingly being replaced by decentralized structures at all levels such as generation, transmission, distribution, and demand in recent years. With this regard, the balancing and ancillary services are experiencing a transition as well. The system-wide services such as balancing and frequency control, which were previously provided by bulk generation units, are in the future increasingly provided by the flexible energy resources (FERs).

Demand-side FERs refer to storage-based resources as well as load-based resources which are capable of changing their power in order to contribute to different kinds of local and system-wide flexibility services. There exist various types of appliances which could participate in smart homes'/communities' energy management or demand response programs by curtailing or shifting their consumption over time [28]. However, the focus of this section is on the FERs that could provide power-based flexibility services to local and system-wide networks. The output power of these resources can be controlled and these resources mostly can have a rapid response to the changes when it is required.

There are different types of FERs introduced so far. However, the FERs which are more commonly utilized in smart homes/communities are battery energy storage system (BESS), plug-in electric vehicle (PEV) and thermostatically controlled load (TCL). A general categorization of the most common FERs in smart homes/communities is presented in Figure 2.4. These resources will be explained in the following subsections.

### 2.3.1   Battery Energy Storage System

BESSs as one of the most prevailing FERs in electrical systems have become quite popular in smart homes/communities, especially for those households that are equipped with small-scale generation units such as solar panels. BESS could be utilized efficiently in smart homes or communities for energy management purposes

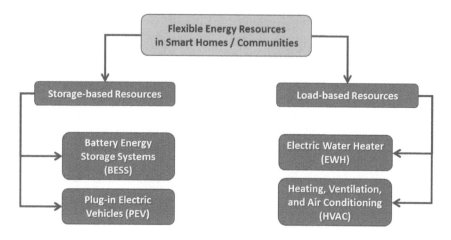

**FIGURE 2.4**  Overview of the most common FERs in smart homes/communities.

as well as in flexibility enhancement of the whole community in an aggregated manner. Accordingly, the small-scale BESS in smart homes could contribute to the cost reduction of the house by discharging its stored energy to the house when it is required. Moreover, the BESSs in a community could be aggregated and controlled to provide flexibility services to the upstream grids (i.e., transmission system-level or distribution system-level networks) for fulfilling flexibility needs of networks. In return, the households or community members can take the advantage of selling energy/flexibility to different entities. Equations (2.1) and (2.2) indicate the relation between the stored energy in the BESS and its power rate when it is in charging and discharging mode, respectively.

$$E(t) = E(t-1) + \gamma P^{ch}.T \qquad (2.1)$$

$$E(t) = E(t-1) - \frac{P^{dch}}{\gamma}.T \qquad (2.2)$$

Equation (2.1) states that the stored energy in the BESS at each time step is directly related to the stored energy at the previous time step in addition to the multiplied values of power rate, efficiency, and the duration of charging considering the efficiency of BESS. In contrast, the stored energy after a period of discharging would be equal to the stored energy in the previous time step minus the discharged energy in the discharging period considering the efficiency of BESS, as in Eq. (2.2). It has to be mentioned that the power rate of BESS could be either constant or adjustable. It is obvious that the adjustable power rate of BESS could be more beneficial than the constant power rate when it comes to flexibility provision. Thus, the power output and its direction (charging or discharging) should be determined by the energy management system. However, the stored energy in the BESS should remain within its limit. This limitation can be denoted by a simple constraint denoted by (2.3).

$$E^{\min} \le E(t) \le E^{\max} \qquad (2.3)$$

## 2.3.2  PLUG-IN ELECTRIC VEHICLE

PEVs have been the most prevailing FER among the smart homes' dwellers in recent years. The presence of PEVs could affect the operation of the electrical network adversely. However, the optimized operation and charging strategy of these resources could help the networks in terms of flexibility provision. In this manner, a smart home can schedule charging of the PEVs through its HEMS. A community manager can also control the charging behavior of its EVs through its energy management system. PEVs are potent resources which can provide the operators with flexibility services. In this way, their charging pattern can be modified according to the system flexibility needs [29].

The equation that represents the operation of the PEV's battery is the same as BESS's equation. Regarding the operation of PEVs in smart homes/communities, the constraints related to the stored energy in the PEVs' battery must be taken into account in the problem formulation as in (2.4):

$$E^{\text{PEV,min}} \le E^{\text{PEV}}(t) \le E^{\text{PEV,max}} \tag{2.4}$$

Constraint (2.4) limits the upper and lower bounds of stored energy in the PEVs' battery in order to prevent rapid depreciation of the battery since limiting the cycles of charging/discharging prevent shortening the life of PEV's battery. In constraint (2.4), $E^{\text{PEV}}(t)$ is the stored energy in the PEV's battery at time slot $t$ whilst $E^{\text{PEV,min}}$ and $E^{\text{PEV,max}}$ are the minimum and maximum desired level of stored energy in the PEV's battery, respectively.

In contrast, models of PEVs have been introduced in recent years that have the capability of vehicle-to-grid (V2G) energy flow as well. This means that PEVs could contribute to upward flexibility provision to the network by injecting the stored energy in their batteries back to the grid in the critical moments. It is noticeable that the V2G operation mode of the PEVs is along with a depreciation in their batteries, which should be considered in the V2G operation mode [30, 31].

Moreover, the charging infrastructure of PEVs may vary in terms of power rate and technology. The power of charging in these charging infrastructures could be either constant or dynamic [29]. There have been introduced some domestic PEVs' charging piles recently that have the ability of changing the charging/discharging power rates. These charging/discharging method are known as dynamic charging/discharging. The benefit of dynamic power rates for PEVs is in their capability of flexibility provision in a shorter period of time. With this regard, the constraints related to dynamic charging/discharging of PEVs also must be considered in flexibility provision problems as in (2.5) and (2.6):

$$0 \le P^{ch,\text{PEV}}(t) \le P^{ch,\text{max}} \tag{2.5}$$

$$0 \le P^{dch,\text{PEV}} \le P^{dch,\text{max}} \tag{2.6}$$

## 2.3.3  THERMOSTATICALLY CONTROLLED LOAD

Another type of FERs that could be found almost in every smart home is TCL. These FERs are, in fact, demand-based FERs which are capable of changing their power consumption through a thermostat and a command signal. In other words, TCL follows the variation in temperature on one hand, and is responsive to the receiving command signals on the other hand. The input command signal to the TCLs could be based on different purposes such as demand response programs, flexibility provision, or even energy management inside the smart homes/communities.

There are different types of TCLs inside smart homes such as electric water heater (EWH) and heating, ventilation, and air conditioning (HVAC) systems. These TCLs could effectively contribute to the issue of flexibility provision from the demand side. It has to be mentioned that these resources are not as fast as storage-based flexible resources when it comes to responsiveness. Therefore, they might not be capable of participating in all kinds of flexibility services. However, since the TCLs are responsible for a great portion of demand-side energy consumption, utilization of their flexibility could be quite beneficial in electricity network service provision [32].

One of the most popular flexible resources amongst TCLs is EWH. This resource is a load that its power consumption could be changed over time by changing the status of the device between ON and OFF in order to provide flexibility services. Accordingly, when the device changes its status from ON to OFF, its power consumption is curtailed and changes to zero, which could be realized as upward flexibility service to the grid. In contrast, when the device changes its status from OFF to ON, its power consumption increases to the nominal rate, which could be realized as downward flexibility service to the grid. However, there have been introduced some kinds of EWH with cutting-edge technologies in recent years, in which their power consumption is adjustable as well.

Another type of TCL which could be found in smart homes is HVAC. This resource is also a load that its power consumption could be changed for flexibility services. In recent decades, the brand new models of HVAC are introduced that are more flexible. These HVACs could adjust their power consumption to an exact value by using power electronic devices (i.e., inverter). This ability could be quite advantageous when a specific amount of flexibility is required from the smart homes or communities.

## 2.4  FLEXIBILITY PROVISION BY SMART HOMES AND COMMUNITIES

As previously mentioned, smart homes have some controllable appliances and storage-based resources as FERs, enabling them to provide flexibility services for the system operators, including DSOs and TSOs. DSOs can procure local services from smart homes. These services can be related to voltage control and congestion management of the distribution network. However, for providing TSO-level services, smart homes need to be aggregated in order to be able to provide system-wide services. System-wide services are mainly referred to those utilized to control the frequency of the system. Figure 2.5 illustrates the potent flexibility services that can be procured from smart homes and communities.

### 2.4.1  LOCAL FLEXIBILITY PROVISION

Distribution feeders in rural areas, especially those within weak electrical systems and the high number of customers, are often operated close to their voltage and thermal

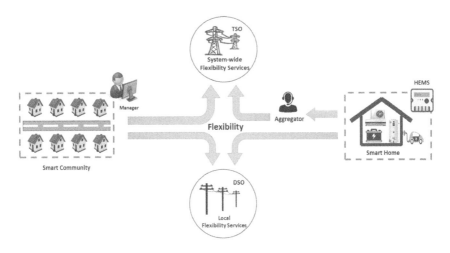

**FIGURE 2.5** Potent services that can be provided by smart homes and a community.

limits. The excessive number of distributed generation units connected to distribution networks may cause violation of the physical capacity constraints. These capacity constraints can be defined as $P^{\text{generation}} - P^{\text{load}} < P^{\text{max}}$ and $P^{\text{load}} - P^{\text{generation}} < P^{\text{max}}$, where $P^{\text{generation}}$ is the power produced and $P^{\text{load}}$ is the amount consumed. Accordingly, with high number of DGs, constraint $P^{\text{generation}} - P^{\text{load}} < P^{\text{max}}$ may be violated, which in turn jeopardizes the security of the distribution network. Another situation can happen for the network with a huge number of electric vehicles which are charging simultaneously. In this situation, constraint $P^{\text{load}} - P^{\text{generation}} < P^{\text{max}}$ regarding the feeder thermal capacity may be exceeded. In addition to capacity constraints, the future DSOs should deal with serious challenges related to voltage control since DGs which are located close to the loads may increase voltage at these nodes if simultaneously generation is very high and load is very low [33].

Currently, DSOs perform some costly actions in order to alleviate the congestion of feeders and control voltages in the distribution network. Changing the set points of on-load tap changer (OLTC) transformers, re-dispatching the generation units, utilizing voltage regulators, capacitor banks, and static voltage compensators are the actions taken by the current DSOs in order to operate the network. However, these methods can be costly and may fail to effectively operate the distribution networks with high injection of intermittent power [34].

Demand-side flexible energy resources such as smart homes are able to provide an alternative solution instead or in addition to the current DSO actions. In fact, smart homes have the capability to reshape their consumption and curtail their production according to the requests sent by the DSO. In return, they will receive monetary benefits.

In order to find the optimal amount of flexibility, DSO needs to solve an optimization problem taking into account the power flow equations in the distribution network. Afterward, the DSO finds the amount of downward and upward flexibility which is required for each node. These upward and downward flexibility need can be provided by the flexible energy resources located at those specific nodes. Smart homes and residential energy communities are thus the potential flexible resources capable of providing flexibility services. They can provide the required flexibility through controlling their appliances or utilizing their storage-based resources. If the DSO needs downward flexibility, the smart homes can charge their EVs and batteries or switch off their appliances such as washing machine. If the DSO requests for upward flexibility, it can discharge its batteries or curtail its curtailable loads.

An energy community can consist of a group of smart homes that voluntarily join the community [29]. The members of a community can be smart consumers or smart pro-active consumers which are called prosumers. Energy communities might have some shared resources in order to increase flexibility and self-sufficiency of the community such as a bulk energy storage system, a PV system, as well as a wind turbine. The energy community can be also a potent resource for providing local flexibility services. In this way, the manager of the community can control the resources within the community aiming to maximize the total profits of the members.

### 2.4.2 SYSTEM-WIDE FLEXIBILITY PROVISION

The volatility and uncertainty resulted from intermittent renewable-based power injected in power systems has decreased the stability and flexibility of the systems. However, the TSO needs to increase the flexibility of the system through adjusting the operating point of the system and accommodate the variations and fluctuations resulted from uncertainties and variability of loads and renewable-based generation resources. The flexibility of the system is enhanced by using flexible energy resources [25]. Flexible energy resources can be located at different levels of the power system, including DSO-level (those located at distribution networks) as well as TSO-level (those connected to transmission networks). Currently, conventional generators are the only flexible energy resources utilized to provide system-wide flexibility services. System-wide flexibility services are mainly referred to as the services related to controlling the frequency of the whole system.

In order to increase the flexibility of power system, flexible energy resources connected to distribution networks should also be able to provide system-wide flexibility services. In this regard, the aggregated smart homes and distribution-network-located energy communities are potent resources which can provide the TSO with frequency services.

Frequency services mostly include reserves that help TSOs to maintain the frequency of the whole system at a predefined threshold. The reserve services can be different based on the characteristic of the system and the current state of the system.

In Finland, as an example of Nordic countries, frequency reserve services are divided into primary reserves that are called "frequency containment reserve" (FCR), the secondary reserves that are called "automatic frequency restoration reserve" (aFRR), and tertiary reserves that are named as "manual frequency restoration reserve" (mFRR). These reserve services are requested based on the need of the system. However, the primary reserve, FCR, should be procured all the time. This reserve is required to automatically respond to spontaneous frequency deviations. FCR may be adopted only for normal operation of the system which is called (FCR-N), or be deployed in disturbance condition which is called (FCR-D). Moreover, the aFRR is used to restore the frequency automatically. However, the manual FRR is just applied in case of outages, or some situations associated with unexpected sustained activation of aFRR [35]. In addition to the mentioned services, a new kind of flexibility service was introduced in Finland which is called "fast frequency reserve" (FFR). This kind of service is deployed to handle rapid frequency fluctuations in low inertia situations [29].

As previously mentioned, currently, fuel-based generators are the main resources providing reserves for the TSO. However, distribution-network-connected resources such as aggregated smart homes and communities are also capable to provide different types of reserves and frequency services. With the growing production of renewable-based resources in the future, the participation of distribution-network-connected resources as well as transmission network resources is of vital necessity in order to increase the flexibility of the system.

The flexibility of smart homes can be aggregated and be utilized to provide different types of reserve services. Each reserve service has its own reserve market with its specific technical consideration. Accordingly, a smart home selected as an FCR provider should have a resource or a combination of resources that can provide upward and downward flexibility. The direction of the flexibility is mainly determined in real time based on the instant need of the TSO. In other words, the TSO chooses the type of flexibility service that will be activated in real time. This request can be sent to the resources through its control system. The resource (e.g., the smart home) should be able to automatically react to the request of the TSO.

The HEMS of a smart home can choose the most beneficial services that the home can provide for the grid. In this regard, there exists a variety of system-wide and local services that the smart home is able to provide for the TSO and DSO. The HEMS can run an optimization problem to select the services that help the household maximize its profits (or minimize its costs). However, the technical constraints of each service should be taken into consideration as well. It is worth mentioning that a household, itself, cannot separately participate in reserve markets since it does not have enough capacity to take part solely in the market. Thus, the smart homes are required to be aggregated by an aggregator so as to be able to contribute to the provision of system-wide services.

When it comes to an energy community, a community manager via a centralized energy management system controls the flexible resources of the community. These flexible resources can be the shared resources which are centrally controlled or those which belong to the household as members of the community. In this way, the community manager aims to maximize the profits of all members by choosing the most appropriate flexibility services that can be provided for DSO and TSO.

## 2.5 CASE STUDY

In the case study, we consider an energy community which consists of two 200kWh/50kW batteries as well as five similar electric vehicles with a capacity of 40 kWh as flexible energy resources. PEVs are considered to be charged with the constant power which is equal to 5 kW. It is assumed that PEVs should be charged just early in the morning, i.e., from 1:00 to 8:00.

The community aims to provide upward and downward services for the grid operators. Thus, the community is assumed to find the capacity that can be offered for grid services. In this regard, the objective function can be maximizing the profits achieved from the provision of upward and downward flexibility.

$$\max_{\text{Flex}_t^{up}, \text{Flex}_t^{dn}} \pi_t^{\text{flex},up} \text{Flex}_t^{up} + \pi_t^{\text{flex},dn} \text{Flex}_t^{dn} \tag{2.7}$$

Where, $\text{Flex}_t^{up}$ is the upward capacity of the community devoted to upward flexibility services at time slot $t$ and $\text{Flex}_t^{dn}$ denotes the downward flexible capacity of the community devoted to downward flexibility services at time slot $t$. In addition, $\pi_t^{\text{flex},up}$ and $\pi_t^{\text{flex},dn}$ are the prices of upward and downward flexibility, respectively. Regarding the proposed case study, the prices of upward and downward flexibility are extracted from balancing energy markets of Finland, similar to those considered in ref. [29].

Charging capacities of batteries and PEVs are considered as downward capacities of the community while discharging batteries are regarded as the upward flexible capacity of the community. The related balancing constraints are denoted with (2.8) and (2.9):

$$\sum_b P_{b,t}^{ch} + N_t^{ev} P^{ev} = \text{Flex}_t^{dn} \tag{2.8}$$

$$\sum_b P_{b,t}^{dis} = \text{Flex}_t^{up} \tag{2.9}$$

In constraint (2.8), $N_t^{ev}$ denotes the number of PEVs which are being charged at time slot $t$. $P^{ev}$ is the constant power used for charging EVs. In addition, $P_{b,t}^{ch}$ is the charging power of battery $b$ at time slot $t$. Accordingly, constraint (2.8) states that the downward flexibility ($\text{Flex}_t^{dn}$) should be provided by charging power of batteries and PEVs of the community. However, according to constraint (2.9), only discharging power of batteries is taken into account as upward flexible capacity of the community, which is denoted by $\text{Flex}_t^{up}$.

Moreover, the constraints related to the charging/discharging power and energy of batteries and PEVs were introduced in the previous section. As a result, the optimization problem consists of an objective function (2.7) as well as constraints (2.8), (2.9), and (2.1)–(2.6). The obtained optimization problem is a simple linear programming problem which can be easily solved.

According to the above-mentioned equations, the upward flexible and downward flexible capacities of the community for 24 hours are estimated based on its flexible energy resources including its PEVs and batteries. The obtained upward and downward flexibility that the community will provide is depicted in Figure 2.6. It is worth mentioning that the case study aims to provide TSO-level flexibility services. However, this flexibility provision method can be also utilized for DSO-level flexibility. As can be seen in the figure, the community can provide both upward and downward flexibility in time slots 2:00–6:00, 8:00–16:00, and 19:00 using its batteries and PEVs. However, in the other time slots, the community is just able to provide either upward or downward flexibility.

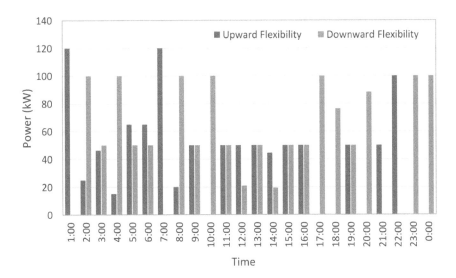

**FIGURE 2.6** The amount of upward and downward flexible capacities provided by the considered community.

The variation in the state-of-charge (SOC) of batteries and PEVs of the community are illustrated in Figures 2.7 and 2.8, respectively.

Figure 2.7 states that the variation of SOC for two batteries is in opposite directions in the most of time slots. It means that, during time slots, battery 1 is charging and battery 2 is discharging. In this way, the community would be able to provide both upward and downward flexibilities simultaneously.

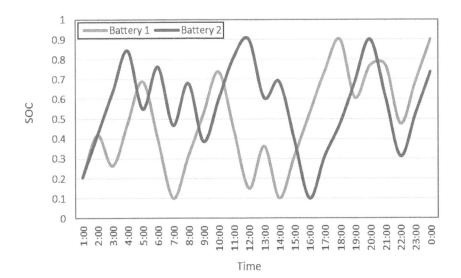

**FIGURE 2.7** The SOC of batteries in the community.

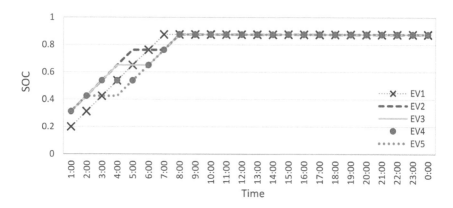

**FIGURE 2.8** The SOC of PEVs' batteries in the community.

Figure 2.8 shows the SOC variation of the batteries of PEVs. The downward flexible capacities of PEVs are mainly deployed in early morning to satisfy the constraint applied by the PEV owners since after these hours, the PEV owners do not want their vehicles to be charged. Thus, their flexibility capabilities are highly exploited in our model.

## 2.6  SUMMARY

The impacts of decentralization as well as the high penetration of renewable energy resources into power systems have created the necessity of changing the way of network management for power system operators in recent years. Although the effects of renewables at all levels of the system can be seen, it is more likely to become an issue, especially in distribution networks, due to the ever-increasing utilization of small-scale renewable generation units (e.g., domestic PV panels).

Distribution-network-located flexible energy resources such as smart homes and energy communities can increase flexibility of power system. Smart homes and communities can reshape their consumption and production through their flexible energy resources, including battery energy storage systems, electric vehicles, or their other controllable appliances. By utilizing these resources, they are able to contribute to enhancing the grid's flexibility. In this way, for example, PEVs and batteries should be charged if the system needs downward flexibility, while the batteries should be discharged during time slots when the system operator requests for upward flexibility.

Different kinds of DR programs can incentivize customers to actively respond to the system requests. These programs can indirectly control the flexible energy resources such as direct load programs or can indirectly affect the load of the customers. However, in order to receive an acceptable response, customers need to be equipped with decent ICT technologies and energy management systems.

Both customers and system operators can benefit from DR programs and flexibility services procured from demand-side resources. DSOs can procure local services from smart homes and energy communities connected to distribution networks. These services can be related to voltage control and congestion management of the distribution network. However, for providing TSO-level services, smart homes need to be aggregated in order to be able to provide system-wide services. System-wide services are mainly referred to those utilized to control the frequency of the system. In this way, the resources need to be automatically controlled.

## REFERENCES

[1] G. Migliavacca, M. Rossi, H. Gerard, M. Džamarija, S. Horsmanheimo, C. Madina, I. Kockar, G. Leclecq, M. Marroquin, and H. Svendsen, "TSO-DSO coordination and market architectures for an integrated ancillary services acquisition: the view of the SmartNet project," *CIGRE*, vol. 5, no. 306, pp. 1–10, 2018.

[2] H. Khajeh, M. Shafie-khah, and H. Laaksonen, "Blockchain-based demand response using prosumer scheduling," *Blockchain-based Smart Grids*, Elsevier, 2020, pp. 131–144.

[3] T. K. Chau, S. S. Yu, T. Fernando, and H. H.-C. Iu, "Demand-side regulation provision from industrial loads integrated with solar PV panels and energy storage system for ancillary services," *IEEE Trans. Industr. Inform.*, vol. 14, no. 11, pp. 5038–5049, 2017.

[4] P. H. Divshali and C. Evens, "Stochastic bidding strategy for electrical vehicle charging stations to participate in frequency containment reserves markets," *IET Gener. Transm. Distrib.*, vol. 14, no. 13, pp. 2566–2572, 2020.

[5] K. Oikonomou, M. Parvania, and R. Khatami, "Coordinated deliverable energy flexibility and regulation capacity of distribution networks," *Int. J. Electr. Power Energy Syst.*, vol. 123, p. 106219, 2020.

[6] J. Ali, S. Massucco, and F. Silvestro, "Distribution level aggregator platform for DSO support—Integration of storage, demand response, and renewables," *Front. Energy Res.*, vol. 7, p. 36, 2019.

[7] F. Lezama, J. Soares, B. Canizes, and Z. Vale, "Flexibility management model of home appliances to support DSO requests in smart grids," *Sustain. Cities Soc.*, vol. 55, p. 102048, 2020.

[8] V. Davatgaran, M. Saniei, and S. S. Mortazavi, "Smart distribution system management considering electrical and thermal demand response of energy hubs," *Energy*, vol. 169, pp. 38–49, 2019.

[9] A. Abessi, A. Zakariazadeh, V. Vahidinasab, M. S. Ghazizadeh, and K. Mehran, "End-user participation in a collaborative distributed voltage control and demand response programme," *IET Gener. Transm. Distrib.*, vol. 12, no. 12, pp. 3079–3085, 2018.

[10] M. Beaudin and H. Zareipour, "Home energy management systems: A review of modelling and complexity," *Renew. Sustain. Energy Rev.*, vol. 45, pp. 318–335, 2015.

[11] K. O. Aduda, T. Labeodan, W. Zeiler, G. Boxem, and Y. Zhao, "Demand side flexibility: Potentials and building performance implications," *Sustain. Cities Soc.*, vol. 22, pp. 146–163, 2016.

[12] J. Han and M. A. Piette, "Solutions for summer electric power shortages: Demand response and its applications in air conditioning and refrigerating systems," 2008.

[13] H. Khajeh, A. A. Foroud, and H. Firoozi, "Robust bidding strategies and scheduling of a price-maker microgrid aggregator participating in a pool-based electricity market," *IET Gener. Transm. Distrib.*, vol. 13, no. 4, pp. 468–477, 2019.

[14] P. Palensky and D. Dietrich, "Demand side management: Demand response, intelligent energy systems, and smart loads," *IEEE Trans. Industr. Inform.*, vol. 7, no. 3, pp. 381–388, 2011.

[15] W. W. Hogan, "Time-of-use rates and real-time prices," *John F. Kennedy Sch. Gov. Harvard* University, 2014.

[16] M. H. Albadi and E. F. El-Saadany, "A summary of demand response in electricity markets," *Electr. Pow. Syst. Res.*, vol. 78, no. 11, pp. 1989–1996, 2008.

[17] X. Yan, Y. Ozturk, Z. Hu, and Y. Song, "A review on price-driven residential demand response," *Renew. Sustain. Energy Rev.*, vol. 96, pp. 411–419, 2018.

[18] V. Venizelou, N. Philippou, M. Hadjipanayi, G. Makrides, V. Efthymiou, and G. E. Georghiou, "Development of a novel time-of-use tariff algorithm for residential prosumer price-based demand side management," *Energy*, vol. 142, pp. 633–646, 2018.

[19] H. T. Haider, O. H. See, and W. Elmenreich, "A review of residential demand response of smart grid," *Renew. Sustain. Energy Rev.*, vol. 59, pp. 166–178, 2016.

[20] F. Shariatzadeh, P. Mandal, and A. K. Srivastava, "Demand response for sustainable energy systems: A review, application and implementation strategy," *Renew. Sustain. Energy Rev.*, vol. 45, pp. 343–350, 2015.

[21] A. Conteh, M. E. Lotfy, K. M. Kipngetich, T. Senjyu, P. Mandal, and S. Chakraborty, "An economic analysis of demand side management considering interruptible load and renewable energy integration: A case study of Freetown Sierra Leone," *Sustainability*, vol. 11, no. 10, p. 2828, 2019.

[22] H. Firoozi and H. Khajeh, "Optimal day-ahead scheduling of distributed generations and controllable appliances in microgrid," in *2016 Smart Grids Conference (SGC)*, 2016, pp. 1–6.

[23] M. H. Imani, P. Niknejad, and M. R. Barzegaran, "The impact of customers' participation level and various incentive values on implementing emergency demand response program in microgrid operation," *Int. J. Electr. Power Energy Syst.*, vol. 96, pp. 114–125, 2018.

[24] G. Strbac, "Demand side management: Benefits and challenges," *Energy Policy*, vol. 36, no. 12, pp. 4419–4426, 2008.

[25] H. Khajeh, H. Laaksonen, A. S. Gazafroud, and M. Shafie-Khah, "Towards flexibility trading at TSO-DSO-customer levels: A review," *Energies*, vol. 13, no. 1, pp. 165, 2020.

[26] M. H. Albadi and E. F. El-Saadany, "Demand response in electricity markets: An overview," in *2007 IEEE power engineering society general meeting*, 2007, pp. 1–5.

[27] A. Asadinejad and K. Tomsovic, "Optimal use of incentive and price based demand response to reduce costs and price volatility," *Electr. Power Syst. Res.*, vol. 144, pp. 215–223, 2017.

[28] H. Khajeh, H. Firoozi, H. Laaksonen, and M. Shafie-khah, "A New Local Market Structure for Meeting Customer-Level Flexibility Needs," in *2020 International Conference on Smart Energy Systems and Technologies (SEST)*, 2020, pp. 1–6.

[29] H. Firoozi, H. Khajeh, and H. Laaksonen, "Optimized Operation of Local Energy Community Providing Frequency Restoration Reserve," *IEEE Access*, vol. 8, pp. 180558–180575, 2020.

[30] M. A. Quddus, M. Kabli, and M. Marufuzzaman, "Modeling electric vehicle charging station expansion with an integration of renewable energy and vehicle-to-grid sources," *Transp. Res. Part E Logist. Transp. Rev.*, vol. 128, pp. 251–279, 2019.

[31] H. Mehrjerdi and E. Rakhshani, "Vehicle-to-grid technology for cost reduction and uncertainty management integrated with solar power," *J. Clean. Prod.*, vol. 229, pp. 463–469, 2019.

[32] W. Mendieta and C. A. Cañizares, "Primary frequency control in isolated microgrids using thermostatically controllable loads," *IEEE Trans. Smart Grid*, 2020.

[33] H. Zoeller, M. Reischboeck, and S. Henselmeyer, "Managing volatility in distribution networks with active network management," 2016.

[34] I. Wasiak, R. Pawelek, and R. Mienski, "Energy storage application in low-voltage microgrids for energy management and power quality improvement," *IET Gener. Transm. Distrib.*, vol. 8, no. 3, pp. 463–472, 2013.

[35] G. De Zotti, S. A. Pourmousavi, H. Madsen, and N. K. Poulsen, "Ancillary services 4.0: A top-to-bottom control-based approach for solving ancillary services problems in smart grids," *IEEE Access*, vol. 6, pp. 11694–11706, 2018.

# 3 A System-of-Systems Planning Platform for Enabling Flexibility Provision at Distribution Level

*H. Arasteh[1], S. Bahramara[2], Z. Kaheh[3],*
*S. M. Hashemi[1], V. Vahidinasab[4],*
*P. Siano[5], and M. S. Sepasian[6]*
[1]Niroo Research Institute, Tehran, Iran
[2]Department of Electrical Engineering,
Sanandaj Branch, Islamic Azad University, Sanandaj, Iran
[3]Tarbiat Modares University, Tehran, Iran
[4]Shahid Beheshti University, Tehran, Iran and Newcastle
University, Newcastle upon Tyne, United Kingdom
[5]University of Salerno, Fisciano (SA), Italy
[6]Shahid Beheshti University, Tehran, Iran

## CONTENTS

## 3.1    INTRODUCTION

Decarbonization policies have globally changed the landscape of electrical energy industry and necessitate rapid integration of renewable energy sources (RESs). In such an atmosphere, significant decrement in renewable technologies has caused rapid growth in the share of these intermittent but clean and affordable RESs in the energy production basket at all levels of the power sector, especially the distribution level. Several countries such as the UK, Germany, Sweden, and Denmark have targets to decarbonize their electricity systems by 2050. This changing paradigm in energy production calls for a new planning approach to respond to these intermit generations by flexibility provision. The variable power generation of the RESs increases the ramp rate of the system's net load. For instance, the simultaneous growth in demand and the reduction in the power generation of photovoltaic (PV) systems around the evening causes 13 GW ramp rate requirements over three hours for the California independent system operator (CAISO). Indeed, because of the uncertain generation pattern of RESs, the way to handle the rapidly increasing uncertainties and variability in power system planning and operation should be studied. In such a framework, the independent system operator (ISO) needs flexible ramp products (FRPs) to meet its required ramp. Flexibility is introduced as one of the main concerns of future renewable-dominated power systems. Electrical energy storage systems (ESSs) and demand response (DR) programs are introduced as the important energy resources in distribution systems to provide the FRPs to the ISO. Therefore, the distribution system expansion planning (DEP) problem changes in the presence of RESs and flexible energy resources. For this purpose, new decision-making frameworks are required for the DEP problem to meet demand in future.

This chapter aims to study the problem of DEP considering the flexibility providers. For this purpose, after introducing the traditional framework for the DEP problem, the effect of the integration of the flexibility resources on this problem is described in detail. In this study, DR providers (DRPs) are introduced as the flexibility service providers, which will play an important role besides other independent entities to satisfy the increasing electrical demand. In the presence of the DRPs, the objective function, technical constraints, load models, and uncertain parameters changed so that new approaches are required to model this framework. Moreover, there are different independent and autonomous entities in distribution systems whose behaviors should be modeled in the DEP problem. It means several independent entities are playing their roles in the distribution systems with their objectives (that may conflict with the objectives of other players), while they can have interoperability with each other. Hence, an efficient modeling architecture is essential to optimally model such interactions. In this chapter, a system-of-systems (SoS) approach is proposed as an efficient framework to model the behavior of these independent entities. The SoS is an approach for modeling the collaboration of different entities with distinct objective functions, which are able to cooperate. Independent systems under the SoS framework aim at achieving their own goals. The outcome of the SoS is an optimum solution that will determine the strategies of all the entities in an optimal manner.

The rest of the chapter is organized as follows. An overview of the DEP problem and the flexibility in the power system is presented in section 3.2. The traditional approaches for the DEP problem are described in section 3.3. The DEP problem in the presence of flexible energy resources is described in section 3.4. Proposing the SoS framework to model the DEP problem is the aim of section 3.5. Finally, section 3.6 presents the status quo and outlook.

## 3.2   OVERVIEW

The power system flexibility is an emerging concept affecting the system's planning and operational aspects. Although the extension of different clean energy resources through the power systems is very attractive in the economic and environmental aspects, it creates serious problems for the planners and operators of system. Increasing the penetration of the wind turbines (WTs) and PV panels as the main clean energy resources raises a question: How can it be possible to guarantee the secure operation of the system when a high amount of load is supplied by the uncertain generation units? In the bulk power system, providing fast start units and units with high ramping performance is a usual solution to create enough flexibility against the uncertainties of wind and PV resources. In the power distribution systems, different flexibility resources such as ESSs, DR resources, and distributed generation (DG) technologies are capable to provide flexibility. The expansion of power distribution system considering flexibility indices is very different from the conventional expansion viewpoints. Different flexibility providers have different performances, and the system planner should economically manage these resources to meet the desired flexibility level of the distribution system.

### 3.2.1   Overview of the Electric Distribution Network Expansion Planning

Power distribution system planning is a long-term decision-making to determine the expansion procedure of the resources and network. Three main planning methods are used in DEP that are dynamic, static, and pseudo-dynamic methods [1]. In the dynamic approach, expansion planning of the system in all years is performed simultaneously. In the static expansion method, only the horizon year is considered in the system expansion procedure, and there are no exact expansion decisions for the other years. In the pseudo-dynamic method, the expansion planning problem is separately solved for different years. Although the pseudo-dynamic approach is not as optimal as the dynamic approach, its computational burden is lower.

The power distribution expansion problem is a complex optimization model. So, in addition to the usual mathematical programming approaches, methods such as genetic algorithm (GA) [2, 3], learning automata-based algorithm (LAA) [4], immune system inspired algorithm (ISIA) [5], improved harmony search algorithm [6], and multi-objective particle swarm optimization (MOPSO) [7] have been used to solve this optimization model.

The distribution company (DISCO) is the main planner of the system and its goal is to provide a reliable network with the minimum cost. However, different goals can be considered as the objective function of DEP. In addition to the cost-related objectives, minimum power loss [8], minimum voltage drops [9], minimum emission [10], and maximum reliability of the systems [11, 12] are some of the popular objective functions of DEP. These are used in single-objective or multi-objective optimization frameworks. Although the power loss can be directly included in the objective function, usually its related cost is considered along with the other cost terms. The conventional generation technologies that consume fossil fuels are producing pollutants. Considering the international rules such as the Kyoto protocol, emission of the dangerous pollutants should be constrained in the power generation portfolio. Renewable power generation technologies are an appropriate solution for this purpose. Adding pollution penalty to the objective function may increase the renewable capacity in the system expansion problem. The distribution system should benefit from a high-reliability level to ensure the continuity of load supplying. Usually, the objective function includes the reliability cost using the value of lost load (VoLL) and the load shedding variable. When the system planner wants to consider all of the mentioned objectives, he/she should use a multi-objective framework in which different methods can be applied to prioritize objectives [13, 14].

Power distribution system expansion planning is a complex optimization problem, including integer and continuous decision variables and non-linear models. DEP includes short-term behavior of the system during the long-term horizon of the expansion planning by modeling the operational constraints of the components and network. The permissible loading range of the feeders, transformers, DGs, and ESSs are constrained in DEP. The system reconfiguration plans can be modeled in the planning problem of the distribution systems [15]. Also, bus voltages and line flows are calculated using the power flow models. DEP is a large-scale optimization problem, including a lot of decision variables of the system expansion during a long-term horizon. Therefore, the usual formulation of AC power flow, which is nonlinear and non-convex, is not appropriate for DEP. The methods such as linear distflow [16] and second-order cone programming (SOCP) [17] convexify the AC power flow equations and can be employed in DEP. In addition to the technical constraints of power distribution systems, increasing interconnection with the other systems more complicates the DEP. In other words, in the concept of smart city [18], different sectors including electricity infrastructure, transportation systems, and data networks should be coordinated. For example, increasing uptake of electric vehicles (EVs) as one of the effective components of the distribution system calls for a distinguished approach to DEP. Increasing penetration of EVs interconnects DEP with the traffic management system [19], and the EV charging stations should be located considering both the traffic flow and the distribution system conditions.

Although DISCO is mainly responsible for system planning, the other entities also participate in this process. The private investors (PIs) are the profit maximizer entities investing to install different components of the system. DRPs are the effective entities that employ demand-side resources and integrate them to provide the required services to the distribution system [20]. In practice, system expansion is the result of interaction of different entities [21]. In such a condition, the centralized optimization, in which DISCO independently makes decisions of the DEP, is replaced by the decentralized programming. In decentralized programming, several independent agents with different goals participate with each other to gain their desired profits [22]–[24]. Methods such as multi-agent systems [25] are applied for such structures.

System planning is a long-term decision-making problem containing many parameters. Most of the input parameters of the scheduling models are provided by the forecasting methods (parameters such as load profile, electricity price, output power of WTs, and PV panels). According to the mentioned uncertainties of input data, an appropriate optimization model should be applied in the decision-making process to ensure that the determined decision variables comply with the actual condition of the system. Methods such as robust optimization [26], stochastic programming, information gap decision theory [27], and fuzzy logic are popular algorithms for modeling the uncertainties.

### 3.2.2 OVERVIEW OF THE POWER SYSTEM FLEXIBILITY

According to ref. [28], power system flexibility is the ability to manage uncertainty and variability of the load and supply during different time-scales in a reliable and cost-efficient manner. Load variations during the operating hours should be followed by power resources. In the case of a severe change of the load, only some facilities such as high ramp and fast start generation units and demand-side resources can follow the variations. In the systems with high penetration of the wind and solar-based generation units, the net load, which is calculated by subtracting the renewable generation from the total load, has a severe variation during the operation horizon. So, such a system requires high flexibility. The system flexibility is affected by supply-side resources, demand-side resources, and infrastructure-related aspects.

From the viewpoint of flexibility, generation units, as the supply-side resources, are divided into base-load units and load-following units [29]. Base-load power plants usually produce their nominal power during the operation horizon. Ramping or shutting down of the base-load power plants is undesired from the economic and technical viewpoints. In return, the load-following power plants balance the supply and demand during the operation horizon. They have a high ramp rate and can be rapidly started. In the power systems with high penetration of renewable resources, system stability is very challenging [30]. So, many ancillary services are employed to support the system operation during different conditions, to increase the system flexibility in several time-scales (from one millisecond to several months [31]). For example, the primary and secondary frequency control resources guarantee the balance of load and supply in the time-scales from seconds to minutes. Any power mismatch between demand and supply is covered by the spinning reserve of the online generation units. Also, the non-spinning reserve is provided by the fast start generation technologies to support the demand-supply power balance. The other resources of flexibility are the energy storage units. They store the generated energy of the renewable resources during the off-peak hours and deliver it during the peak periods. Besides, they can be strategically charged and discharged targeting the economic goals and benefiting from the electricity price differences in the operation hours.

DR programs integrate the responsive loads to participate in the system operation and change the load pattern [32]. DR includes several programs such as load shifting, valley filling, load growth, and peak shaving [33, 34]. DR can effectively increase the flexibility of system through the programs of load shifting and peak shaving. In addition to the supply-side and demand-side resources, flexibility can be provided by other methods. For example, local supplying of loads using microgrids (MGs), which are the small power systems containing DGs and ESSs, increases the system flexibility from structural aspects [35]. Also, using the remedial reconfiguration to employ the adjacent feeders in critical conditions reduces the load downtime, and improves flexibility [36]. Although the remedial reconfiguration is not considered as a short-term decision variable of the normal operation, it can be considered as a flexibility resource since it is an appropriate solution to deal with the unexpected outage of the lines as an uncertain factor. In such conditions, the interrupted loads are restored by a switching action on the adjacent feeder. In some researches, some of the loads' vulnerability indices have been introduced against the power interruption. In critical conditions, the loads can be coordinately switched on and off based on their vulnerability indices in order to minimize the load damage [37]. This switching action can be considered as a flexibility resource because it can be a good solution to deal with the unexpected outage of the feeders or transformers. Another structure-related solution for flexibility is using multi-carrier energy system (MCES) [38] that simultaneously improves the flexibility from the technical and economic viewpoints. MCES, containing several energy carriers, can economically and reliably supply the load. MCES is composed of a number of energy hubs (EHs). An EH contains several components that converting different energy carriers into each other. For example, combined heat and power (CHP) units convert the natural gas to electrical and thermal energy carriers to serve the related loads. Therefore, when the electrical network is not available, the CHP unit can serve the electrical load by consuming natural gas. So, such a structure will increase flexibility.

Including a flexibility concept in the optimization problems of power system operation and planning requires flexibility indices. The distance between the system variables and their upper and lower bounds can be considered as a flexibility index [36]. Depending on the ramping requirements of the system and the ramping capability of the resources, it may be impossible to follow the variations of the net load during some of the operation states. Such an analysis can be used to define a flexibility metric [39]. Renewable energy curtailment can be considered as another flexibility measure. Inflexible systems have to curtail renewable power in situations such as off-peak periods because they cannot shut down the base-load power plants. Although it is a technically feasible solution, it is undesired in the economic viewpoint [40, 41]. The flexibility concept can be analyzed in comparison to the other known concepts such as system security. A secure operation guarantees that the system can tolerate the single or double contingencies of the lines and generation units. Usually, some of the preventive and corrective actions are performed to provide security [42]. The preventive actions move the normal operating point of the system away from the

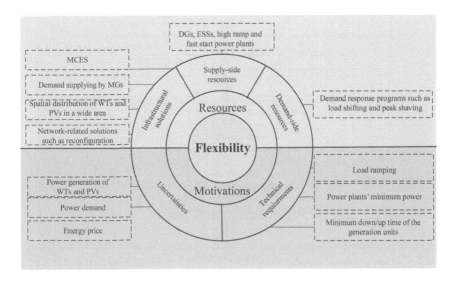

**FIGURE 3.1** Motivations and resources of the power system flexibility.

optimal condition. By increasing flexibility, the system operator can secure the system using corrective actions [43].

In the conventional structure of the long-term planning of the power systems, the systems' short-term behaviors have been modeled by the simple operational states, in which, a single load level [2] or several load levels are selected based on the load duration curve (LDC) [44, 45]. This structure is not appropriate to consider the flexibility constraints such as the ramping capability because it removes the chronological order of the operation states. In return, in some studies, some daily load profiles are considered instead of only one value of the load, and then the time duration of occurring the load profiles are presented [46]. By using the representative days, containing the hourly profile of the load, the planning horizon is divided into small time sections in which short-term flexibility constraints are considered [47]–[49]. Although the system requires higher flexibility in case of high penetration of WTs and PV panels, the spatial distribution of these resources in a wide area effectively reduces the rapid change of their output power [50].

A graphical overview of the motivations, resources, and features of the power system flexibility is presented in Figure 3.1.

## 3.3 TRADITIONAL APPROACHES TO DISTRIBUTION EXPANSION PLANNING

The traditional DEP problem aims to meet the demand in a specific time horizon in the future by installing new substations and feeders, reinforcing and/or replacing the existing substations and feeders. In this section, the traditional DEP problem is described and a brief review is done on different models and solution approaches which are used in the literature.

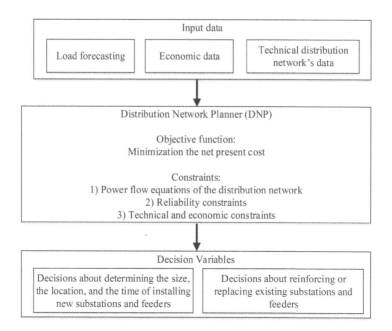

**FIGURE 3.2** Decision-making problem of the traditional DEP problem.

### 3.3.1 DESCRIPTION OF TRADITIONAL DISTRIBUTION EXPANSION PLANNING PROBLEM

The decision-making problem in the traditional DEP problem is shown in Figure 3.2. The required input data of this problem include (1) load forecasting: the amount of forecasted load for the specific period in the future, (2) economic data: the investment, replacement, operation, and maintenance cost of substations and feeders, and (3) technical distribution network's data including the power flow required data. The main objective function of the DEP problem is the net present cost (NPC) which is solved considering the technical constraints of the distribution network. The decision variables of the DEP problem which are obtained from solving the problem are the optimal size and location of the new substations and feeders as well as the decisions on reinforcing and/or replacing the existing substations and feeders.

### 3.3.2 MATHEMATICAL MODELING

The objective function of the traditional DEP problem consists of several terms, including the investment cost of the new substations and feeders, the reinforcement and replacement cost of the existing substations and feeders, the energy loss cost, the operation and maintenance cost, the cost of power purchased from the main grid, and the reliability cost. All or some of these costs are considered in the objective function of the DEP problem in the literature. From the objective function point of view, two types of problems are proposed in literature consisting of single-objective and multi-objective forms. In the single-objective problems, NPC is mainly used in the DEP problem. Moreover, in the multi-objective problems, NPC, reliability, and power loss are used. The aim of this sub-section is to propose a simple model for the DEP problem which is used in the previous studies. This model aims to install new substations and feeders as well as to reinforce the existing substations and feeders in the distribution network to meet the active peak load at a specific time in the future as proposed in Eqs. (3.1)–(3.10).

$$\text{Minimize NPC} = C^F + C^S + C^L + C^P \tag{3.1}$$

where, $C^F$, $C^S$, $C^L$, and $C^P$ are the cost of upgrading the feeders, the cost of upgrading the substations, the energy loss cost, and the cost of purchased energy from the main grid. In the following equations, $\varsigma^\upsilon(y) = \left[ \dfrac{d}{1-(1+d)^{-n(.)}} \displaystyle\sum_{k=y}^{Y} \dfrac{1}{(1+d)^k} \right]$
is the factor to convert the investment costs (related to the "$\upsilon^{th}$" action) in different years to the first year by considering the lifetime of the components. Here, $d$ is the discount rate, and $n(.)$ is the lifetime of the component "$(.)$".

$$C^F = \sum_{y=1}^{Y} \left\{ \sum_{cf}^{CF} \sum_{x}^{X} \left( \varsigma^i(y)\rho^i(cf,x,y)u^i(cf,x,y) + \varsigma^r(y)\rho^r(cf,x,y)u^r(cf,x,y) \right) \right\} \tag{3.2}$$

where, $y$ is the index of the years, $Y$ indicates the horizon planning year, $cf/CF$ are the index/set of the candidate feeders to be upgraded, $x$ is the index of feeder types, $X$ is the set of feeder types, $\rho^i(cf,x,y)$ is the cost to install the "$x^{th}$" feeder in year "$y$" to upgrade the "$cf^{th}$" feeder, $\rho^r(cf,x,y)$ is the cost to reinforce the "$cf^{th}$" feeder in year "$y$" by replacing the existing feeders by the feeder with the type of "$x$". Moreover, $u^i(cf,x,y)$ is an integer variable that is 1, if the candidate feeders "$cf$" is upgraded by installing the feeder with the type of "$x$" in year "$y$"; otherwise, it is 0. Also, $u^r(cf,x,y)$ is another integer variable that is 1, if the candidate feeders "$cf$" is reinforced with the feeder with the type of "$x$" in year "$y$"; otherwise, it is 0.

$$C^S = \sum_{y=1}^{Y} \left\{ \varsigma^S(y) \left( \sum_{s}^{S} \rho_s(y) \right) \right\} \tag{3.3}$$

where, $s$ and $S$ are the index and set of the candidate substations (new substations to be installed or existing substations to be upgraded), and $\rho_{s(y)}$ is the required investment costs in substation "$sth$" in year "$y$".

$$C^L = \sum_{y=1}^{Y} \left\{ \frac{1}{(1+d)^y} \left( 8760\,\text{LSF}(y)\,P_{\text{loss}}(y)\,\rho_M(y) \right) \right\} \tag{3.4}$$

where, $\rho_M(y)$ is the cost of energy losses in year "$y$". It is assumed that $P_{\text{loss}}(y)$ is the power loss of the system in the peak load in year "$y$" which is determined using Eq. (3.5). The power loss in the peak load is multiplied in the loss factor (LSF) to obtain the average power loss of the network and the resulted term is multiplied by 8760 to obtain the total energy loss in one year.

$$P_{\text{loss}}(y) = \sum_{\substack{i \in N \\ i \neq j}} \sum_{\substack{j \in N \\ i \neq j}} R_{ij}(y) I_{ij}^2(y) \quad , \quad I_{ij}(y) = \frac{V_i(y) - V_j(y)}{Z_{ij}(y)} \tag{3.5}$$

In Eq. (3.5), $i$ and $j$ are the indices of nodes, $N$ is the set of nodes, $V_i(y)$ and $V_j(y)$ are the voltage of nodes at peak load in year "$y$", $I_{ij}(y)$ is the feeder current from node $i$ to $j$, $Z_{ij}(y)$ and $R_{ij}(y)$ are the impedance and resistance of the line connecting node $i$ to $j$.

The cost of the purchased energy from the main grid could be formulated as Eq. (3.6).

$$C^P = \sum_{y=1}^{Y} \left\{ \frac{1}{(1+d)^y} \left( 8760 \, LF(y) \, P_h(y) \, \rho_e(y) \right) \right\} \tag{3.6}$$

where, $\rho_e(y)$ is the average price of purchasing energy from the market in the year "$y$", and $P_h(y)$ is the purchased power in the peak load in year "$y$" from the existing substation $h$. $LF$ is used to show the load factor. In Eq. (3.6), the purchased power in the peak period is multiplied by the load factor ($LF$) to obtain the average power purchased from the upstream and the resulted term is multiplied by 8760 to obtain the total purchased energy.

The proposed objective function of the DEP problem is solved considering several technical constraints which are described as follows:

- Voltage limitations: Equation (3.7) is used to model the voltage limitations of the distribution network. In this equation, $V_i^{\min}$ and $V_i^{\max}$ are the minimum and the maximum acceptable voltages of node $i$.

$$V_i^{\min} \leq V_i(y) \leq V_i^{\max} \qquad \forall i, y \tag{3.7}$$

- Distribution substation capacity: The amount of purchasing power from the main grid is limited to the maximum capacity of an existing substation in the distribution network as modeled in Eq. (3.8). $P_h^{\max}(y)$ is the maximum capacity of the existing substation in year "$y$".

$$P_h(y) \leq P_h^{\max}(y) \qquad \forall h, y \tag{3.8}$$

- Thermal capacity of distribution feeders: The thermal capacity limitation of the distribution feeders is formulated by Eq. (3.9). In this equation, $I_{ij}(y)$ and $I_{ij}^{\max}(y)$ are the amount of current flow in feeder connecting node $i$ to node $j$ and its maximum value, respectively.

$$\left| I_{ij}(y) \right| \leq I_{ij}^{\max}(y) \qquad \forall\, i, j, y \atop i \neq j \tag{3.9}$$

**TABLE 3.1**

**Comparison of Different Studies on the Traditional DEP Problem**

| Reference | Static/ Dynamic | Objective Function | Type of Load Modeling | Uncertainties | Test System | Solving Approach |
|-----------|-----------------|--------------------|-----------------------|---------------|-------------|------------------|
| [2] | Static | Cost | One level | No | 24-node | GA |
| [5] | Static | Cost | One level | Demand Energy tax | 23-node | ISIA |
| [4] | Static | Cost | One level | No | 62-node | LAA |
| [3] | Dynamic | Cost | One level | No | 16-node | GA |

- Active power balance limitation: The active power balance constraint in the distribution network's buses is modeled as Eq. (3.10) where the amount of power flow in feeder connecting node $i$ to node $j$ ($P_{ij}(y)$) is described in Eq. (3.11) [51] (it should be noted that $P_{ij}(y)$ denotes the transmitted power from bus "$j$" to "$i$", and $P_{ji}(y)$ is the transmitted power from bus "$j$" to "$i$"). The amount of the peak load is illustrated using $D_i(y)$ in this equation.

$$\left( \sum_{\substack{j \\ j \neq i}} \left\{ P_{ij}(y) - P_{\text{loss}}(y) \right\} - \sum_{\substack{j \\ j \neq i}} P_{ji}(y) \right) = D_i(y) \qquad \forall\, i,j,y \atop i \neq j \qquad (3.10)$$

$$P_{ij}(y) = \left( \frac{R_{ij}(y)}{Z_{ij}^2(y)} \right) (V_i^2(y) - V_j^2(y)) \qquad \forall\, i,j,y \atop i \neq j \qquad (3.11)$$

- Radial structure limitation: Regarding the solution methodology of the problem, the radial structure of the distribution network is satisfied.

### 3.3.3 BRIEF OVERVIEW OF THE PREVIOUS STUDIES

The studies which are done on the traditional DEP are reviewed from different aspects in Table 3.1. As described in this table, the main objective function of this problem is the total cost of system. Since the DEP problem is done to meet the peak demand of the distribution network in a specific period in the future, the demand is considered as the main parameter in this problem. In these studies, the peak load is only considered in the problem as proposed in ref. [2]. This type of load modeling is named as one level model in the literature. The demand and the market energy price are two main uncertain parameters in the traditional DEP problem.

## 3.4 INTEGRATION OF FLEXIBILITY RESOURCES INTO DISTRIBUTION EXPANSION PLANNING PROBLEM

Although the high penetration of the RESs in the power system can facilitate the transition of these systems to low-carbon ones, their presence creates several challenges to the power system decision-makers. The uncertain and intermittent output power of the RESs, especially the WTs and PV systems, are the two main characteristics of these resources which must be managed in the power systems. The intermittent behavior of the output power of the WT and PV systems creates a severe ramp rate in the net load of the system as reported by the CAISO. Regarding the high amount of installed capacity of the WTs and especially the PV systems in the distribution networks, the net load of the distribution networks has a high amount of ramp up and ramp down which needs to be managed. To manage these high ramps, flexible energy sources such as ESSs, DGs, and DRPs can be employed. This section aims to investigate the DEP problem in the presence of flexible energy sources.

### 3.4.1 DESCRIPTION OF DISTRIBUTION EXPANSION PLANNING PROBLEM WITH FLEXIBLE RESOURCES

The DEP problem in the presence of flexible energy sources has changed in comparison with the traditional one as described in Figure 3.3. As shown in this figure, the technical and economic data of the flexible energy sources are added to the previously required input data of the DNP problem. Moreover, in this DEP problem,

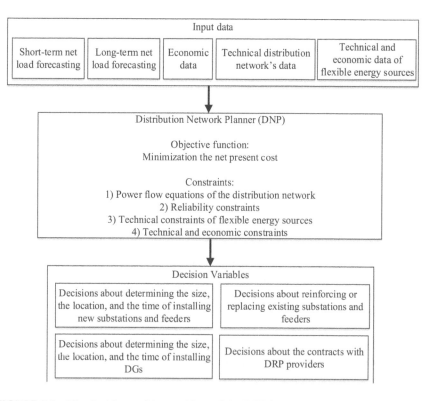

**FIGURE 3.3** The decision-making problem of the DEP in the presence of flexible energy sources.

the forecasted amount of the net load of the distribution network should be used where the net load is the amount of the actual demand of the distribution network minus the power generation of the RESs. For this purpose, two types of net load forecasting are required, i.e., long-term and short-term. The long-term net load forecasting considers long-term changes in load and generation of the RESs (here, the annual peak load of the system should be satisfied through the DEP problem). However, the short-term net load forecasting considers short-term variations in load and generation (here, the flexibility requirements should be satisfied). The main objective function of the DEP is the NPC which will be solved considering the technical constraints of the distribution network and flexible energy sources. The decision variables of the DEP problem in the presence of flexible energy sources are the optimal size and location of the new substations and feeders, the decisions on the reinforcing and/or replacing the existing substations and feeders, the optimal size and location of the DGs and electrical ESSs, and the optimal contracts with the DRPs.

### 3.4.2 Problem Formulation

In this sub-section, a simple formulation is presented for the static DEP problem in the presence of flexible energy sources as Eqs. (3.12)–(3.17) to install new substations, new feeders, DGs, and optimal contracts with DRPs.

$$\text{Minimize NPC} = C^f + C^S + C^L + C^P + C^{DG} + C^{DRP} \tag{3.12}$$

where, $C^{DG}$ and $C^{DRP}$ are the installation and operation costs of the DGs, and the cost of implementing the DRPs. These terms are described in detail as follows. DG installation and operation costs could be formulated as Eq. (3.13).

$$C^{DG} = \sum_{y=1}^{Y} \varsigma^{DG}(y) \sum_{i \in N} \sum_{k}^{K} \rho_k^{\text{inv}} P_{i,k}^{\text{max}}(y) + \sum_{y} \frac{1}{(1+d)^y} \sum_{i \in N} \sum_{k}^{K} 8760 LF(y) \rho_k^o p_{i,k}(y) \tag{3.13}$$

where, $k$ and $K$ are the index and set of the DGs, $\rho_k^{\text{inv}}$ is the investment cost of the "$k$th" DG, and $P_{i,k}^{\text{max}}(y)$ is the capacity of the DGs which are installed in year "$y$", $\rho_k^o$ is the annual operation cost of the "$k$th" DG, and $p_{i,k}(y)$ is the power generation of the DGs in peak load in year "$y$".

DR implementation costs could be explained as Eq. (3.14).

$$C^{DRP} = \sum_{y} \frac{1}{(1+d)^y} \sum_{b}^{B} \rho_b^{DRP}(y) p_{i,b}(y) \tag{3.14}$$

where, $b$ and $B$ are the index and set of the DRPs, $\rho_b^{DRP}(y)$ is the price of providing DR in year "$y$", and $p_{i,b}(y)$ is the amount of DR provided in year "$y$". Please note that if DR prices are different in various periods (based on the DR contracts), Eq. (3.14) should be modified to consider different DR prices.

The previous technical constraints, e.g. voltage limitations, distribution substation capacity, thermal capacity of feeders, and radial structure limitation are also used in this problem. Moreover, other constraints are as follows:

- DG capacity constraint: The power generation of the DG is limited to its maximum value as modeled in Eq. (3.15).

$$p_{i,k}(y) \leq P_{i,k}^{\max}(y) \qquad\qquad \forall i,k,y \qquad\qquad (3.15)$$

- DRP constraint: The amount of the DR provided for the system is limited to the maximum value of the DR that can be provided by the DRPs as shown in Eq. (3.16). In this equation, $P_{i,b}^{\max}(y)$ is the maximum value of DR that can be provided at bus $i$ in year "$y$".

$$p_{i,b}(y) \leq P_{i,b}^{\max}(y) \qquad\qquad \forall i,b,y \qquad\qquad (3.16)$$

- Active power balance limitation: The active power balance constraint in the distribution network's buses in the presence of the DGs and DRPs is modeled as Eq. (3.17).

$$\left( \sum_{\substack{j \\ j\neq i}} \left\{ P_{ij}(y) - P_{\text{loss}}(y) \right\} - \sum_{\substack{j \\ j\neq i}} P_{ji}(y) \right) + \sum_{k}^{K} p_{i,k}(y) + \\ \sum_{b}^{B} p_{i,b}(y) = D_i(y) \qquad \underset{i\neq j}{\forall\, i,j,y} \qquad\qquad (3.17)$$

The proposed model in this section could be developed to consider the electrical ESS using the proposed approach in ref. [46]. Moreover, the DEP problem can be developed to supply the required reactive power of the distribution network considering the reactive planning problem as described in ref. [52].

### 3.4.3 Overview of the Previous Studies

The studies which are done on the DEP in the presence of flexible energy sources are reviewed from different aspects in Table 3.2. As shown in this table, most of these studies used a multi-objective function to solve the DEP problem in the presence of flexible energy sources. In this problem, three main uncertain parameters consist of energy market price, demand, and output power of RESs, as shown in Table 3.3. In ref. [45], multi-level demand is considered, as shown in Table 3.3. As shown in this table, three load levels are defined where the percentage of the peak load and the amount of time duration of each level is determined. Also, in some studies, some daily load profiles are considered instead of only one value of the load, and then the time duration of occurring the load profiles is presented [46].

**TABLE 3.2**

**Comparison of Different Studies on the DEP Problem in the Presence of Flexible Energy Sources**

| Reference | Static/ Dynamic | Objective Function | Type of Flexible Sources | Type of Load | Uncertain Parameters | Test System | Solving Approach |
|---|---|---|---|---|---|---|---|
| [46] | Static | Cost | DG | Multi-level | No | 9-node | GA |
| [6] | Dynamic | Cost emission power loss | DG | One level | Demand energy price | 69-node | IHSA |
| [53] | Dynamic | Cost emission power loss | DG | One level | Demand energy price RESs | 69-node | NSIHSA-II |
| [46] | Dynamic | Cost | DG PEVs | Daily load profiles | No | 24-node | CPLEX 12 |
| [11] | Dynamic | Cost reliability | DG DRP | One level | Demand RESs | 118-node | MOPSO |

**TABLE 3.3**

**Multi-Level Modeling of Demand in the DEP Problem [45]**

| Load Level | Percentage of Peak Load (%) | Duration [h] |
|---|---|---|
| 1 (high) | 100 | 1500 |
| 2 (normal) | 70 | 5000 |
| 3 (low) | 50 | 2260 |

## 3.5 A SYSTEM-OF-SYSTEMS PLANNING APPROACH FOR THE INCLUSION OF FLEXIBILITY PROVIDERS

### 3.5.1 DISTRIBUTION EXPANSION PLANNING AS A SYSTEM-OF-SYSTEMS

Nowadays, due to the increase in the complexity of systems, traditional approaches for solving these systems are not efficient enough [54]. The complexity of problems is one of the main drivers to use other frameworks such as SoS [55]. Having a larger number of more complex and integrated systems, the researchers have highly focused on studying different aspects of such systems. Gradually, the SoS is known as an economic and strategic way for improving the performance of these systems and several definitions have been reported to explain the concept of the SoS [56]. However, generally, the SoS is defined as a system consisting of different components that are integrated for a common goal. Therefore, the SoS includes several systems with the following specifications [57]:

- Independent from the operational viewpoint: It means that each system is able to be operated individually or under the SoS framework.
- Independent from the managerial viewpoint: It means that each system is operated based on its independent objective.

The SoS integrates different independent and heterogeneous systems with distinct objective functions into a more complex system in a way that its performance is better than the performance of the separated systems. Determination of the optimal solution of the SoS approach is an important task that will specify the behaviour of its sub-systems [58].

In the electric power delivery chain system, several entities have emerged. On one hand, such entities are independent and autonomous players who have their objectives that might be different or even conflicting with the goals of other players. On the other hand, these heterogeneous constituents have interoperability and could exchange limited data.

Under such a situation, a comprehensive approach is required to efficiently model the players' interactions and satisfy their goals.

As pre-mentioned, nowadays, researchers are highly attracted toward the SoS approach because of its managerial advantages as well as its good performance to model the interactions of different entities. Due to the presence of different players in the electric distribution networks, their decisions could be considered using this framework. The SoS approach is an interesting way to improve the managerial aspects and overcome the relevant challenges [59]. The SoS contains different independent constituents that cooperate to satisfy their own goals [60]. The SoS approach has the following main features [61]–[63]:

1. Autonomy: It means that the constituents of the SoS are able to decide and act as an autonomous entity.
2. Belonging: The autonomous entities should decide to be completely autonomous or join a collective framework where they share some of their variables with others. Belonging means that the constituents of the SoS can mitigate their risk levels by joining the SoS framework and have more secure relationships.
3. Diversity: It means that under the SoS umbrella, various entities could be aggregated to fulfill the social function. In other words, it is assumed that the SoS aims to increase the number of its constituents by aggregating them using cooperation and collaboration.
4. Connectivity: It means that the sub-systems of the SoS have a relationship with each other for enhancing the SoS capabilities.
5. Emergence: A feedback for the autonomy and diversity to control the autonomy and heterogeneity for providing collaboration and functionalities.

Although the sub-systems of the SoS aim to optimize its objective function, each of them should cooperate with others and share some variables. The operating point of the constituents will be determined through the final solution of the SoS. Since the final solution of the SoS will determine the operating point of all its constituents, all the constraints, objective functions, and behaviours of the sub-systems should be considered by the SoS. It should be mentioned that limited data could be transferred between the players.

Here, the following entities are considered as the autonomous players in the electric distribution systems: DISCO, PI, and DRP. Indeed, Since the PI, DRP, and DISCO are different autonomous systems with distinct objective functions, the SoS could be employed to model them.

The goal of DISCO is to solve a DEP problem to minimize the cost terms and maximize the reliability levels, while PI and DRP aim to maximize their profits. PI invests in distributed energy resources (DERs) to sell electric energy and gain economic benefits. Moreover, DRP is an independent entity that should aggregate the DR potentials from the customers' sides and sell the available DR capacities to DISCO. Indeed, a DRP should negotiate with the DISCO for selling the aggregated DR capacities through the DR contracts to maximize its profit. Therefore, DISCO, PI, and DRP are heterogeneous and autonomous players with separate objectives, but they have interoperability and are able to share some variables. A DISCO should optimize a multi-objective problem (to minimize its cost terms and maximize the reliability level), but the PI and DRP aim to maximize their profit functions.

DISCO should offer some guaranteed purchasing prices to PIs in order to convince them to invest in DG planning. Moreover, it should negotiate with DRP to sign DR contracts to purchase the guaranteed available DR capacities.

As mentioned above, the SoS framework contains several independent and heterogeneous constituents with individual and in some cases conflicting goals, which have interoperability and cooperate for a common aim [61]. In the SoS model, not only the constant parameters and decision variables are needed (that are generally required to model any constituent) but also adaptive parameters and shared variables are needed [58]. Adaptive parameters are constant which would be determined for one constituent from other constituents. As an instance, as DISCO's adaptive parameter, it could determine the permissible penetration level of DERs in each bus, while as the PI's adaptive parameter, it could specify the possible investments in each year. Shared variables are common among two independent players, and they reflect the effects of different conditions of autonomous entities on each other. As an instance, one of the shared variables between PI and DISCO is the guaranteed purchasing prices of DERs. Similarly, DR prices are the shared variables between DISCO and DRP.

According to the data exchange among the entities, each entity could be defined as "Origin" or "Client" [58]. The client is an entity who has sent a signal to receive some data from another entity. Moreover, the origin is an entity that has received the signals to share some data about its adaptive parameters or shared variables. As an instance, if the PI sends signals to DISCO to know the electricity price data at each bus, it would be defined as the client of DISCO (in this example, PI is a Client and DISCO is an Origin).

Therefore, DISCO, PI, and DRP are autonomous and heterogeneous entities that are gathered using the SoS, intending to achieve an optimum solution to determine the expansion planning of DERs (by the PI), and the DR contracts among the DRP and DISCO. PI and DRP are commercial entities that play their roles to maximize their expected profits. However, the DISCO aims to upgrade the distribution system in the most economic and reliable manner.

Consequently, through such a framework, (1) DISCO should specify its adaptive parameters and submit them to PI or DRP (e.g., the permissible penetration level of DERs); (2) PI and DRP should determine their adaptive parameters and submit them to DISCO (e.g., the possible DER investments by PI, and the maximum available capacity of DR by DRP); (3) DISCO specifies the guaranteed purchasing prices for PI, as well as the DR prices for DRP; and (4) PI determines the guaranteed power generation, and DRP determines the guaranteed available DR capacity, and then they should submit these signals to DISCO. It should be mentioned that DISCO will also submit the suggested prices for purchasing power from PI, and DR from DRP. Then, based on the received data, PI and DRP will determine their strategies. Regarding the behaviors of PI and DRP, DISCO could modify its offers. This is a repetitive procedure that should be continued until the convergence of the results to a common point. This solution is the optimum point of the SoS that will determine the strategies of all heterogeneous agents.

### 3.5.2  SoS-based Distribution Expansion Planning Framework

As mentioned, the SoS is introduced here to model the behavior of PI, DRP, and DISCO. The solution of the proposed model will determine the decision of PI to install DGs, as well as the strategy of DRP to provide the guaranteed amount of DR. In such a framework, DISCO, PI and DRP are independent and heterogeneous systems. Therefore, the SoS framework is employed to model the potential of DERs (by modeling the behavior of a PI), the capacity of DR programs (by modeling the strategy of a DRP), and also the DEP problem (from the DISCO's point of view). The SoS is introduced as a suitable concept for modeling the independent behavior of different heterogeneous agents with distinct goals which are able to work together and share some data. In the proposed model, the objective of DISCO is to solve a DEP problem in the most economic and reliable way, while PI and DRP are commercial agents to maximize their profit functions. DISCO should offer interesting purchasing prices for persuading the PI to invest in DG planning. Also, DISCO should negotiate with DRP (as a market player to integrate the DR potentials from the customer side) to sign a contract to purchase the guaranteed DR capacities. It should be noted that each entity should cope with some uncertainties in its problem. Since the SoS constituents are independent players, they are able to freely determine their acceptable risk levels and decide about their own desired approaches to model the uncertainties. Hence, from the players' viewpoints, the acceptable risk levels and the methods to model the uncertainties could be different.

As mentioned, since the SoS will determine the operating points of all its constituents, objective functions and constraints of the players should be considered. The objective functions of DISCO could be written as follows:

- To minimize the planning and operation cost terms.
- To maximize the reliability level.

The economic objective function is to minimize the planning and operation cost that will include the following terms:

- Upgrading costs of the feeders.
- Cost of energy losses.

- Electricity purchasing prices from the upstream network.
- Guaranteed purchasing prices from PI.
- Cost of DR contract with DRP.

Moreover, the second objective function is to maximize the reliability level of the system. Different reliability indexes could be considered as the objective function such as expected energy not supplied (EENS). It should be noted that other objective functions could also be considered (e.g., emission rate).

From DISCO's point of view, the following constraints should be satisfied:

- Radiality of the distribution system (the distribution networks should be operated radially and no islanding should occur).
- Minimum and maximum voltage levels.
- Maximum capacity of feeders.
- Load balance.
- Financial constraints (financial limitations to upgrade the distribution system).

PI is a commercial agent to maximize its profit. The following terms should be modeled through the objective function of PI:

- Revenue by selling power.
- Operation and maintenance (O&M) costs.
- Investment costs.
- Salvage values.

PI should be coped with different constraints such as:

- Maximum size of each DER type in each bus.
- Permissible DERs' generation in each bus.
- Total penetration level of DERs in the network.
- Investment restrictions.
- Pollution emission.

Also, DRP is another commercial agent which wants to maximize its profit function. The following terms could be considered as the terms of its objective function:

- Revenue from the DR contracts by DISCO (to enable a specified amount of DR).
- Costs to enable DR (to persuade the customers to be responsive when they are called by DR).

The magnitude of DR could be considered as one of its constraints. Other constraints could also be considered, e.g. DR frequency and duration.

Three types of data could be transferred among PI, DRP, and DISCO, as follows:

1. DISCO will determine the adaptive parameters (such as the penetration level of DGs and the maximum permissible pollution rate).

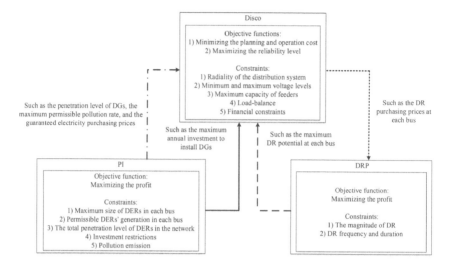

**FIGURE 3.4**   Information flow among DISCO, PI, and DRP in the proposed SoS framework.

2. PI and DRP will determine their adaptive parameters (such as the maximum annual investment to install DGs by PI, and the maximum DR potential at each bus by DRP);

3. DISCO will determine the guaranteed electricity purchasing prices, as well as the DR purchasing prices at each bus and send them to PI and DRP, respectively. Also, based on the received data from DISCO, PI will determine its decision to invest in DGs and submit it to DISCO. Similarly, in response to the received data from DISCO, DRP will determine the DR potentials at each bus and sent them to DISCO.

The data exchange procedure is illustrated in Figure 3.4. Also, Figure 3.5 schematically shows the proposed SoS framework to model the interactions of the SoS sub-systems.

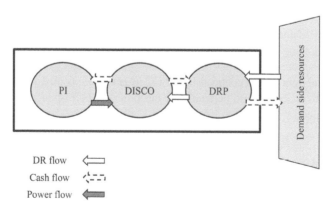

**FIGURE 3.5**   Constituents' interactions through the SoS framework [11].

The proposed framework should be solved to find the solution of the SoS that will specify the strategies of its constituents. More details about the formulations and the optimization procedures could be found in ref. [11].

## 3.6  CONCLUSION AND OUTLOOK

This chapter addresses the planning problem of the distribution systems in the presence of flexibility providers as the emerging market players. To this end, first, a survey is provided regarding the DEP and flexibility problems, and their concepts and necessities are clarified. Then, the conventional DEP problem is introduced and its regular formulations are presented. Also, the integration of flexibility resources in the DEP problem is described that shows the potential changes in the planning criteria of the future distribution systems.

By considering the emergence of autonomous players in the future distribution systems, this chapter proposes a comprehensive and efficient concept to model the independent behavior of different heterogeneous entities. Such players have their distinct objective functions, but they have interoperability and are able to work together and share some parts of their data. DISCO, PI, and DRP are introduced as three important entities which can negotiate with each other to determine their strategies. A PI could be considered as one of the providers of flexibility services since it can invest in various types of DGs and is able to provide flexibility requirements. Also, DRPs are specifically introduced as one of the main flexibility service providers that can act in the system in order to integrate the DR potentials from the customer side. Indeed, on one hand, a DRP should persuade customers to participate in DR programmes by modifying their consumption patterns. On the other hand, it should negotiate with DISCO to sign the DR contracts. Based on this contract, DRP should enable the specified amounts of DR capacities and will be responsible to provide it when it is necessary, and DISCO should be committed to pay for the available purchased DR, even if it is called or not. Similarly, PI and DISCO should have an agreement with each other in which DISCO offers the guaranteed purchasing prices to PI to persuade it to invest in the system, and PI will agree to be responsible to provide a pre-determined power from DGs. Therefore, there are three heterogeneous and independent agents in the distribution system with their own distinct goal, but they can share some data and work with each other to achieve a common goal. Details of the SoS approach for handling such cooperation are also addressed in this chapter. As mentioned, not only the general parameters and variables are needed to model the SoS concept, but also two other types of data are necessary, called shared variables and adaptive parameters. Finally, it is concluded that the aggregation of different agents in the DEP problem through an efficient framework such as SoS could effectively enhance the performance of the total system and help planners to achieve the planning aims.

Considering the proposed works in DEP, the following research directions need to be more investigated:

• Modeling the competition among different SoS constituents by considering several DRPs, as well as several PIs.
• Modeling the DEP problem in the presence of several MGs besides DRPs and PIs through the SoS framework.

- Developing the proposed SoS framework to model the local energy and ancillary service markets in the DEP problem.
- Modeling the DEP problem considering the incentive policies to develop flexible energy sources in the distribution systems.

## REFERENCES

[1] V. Vahidinasab, M. Tabarzadi, H. Arasteh, M. I. Alizadeh, M. Mohammad Beigi, H. R. Sheikhzadeh, K. Mehran, M. S. Sepasian, "Overview of electric energy distribution networks expansion planning," *IEEE Access*, vol. 8, pp. 34750–34769, 2020.

[2] A. Samui, S. Singh, T. Ghose, and S. R. Samantaray, "A direct approach to optimal feeder routing for radial distribution system," *IEEE Trans. Power Deliv.*, vol. 27, no. 1, pp. 253–260, 2011.

[3] D. T.-C. Wang, L. F. Ochoa, and G. P. Harrison, "Modified GA and data envelopment analysis for multistage distribution network expansion planning under uncertainty," *IEEE Trans. Power Syst.*, vol. 26, no. 2, pp. 897–904, 2010.

[4] S. M. Mazhari, H. Monsef, and H. Falaghi, "A hybrid heuristic and learning automata-based algorithm for distribution substations siting, sizing and defining the associated service areas," *Int. Trans. Electr. Energy Syst.*, vol. 24, no. 3, pp. 433–456, 2014.

[5] E. G. Carrano, F. G. Guimarães, R. H. Takahashi, O. M. Neto, and F. Campelo, "Electric distribution network expansion under load-evolution uncertainty using an immune system inspired algorithm," *IEEE Trans. Power Syst.*, vol. 22, no. 2, pp. 851–861, 2007.

[6] A. Rastgou, J. Moshtagh, and S. Bahramara, "Improved harmony search algorithm for electrical distribution network expansion planning in the presence of distributed generators," *Energy*, vol. 151, pp. 178–202, 2018.

[7] S. Ganguly, N. C. Sahoo, and D. Das, "Multi-objective particle swarm optimization based on fuzzy-Pareto-dominance for possibilistic planning of electrical distribution systems incorporating distributed generation," *Fuzzy Sets Syst.*, vol. 213, pp. 47–73, 2013.

[8] A. Onlam, D. Yodphet, R. Chatthaworn, C. Surawanitkun, A. Siritaratiwat, and P. Khunkitti, "Power loss minimization and voltage stability improvement in electrical distribution system via network reconfiguration and distributed generation placement using novel adaptive shuffled frogs leaping algorithm," *Energies*, vol. 12, no. 3, p. 553, 2019.

[9] S.-A. Ahmadi, V. Vahidinasab, M. S. Ghazizadeh, K. Mehran, D. Giaouris, and P. Taylor, "Co-optimising distribution network adequacy and security by simultaneous utilisation of network reconfiguration and distributed energy resources," *IET Gener. Transm. Distrib.*, vol. 13, no. 20, pp. 4747–4755, 2019.

[10] O. D. Melgar-Dominguez, M. Pourakbari-Kasmaei, M. Lehtonen, and J. R. S. Mantovani, "An economic-environmental asset planning in electric distribution networks considering carbon emission trading and demand response," *Electr. Power Syst. Res.*, vol. 181, p. 106202, 2020.

[11] H. Arasteh, V. Vahidinasab, M. S. Sepasian, and J. Aghaei, "Stochastic system of systems architecture for adaptive expansion of smart distribution grids," *IEEE Trans. Ind. Inform.*, vol. 15, no. 1, pp. 377–389, 2018.

[12] S. M. Hashemi, V. Vahidinasab, M. S. Ghazizadeh, and J. Aghaei, "Reliability-oriented DG allocation in radial Microgrids equipped with smart consumer switching capability," in *2019 International Conference on Smart Energy Systems and Technologies (SEST)*, 2019, pp. 1–6.

[13] H. Arasteh, M. Kia, V. Vahidinasab, M. Shafie-khah, and J. P. Catalão, "Multiobjective generation and transmission expansion planning of renewable dominated power systems using stochastic normalized normal constraint," *Int. J. Electr. Power Energy Syst.*, vol. 121, p. 106098, 2020.

[14] H. Arasteh, M. S. Sepasian, V. Vahidinasab, and P. Siano, "SoS-based multiobjective distribution system expansion planning," *Electr. Power Syst. Res.*, vol. 141, pp. 392–406, 2016.

[15] H. Arasteh, M. S. Sepasian, and V. Vahidinasab, "An aggregated model for coordinated planning and reconfiguration of electric distribution networks," *Energy*, vol. 94, pp. 786–798, 2016.

[16] P. Šulc, S. Backhaus, and M. Chertkov, "Optimal distributed control of reactive power via the alternating direction method of multipliers," *IEEE Trans. Energy Convers.*, vol. 29, no. 4, pp. 968–977, 2014.

[17] M. Bazrafshan and N. Gatsis, "Decentralized stochastic optimal power flow in radial networks with distributed generation," *IEEE Trans. Smart Grid*, vol. 8, no. 2, pp. 787–801, 2017.

[18] M. H. Amini, H. Arasteh, and P. Siano, "Sustainable smart cities through the lens of complex interdependent infrastructures: panorama and state-of-the-art," *Sustainable Interdependent Networks II*, Springer, 2019, pp. 45–68.

[19] R. Aghapour, M. S. Sepasian, H. Arasteh, V. Vahidinasab, and J. P. Catalão, "Probabilistic planning of electric vehicles charging stations in an integrated electricity-transport system," *Electr. Power Syst. Res.*, vol. 189, p. 106698, 2020.

[20] H. Arasteh, M. Sepasian, and V. Vahidinasab, "Toward a smart distribution system expansion planning by considering demand response resources," *J. Oper. Autom. Power Eng.*, vol. 3, no. 2, pp. 116–130, 2015.

[21] H. Nezamabadi, V. Vahidinasab, S. Salarkheili, V. Hosseinnezhad, and H. Arasteh, "Game theory application for finding optimal operating point of multi-production system under fluctuations of renewable and various load levels," *Integration of Clean and Sustainable Energy Resources and Storage in Multi-Generation Systems*, Springer, 2020, pp. 189–216.

[22] J. Liu, H. Cheng, P. Zeng, L. Yao, C. Shang, and Y. Tian, "Decentralized stochastic optimization based planning of integrated transmission and distribution networks with distributed generation penetration," *Appl. Energy*, vol. 220, pp. 800–813, 2018.

[23] J. Liu, P. Zeng, H. Cheng, S. Zhang, J. Zhang, and X. Zhang, "Decentralized expansion of transmission networks incorporating active distribution networks," in *2019 IEEE PES Asia-Pacific Power and Energy Engineering Conference (APPEEC)*, 2019, pp. 1–6.

[24] E. S. Parizy, H. R. Bahrami, and K. Loparo, "A decentralized three-level optimization scheme for optimal planning of a prosumer nano-grid," *IEEE Trans. Power Syst.*, 2020.

[25] M. Pipattanasomporn, H. Feroze, and S. Rahman, "Multi-agent systems in a distributed smart grid: Design and implementation," in 2009 IEEE/PES Power Systems Conference and Exposition, 2009, pp. 1–8.

[26] N. Olfatinezhad, V. Vahidinasab, M. Ahmadian, H. Arasteh, J. Aghaei, and K. Mehran, "Flexible two-stage robust model for moving the transmission and reactive power infrastructures expansion planning towards power system integration of renewables," *IET Renew. Power Gener.*, vol. 14, no. 11, pp. 1921–1932, 2020.

[27] M. Kia, M. Shafiekhani, H. Arasteh, S. M. Hashemi, M. Shafie-khah, and J. P. S. Catalão, "Short-term operation of microgrids with thermal and electrical loads under different uncertainties using information gap decision theory," *Energy*, p. 118418, 2020.

[28] International Energy Agency, "International Energy Agency. Status of power system transformation 2018 Advanced power plant flexibility," 2018. [Online]. Available: https://www.oecd-ilibrary.org/energy/status-of-power-system-transformation-2018_9789264302006-en.

[29] P. D. Lund, J. Lindgren, J. Mikkola, and J. Salpakari, "Review of energy system flexibility measures to enable high levels of variable renewable electricity," *Renew. Sustain. Energy Rev.*, vol. 45, pp. 785–807, 2015.

[30] A. Tuohy, B. Kaun, and R. Entriken, "Storage and demand-side options for integrating wind power," *Wiley Interdiscip. Rev. Energy Environ.*, vol. 3, no. 1, pp. 93–109, 2014.

[31] B. Kirby, "Ancillary services: Technical and commercial insights," *Retrieved Oct.*, vol. 4, p. 2012, 2007.

[32] H. Arasteh, V. Vahidinasab, M. S. Sepasian, and A. Ghaderi, "Integrated Generation, Transmission and Energy Efficiency Planning," in *2019 International Power System Conference (PSC)*, 2019, pp. 571–578.

[33] C. W. Gellings and W. M. Smith, "Integrating demand-side management into utility planning," *Proc. IEEE*, vol. 77, no. 6, pp. 908–918, 1989.

[34] M. H. Amini, S. Talari, H. Arasteh, N. Mahmoudi, M. Kazemi, A. Abdollahi, V. Bhattacharjee, M. Shafie-Khah, P. Siano, J. P. S. Catalão., "Demand response in future power networks: panorama and state-of-the-art," *Sustainable Interdependent Networks II*, Springer, 2019, pp. 167–191.

[35] X. Liu and B. Su, "Microgrids—an integration of renewable energy technologies," in *2008 China International Conference on Electricity Distribution*, 2008, pp. 1–7.

[36] A. Nikoobakht, J. Aghaei, T. Niknam, H. Farahmand, and M. Korp\aas, "Electric vehicle mobility and optimal grid reconfiguration as flexibility tools in wind integrated power systems," *Int. J. Electr. Power Energy Syst.*, vol. 110, pp. 83–94, 2019.

[37] S. M. Hashemi, V. Vahidinasab, M. Ghazizadeh, and J. Aghaei, "Load control mechanism for operation of microgrids in contingency state," *IET Gener. Transm. Distrib.*, vol. 14, no. 23, pp. 5407–5417, 2020.

[38] D. K. Asl, A. Hamedi, and A. R. Seifi, "Planning, operation and flexibility contribution of multi-carrier energy storage systems in integrated energy systems," *IET Renew. Power Gener.*, vol. 14, no. 3, pp. 408–416, 2019.

[39] E. Lannoye, D. Flynn, and M. O'Malley, "Evaluation of power system flexibility," *IEEE Trans. Power Syst.*, vol. 27, no. 2, pp. 922–931, 2012.

[40] J. Cochran, M. Miller, O. Zinaman, M. Milligan, D. Arent, B. Palmintier, M. O'Malley, S. Mueller, E. Lannoye, A. Tuohy, B. Kujala, M. Sommer, H. Holttinen, J. Kiviluoma, S.K. Soonee, "Flexibility in 21st century power systems," National Renewable Energy Lab. (NREL), Golden, CO (United States), 2014.

[41] M. Emmanuel, K. Doubleday, B. Cakir, M. Marković, and B.-M. Hodge, "A review of power system planning and operational models for flexibility assessment in high solar energy penetration scenarios," *Sol. Energy*, 2020.

[42] V. H. Hinojosa, "Comparing corrective and preventive security-constrained DCOPF problems using linear shift-factors," *Energies*, vol. 13, no. 3, p. 516, 2020.

[43] S. M. Hashemi, V. Vahidinasab, M. S. Ghazizadeh, and J. Aghaei, "Valuing consumer participation in security enhancement of microgrids," *IET Gener. Transm. Distrib.*, 2018.

[44] A. J. Conejo, L. Baringo, S. J. Kazempour, and A. S. Siddiqui, "Investment in electricity generation and transmission," *Cham Zug Switz. Springer Int. Publ.*, vol. 119, 2016.

[45] H. Falaghi, C. Singh, M.-R. Haghifam, and M. Ramezani, "DG integrated multistage distribution system expansion planning," *Int. J. Electr. Power Energy Syst.*, vol. 33, no. 8, pp. 1489–1497, 2011.

[46] M. Abdi-Siab and H. Lesani, "Distribution expansion planning in the presence of plug-in electric vehicle: A bilevel optimization approach," *Int. J. Electr. Power Energy Syst.*, vol. 121, p. 106076, 2020.

[47] P. Nahmmacher, E. Schmid, L. Hirth, and B. Knopf, "Carpe diem: A novel approach to select representative days for long-term power system modeling," *Energy*, vol. 112, pp. 430–442, 2016.

[48] H. Teichgraeber and A. R. Brandt, "Clustering methods to find representative periods for the optimization of energy systems: An initial framework and comparison," *Appl. Energy*, vol. 239, pp. 1283–1293, 2019.

[49] N. Helistö, J. Kiviluoma, and H. Reittu, "Selection of representative slices for generation expansion planning using regular decomposition," *Energy*, p. 118585, 2020.

[50] M. Hand, S. Baldwin, E. DeMeo, J. Reilly, T. Mai, D. Arent, D. Sandor, *Renewable Electricity Futures Study. Volume 1: Exploration of High-penetration Renewable Electricity Futures*, National Renewable Energy Lab. (NREL), Golden, CO (United States), 2012.

[51] S. Bahramara, P. Sheikhahmadi, A. Mazza, G. Chicco, M. Shafie-Khah, and J. P. Catalão, "A risk-based decision framework for the distribution company in mutual interaction with the wholesale day-ahead market and microgrids," *IEEE Trans. Ind. Inform.*, vol. 16, no. 2, pp. 764–778, 2019.

[52] M. Resener, S. Haffner, L. A. Pereira, P. M. Pardalos, and M. J. Ramos, "A comprehensive MILP model for the expansion planning of power distribution systems–Part I: Problem formulation," *Electr. Power Syst. Res.*, vol. 170, pp. 378–384, 2019.

[53] A. Rastgou, S. Bahramara, and J. Moshtagh, "Flexible and robust distribution network expansion planning in the presence of distributed generators," *Int. Trans. Electr. Energy Syst.*, vol. 28, no. 12, p. e2637, 2018.

[54] A. M. Ross, D. H. Rhodes, and D. E. Hastings, "Defining changeability: Reconciling flexibility, adaptability, scalability, modifiability, and robustness for maintaining system lifecycle value," *Syst. Eng.*, vol. 11, no. 3, pp. 246–262, 2008.

[55] B. Heydari and K. Dalili, "Emergence of modularity in system of systems: Complex networks in heterogeneous environments," *IEEE Syst. J.*, vol. 9, no. 1, pp. 223–231, 2013.

[56] J. A. Lane and D. J. Epstein, "What is a system of systems and why should I care?," Department of Industrial and Systems Engineering, University of Southern California.

[57] M. W. Maier, "Architecting principles for systems-of-systems," *Syst. Eng. J. Int. Counc. Syst. Eng.*, vol. 1, no. 4, pp. 267–284, 1998.

[58] A. K. Marvasti, Y. Fu, S. DorMohammadi, and M. Rais-Rohani, "Optimal operation of active distribution grids: A system of systems framework," *IEEE Trans. Smart Grid*, vol. 5, no. 3, pp. 1228–1237, 2014.

[59] B. Ge, K. W. Hipel, K. Yang, and Y. Chen, "A novel executable modeling approach for system-of-systems architecture," *IEEE Syst. J.*, vol. 8, no. 1, pp. 4–13, 2013.

[60] M. Jamshidi, *System of Systems Engineering: Innovations for the Twenty-first Century*, vol. 58, Wiley, 2008.

[61] M. J. DiMario, J. T. Boardman, and B. J. Sauser, "System of systems collaborative formation," *IEEE Syst. J.*, vol. 3, no. 3, pp. 360–368, 2009.

[62] B. Sauser, J. Boardman, and D. Verma, "Systomics: Toward a biology of system of systems," *IEEE Trans. Syst. Man Cybern.-Part Syst. Hum.*, vol. 40, no. 4, pp. 803–814, 2010.

[63] B. Sauser and J. Boardman, "Taking hold of system of systems management," *Eng. Manag. J.*, vol. 20, no. 4, pp. 3–8, 2008.

# 4 Application of Regression Tools for Load Prediction in Distributed Network for Flexible Analysis

*Nils Jakob Johannesen*[1] *and Mohan Lal Kolhe*[1]
[1]University of Agder, Kristiansand, Norway

## CONTENTS

## 4.1   LOAD FLEXIBILITY AND MANAGEMENT

Load flexibility relates to the ability of power system to shift the operation. The flexibility has to respond to the variability and uncertainty of the net load. The increasing penetration of variable renewable generation (RG) increases the need for flexibility in the load demand. A flexible power system can adapt to rapid change in supply and demand. The flexibility of resources is defined by their dynamic capabilities such as ramp time, start-up/shut-down time, operating range (minimum and maximum operating level) as well as minimum up and down times of the energy generation system.

The regression tools can be used to understand the variation and uncertainty in load and supply, as well as to analyze and forecast the expected output. Regression techniques can be used to model the past behavior, to understand and help to predict the future scenarios both on demand and generation.

Flexible electric power system operation is going to help in integrating a mix of energy sources that can respond to the varying demand for electricity. This demand is met with three types of plants typically referred to as baseload (meeting the constant demand), intermediate load (meeting the diurnal changes), and peaking (meeting the peak demand). At very high penetration of RG, a key element of system flexibility is the ability of baseload generators, as well as generators providing operating reserves, to reduce output to very low levels while maintaining system reliability. Although baseload generators are a capital incentive, inexpensive small-unit generators are favored [1–5].

Demand side management is an umbrella term that describes the utility company efforts to improve energy consumption at customer site, the demand side of the meter [6]. Demand response (DR) is the customers' adaptation to alter their normal electricity usage in response to the adjusted electricity prices with grid constraints or other incentives created to decrease energy consumption at times of shortage or when system reliability is at risk [7]. The introduction of advanced metering system in the form of smart energy meters (SEM) allows for an unprecedented granularity in data gathering, and hence unlocking the potential of DR. The SEM implements an advanced measurement infrastructure, a two-way communication between the end-user and the distribution management system. SEM monitors, measures, and reports electric energy load demand in near real-time [8]. Traditionally, utilities have used three types of generating facilities to serve the diurnal and seasonal changes in load demand: Baseload, intermediate load, and peak load plants [9]. A load demand curve for a sample European country shown in Figure 4.1 illustrates typical load demand patterns, where the segments indicate natural threshold level typical for baseload, intermediate load, and peak load. Yearly seasonal load demand of a selected European country is given in Figure 4.2.

The diurnal changes start with a surge demand in the morning when industrial companies commence activity and domestic end-users start their home appliances; it is the first peak in the load demand curve. Following the early morning activities, load demand stabilizes; there is a dip in the load demand creating a valley in the load curve. When the working day is over, another surge load follows when people return to home and start cooking. The last diurnal valley in the load demand curve commences in the night time when people go to bed.

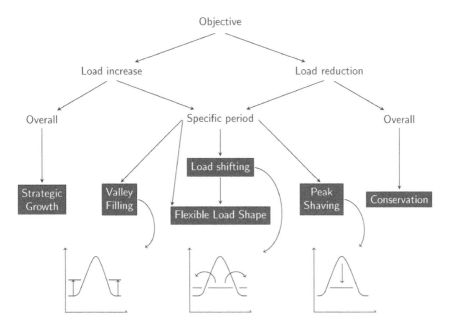

**FIGURE 4.1** The electric load demand curve of a sample European country for one week, indicating level of load curves (ENTSOE-E).

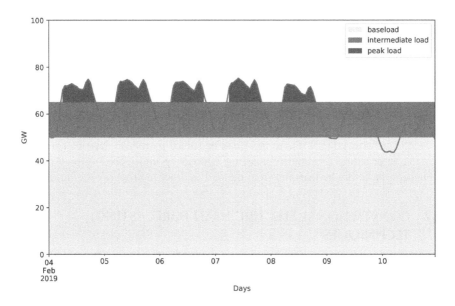

**FIGURE 4.2** The yearly electric load demand curve of a sample European country, depicting seasonal changes (ENTSOE-E).

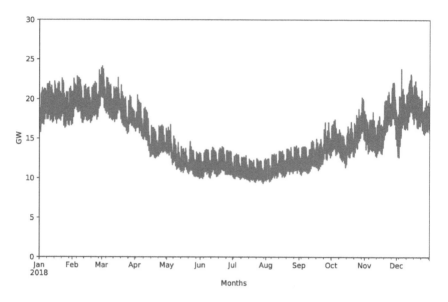

**FIGURE 4.3**   Objective of load shapes [1].

Depending on the operative flexibility of generators, they serve different load demand [10]. Efforts have been done to advance more flexible operation for managing the range between peak power and minimum load. Load cycling has a degenerating effect on units, impairs power production, and leads to frequent breakdowns and unplanned maintenance [11, 12]. Different techniques are used to create a better match between load and supply. Peak clipping or peak load shaving is to reduce the peak demand. Another incentive is to fill up the valleys where demand is low. Load shifting, as seen in Figure 4.3, combines the two previous techniques by shaving of the peak demand and filling the low-demand valleys. Load shifting regime is crucial to development of microgrids within the distributed network. Microgrids are designed without peaking generator, thus reserve their capacity and up to 10% of load is not utilised [13]. These tasks can be solved by robust electric energy load demand forecasting. Demand forecasting is done by understanding how the past influences the future by learning from the past in order to prophesy the future.

## 4.2   CONVENTIONAL ELECTRIC LOAD FORECASTING TECHNIQUES

The electrical load forecasting has been carried out using conventional mathematical techniques. The traditional forecasting techniques are based on linear regression series. Most of them use statistical techniques. A time series is a collected sequence of events, based on the assumption of an inherent structure. The inherent structure is analytically observed trhough means such as autocorrelation, trend, and seasonal behaviour. There are many different scenarios of how these sequences of events are collected and described. The most often used time series techniques are in particular autoregressive moving average (ARMA), autoregressive integrated moving average (ARIMA), autoregressive integrated moving average with exogenous variables (ARIMAX). For stationary processes, ARMA is usually used, and it has been extended to ARIMA for non-stationary processes. ARIMAX is the most natural tool since electrical load generally depends on exogenous variables such as weather and historical time series data. Time series forecasting, its data and analysis will in the future be increasingly important as the availability and scaling of such data is growing through Internet of things (IOTs), the rise of smart cities, and due to the advanced infrastructure metering. The continuous monitoring and data mining will pave the way for adequate time series analysis, both statistical and machine learning techniques, as well as hybrid models will increase.

Time series analysis has traditionally been performed in meteorology, energy, and economics. The era of modern time series analysis started when the Box-Jenkins model was introduced [14]. The Box-Jenkins method has been further developed by the research community to a robust parsimonious ARMA for multivariate forecasting, requiring less human intervention [15]. Additional improvement has been reached with a combined Box-Jenkins econometric approach to forecast monthly peak system load. By observing changes in economic and weather-related variables in a Box-Jenkins time series model, refined forecasts are obtained [16]. It is common for these approaches that they use multiplex mathematical computations and possess a heavy computational burden [17].

Machine learning models seriously contested the classical statistics with the artificial neural networks (ANN) [18]. The neural networks can aid dispatchers deal with uncertain loads [19]. ANN is used with updating network parameters, generating plant control and economic power dispatch problem [20–23]. A typical neural network model, with back propagating, adjusted weights, is presented in Figure 4.4. In the following years during the 1990s, the research on ANN in electric load forecasting was mainly concerned with regional loads in the MW-scale, resembling the load consumption of a medium size European country and including multivariate time series analysis [24–26].

Focus has also been attuned towards case and system dependency of ANN [27], the explainable and interpretative ANN, and the "black box nature" of

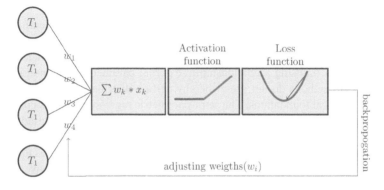

**FIGURE 4.4**   Neural network model with the propagating adjusted weights.

neural networks. This has paved the way for ensembles of trees, linear fits, Support Vector Machines (SVM), and other machine learning models. Some of these models find their origin in the statistics and overlapping with machine learning (see discussion [28]) [29, 30]. Deep learning techniques based on long- or short-term memory and recurrent neural networks have shown promising results for optimal scheduling of microgrids [31]. Also, the convolutional neural networks (CNN) show good results, but need big load schemes in GW-scale to perform well [32].

## 4.3   LEARNING SYSTEMS

Machine learning provides a framework for estimating from the observed data to form an appropriate model in the time dependencies. Machine learning is a subcategory of artificial intelligence and usually divided into two main types, supervised and unsupervised learning. Unsupervised learning is learning without any prior knowledge of the aim of learning, and is also named knowledge discovery. Hence, the unsupervised learning can be state dependent or clustering. For the supervised learning, the aim or the independent variable is known. In supervised learning, data is orchestrated in such a way that it fits the aim.

In supervised learning, $x$ and $y$ are preserved in a train and test set. Here, $D$ is called the training set and $N$ is the number of training examples. Test-set, is preserved for inference purposes. When the inference is performed, the algorithm is normally verified according to a performance metrics. In the predictive or supervised learning approach, the goal is to learn a mapping from inputs $x$ to outputs $y$, given a labeled set of input-output pairs $D = \{(x_i, y_i)\}_{i=1}^{N}$. Given the inputs, $D = \{x_i\}_{i=1}^{N}$, the aim is to recognize patterns in the data [33].

## 4.4   REGRESSION TOOLS

Regression is distinguishable from classification by the response vector ($y$), which is a continuous output of time, whilst in classification, the $y$ vector is categorical. In this sense, the classification is a subdivision of regression [34]. For this reason, regression has been known by machine learning practitioners "learning how to classify among continuous classes" [35].

Regression methods vary from purely statistical methods, machine learning techniques to hybrid models that combine two methods. The regression tools can be parametric, where a particular distribution constitutes the method, either by direct measures or when posing a relationship to external parameters. The non-parametric regression methods do not prescribe any certain distribution, hence regress on pure mathematical foundations. The semi-parametric regression models combine an underlying distribution with a pure mathematical relation. A technique used in many of the regression tools is correlation, either to research the data for their general function, or in multivariate time series where the target vector correlates to external parameters. Correlation is a measurement to how two ranges of data move together. The Pearson Correlation Coefficient ($r$) computes the linear relationship between two variables, in a range from $-1$ to $+1$ [36]. If the relationship is in the proximity of 1, it means that when $x$ increases so does $y$, and at exact linearity, the opposite is true for $-1$, which means that when one variable increases, the other decreases.

$$r = r_{xy} = \frac{n\sum x_i y_i - \sum x_i \sum y_i}{\sqrt{n\sum x_i^2 - (\sum x_i)^2}\sqrt{n\sum y_i^2 - (\sum y_i)^2}} \tag{4.1}$$

Autocorrelation function (ACF) shows how a time series is correlated to its own lagged version at each $lag_k$ [37]:

$$\rho_k(t) = \frac{\sum\limits_{t=1}^{n-k}(x_t - \hat{x})\sum\limits_{t=1}^{n-k}(y_{t+k} - \hat{y})}{\sqrt{\sum\limits_{t=1}^{n-k}(x_t - \hat{x})^2}\sqrt{\sum\limits_{t=1}^{n-k}(y_{t+k} - \hat{y})^2}} \tag{4.2}$$

Cross-correlation can be found when one of the variables is shifted in time ($t$), and can be used to alter the time lags between the variables for a reshaped perspective of the relationship between them. As the times series are cross-correlated, an evaluation of temporal similarity is made [38]:

$$\rho_{xy}(t) = \frac{\sum\limits_{i=1}^{n}(x_i - \hat{x})\sum\limits_{i=1}^{n}(y_{i-t} - \hat{y})}{\sqrt{\sum\limits_{i=1}^{n}(x_i - \hat{x})^2}\sqrt{\sum\limits_{i=1}^{n}(y_{i-t} - \hat{y})^2}} \tag{4.3}$$

Autoregression (AR) is a simple and straightforward regression technique, where past values of the univariate time series are dependent on their own lagged version defined by a parameter weighting of each input, $\phi$, and therefore a parametric model. The current value of $y(t)$ is expressed by previous values of time $y_{t-1}, y_{t-2}, ..., y_{t-p}$. The order of an AR process is defined by the number of past values of $y(t)$ it is regressed on. $AR(p)$ is defined by the last $y_{t-p}$, considered in the process, denoted as:

$$y(t) = \phi_1 y_{t-1} + \phi_2 y_{t-2} + ... + \phi_p y_{t-p} + \varepsilon_t \tag{4.4}$$

where, the error term $\varepsilon_t$ is white noise defined by a constant mean and some unknown fixed variance $\sigma_\varepsilon^2(t)$, a stationary process. The ACF of a white noise process is zero at all lags other than lag zero where it is unity, to indicate that the nature of its process is completely uncorrelated. By using backshift operator ($B$), the previous value of the time series is related to the current value $y_{t-1} = By_t$, and thus, $y_{t-m} = B^m y(t)$, and the error term is explained as:

$$\phi(B)y_t = \varepsilon_t \tag{4.5}$$

An AR process $p$-value is defined by the autocorrelation of residuals of the AR process. If the residuals autocorrelation falls within a confidence interval, normally considered as 95%, the autocorrelation function of the residuals are considered to be white noise. If not, the AR process will still continue to find another parameter, until its residuals satisfy the criteria of white noise. If the current and previous values of a white noise series $\varepsilon_t, \varepsilon_{t_1}$ are expressed linearly, it is known as moving average process (MA), and an equivalent implementation of backshift operator ($B$) would be:

$$y(t) = \theta(B)\varepsilon(t) \tag{4.6}$$

A combination of the two processes is the ARMA. If the mean or covariance of the time series observations change with time, the series is defined as non-stationary, and a differencing process makes it stationary by introducing the $\nabla$ operator, and the AR, MA and ARMA processes are transformed into ARI, IMA or ARIMA process.

### 4.4.1 LINEAR REGRESSION

Another parametric model is multiple linear regression (MLP) that assumes a linear relationship in the training data and to explanatory variables to explain relationship to the response-vector ($y$):

$$y(t) = a_0 + \beta_1 x_1(t) + ... + \beta_n x_n(t) + \varepsilon(t) \tag{4.7}$$

where, $x_1(t),...,x_n(t)$ are independent explanatory variables correlated with the dependent load variable $y(t)$. The independent variables are found through correlation analysis, and coefficient estimation normally found through least square estimation, or iteratively reweighted least squares (IRWLS). All parameters start at 0 and is step-wise improved using backpropagation through a loss function to find appropriate weights, or through finding the intercept $a_0$. Each explanatory variable finding its coefficient based on the covariance and standard deviation of dependent and independent variables is defined as:

$$\beta_x = \frac{\sigma_{xy}}{\sigma_x} \tag{4.8}$$

### 4.4.2 k-NEAREST NEIGHBOR REGRESSION

Opposite to the linear regression (LR) is the k-nearest neighbor (kNN) regressor, which is non-paramteric, relying on its own table look-up and mathematical foundation, and highly non-linear.

$$y_{knn}(x) = \frac{1}{K} \sum_{k=1}^{K} y_k \text{ for } K \text{ nearest neighbors of } x \tag{4.9}$$

The kNN-classifier is illustrated in Figure 4.5, where the left diagram with a small encirclement options for $k = 1$, where simply the nearest neighbor decides the class of prediction, whilst in the right diagram in Figure 4.5, the number of $k$ is increased to more than one [39]. Using $k = 1$ can lead to false prediction, and a

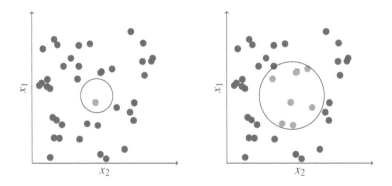

**FIGURE 4.5**   k-nearest neighbor classifying based on the $k$th observation.

set of kNNs is often used. When classifying the dependent variable is categorical, it can easily be made numerical by regression. The kNN regressor makes a regression based on the number of kNNs to minimize false predictions. The model considers a range of different $k$-values to find the optimal value. The kNN regressor needs thorough pre-processing and feature engineering to limit the effect of noise caused by irrelevant features, and is, therefore, dependent on finding the appropriate distance model [34]:

### 4.4.3   DISTANCE

A variety of distances is used in the algorithm. As seen in Equations 4.10, 4.11, 4.12, and 4.13, they are mostly used, since it is easy to intersect by changing the variable $q$. The variable $q$ is also considered to find the optimal value.

#### 4.4.3.1   Manhattan/City Block Distance

$$d(x, y) = \sum_{i=1}^{k} |x_i - y_i| \tag{4.10}$$

#### 4.4.3.2   Euclidean Distance

$$d(x, y) = \sqrt{\sum_{i=1}^{k} (x_i - y_i)^2} \tag{4.11}$$

#### 4.4.3.3   Minkowski Distance

$$d(x, y) = \left( \sum_{i=1}^{k} (|x_i - y_i|)^q \right)^{\frac{1}{q}} \tag{4.12}$$

### 4.4.3.4 Chebychev Distance

$$d(x,y) = \lim_{q \to \infty} \left( \sum_{i=1}^{k} (|x_i - y_i|)^q \right)^{\frac{1}{q}}$$  (4.13)

### 4.4.4 RANDOM FOREST REGRESSION

Random forest (RF) regression is a combination of decision trees, found through recursive partitioning to build a piece-wise linear model. From these tree models, it uses a majority vote for the most popular class. The trees grow dependent on a random vector, and the outputs are numerical scalars [40]. Each leaf on the tree is a linear model constructed for the cases at each node by regression techniques. One sole decision tree encompasses attributes and classes in the data and uses an entropy function and gain function to distinguish its structure. Entropy is known from thermodynamics as a measure of disorder, and later adopted by the information theory. In information theory, entropy is a measure of uncertainty of a variable, and defines a pure classifier [41]. In Equation (4.14), $p$ is positive and $n$ is negative:

$$\text{Entropy}(S) = -p * \log2(p) - n * \log2(n)$$  (4.14)

The entropy function is then used to evaluate the information gathered (gain) of an attribute, and thus to know how to choose the highest gaining attribute as the next branch in the decision tree. The equation yields the expected reduction in entropy, by imposing another branch in the decision tree.

$$\text{Gain}(S,A) = \text{Entropy}(S) - \sum_{v \in \text{Values}(A)} \frac{|S_v|}{|S|} \text{Entropy}(S_v)$$  (4.15)

In Equation (4.15), $A$ are attributes used for splitting the data into subsets ($S$). $S$ is the sum of subsets, and $S_v$ is the value of subsets. Using prior known input/output relationships, the algorithm searches for a model for the best prediction in the training set. The mathematical equations are structured in the algorithm, see Figure 4.6, based on the past knowledge.

### 4.4.4.1 Normalizing

The pre-processing of data is a transformed so that the machine learning algorithm can learn the patterns and generate a sound forecast. In a standard normalization process, input data are transformed with values from zero to one. This is done to make the predictive algorithm more robust [42].

$$\frac{\hat{X} - X_{\min}}{X_{\max} - X_{\min}}$$  (4.16)

$$\frac{\hat{X}}{X_{\text{sum}}}$$  (4.17)

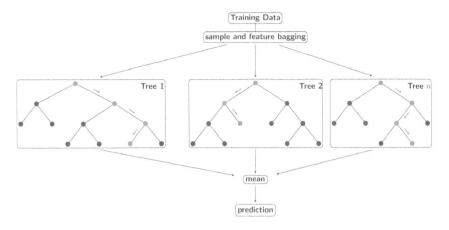

**FIGURE 4.6**   Random forest regression diagram sampling and voting from $n$ trees.

$$\frac{\hat{X}}{X_{\max}} \qquad (4.18)$$

$$\frac{\hat{X} - X_{\text{avg}}}{X_{\max} - X_{\text{avg}}} \qquad (4.19)$$

### 4.4.4.2   Performance Metrics

To evaluate the performance of load forecasting, a performance metric is used, including mean absolute error (MAE), mean absolute percentage error (MAPE), mean squared error (MSE), and symmetric mean absolute percentage error (SMAPE) [43]. They are defined as:

$$\text{MAE} = \frac{1}{n}\sum_{i=1}^{n}|y_i - \hat{y}| \qquad (4.20)$$

$$\text{SMAPE} = \frac{1}{n}\sum_{i=1}^{n}\left(\frac{|y_i - \hat{y}|}{\left(|y_i| + |\hat{y}|\right)/2}\right)*100 \qquad (4.21)$$

$$\text{MSE} = \frac{1}{n}\sum_{i=1}^{n}(y_i - \hat{y})^2 \qquad (4.22)$$

$$\text{MAPE} = \frac{1}{n}\sum_{i=1}^{n}\left|\frac{y_i - \hat{y}}{y_i}\right|*100 \qquad (4.23)$$

### 4.4.5   Visual Inspection

The first thing is to plot the time series of the data shown in Figures 4.7 and 4.8. In these plots, the time series are plotted as univariate time series with $y$-axis

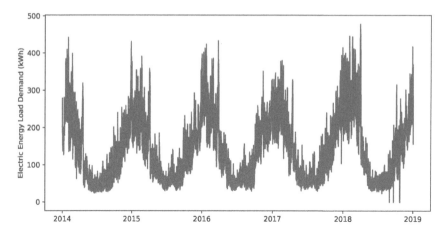

**FIGURE 4.7**   Rural load.

representing the univariate or dependent variable, and *x*-axis being the time axis. By visual inspection, these plots are illustrating the main structure of the time series. Important information such as time span, trends, and cycles are emerging in the figures. When applying intuition to visually inspect these time series, they certainly display some repetitive patterns, as in Figure 4.7, where load pattern seems to be taken a U-wave form that repeats itself over time. Figure 4.8 is much more dense than Figure 4.7, and looks to contain more information.

In some instances, a univariate time series can be explained by itself as is the case for univariate analysis; even then a univariate series can and most likely will be affected by other influences, but remains self-explanatory for this purpose. For the multivariate case where explanatory independent features are added, they are not directly connected to the dependent/response variable such as weather parameters, yet correlation exists to aid the time series analysis.

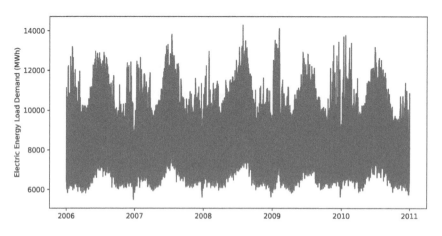

**FIGURE 4.8**   Urban load.

## 4.5 APPLICATIONS OF REGRESSION TECHNIQUES FOR ELECTRIC LOAD FORECASTING

Recent research from 2018 on computational intelligence approaches for energy load forecasting that reviewed more than 50 research papers related to the subject outlines the complexity of demand patterns as potentially influenced by factors such as climate, time periods, holiday or working days and other factors such as social activities, economic factors, including power market policies. Electrical energy demand is influenced by meteorological weather conditions; therefore, it is necessary to include the impact of meteorological weather parameters on electrical energy demand forecasting. The future electrified grid will increasingly depend on renewable intermittent energy sources (solar, wind), and the individual load profiles of such a system will change radically as home appliances include new energy demanding appliances (e.g., heat pump, electric vehicles, and induction stove) [44].

The regression models kNN, LR, and RF are supervised machine learning algorithms with a numerical outcome. The model is trained to find rules for pattern recognition in the input to output relation. The inputs to the model are known as features. Neural networks are the preferred machine learning tool and are known as both feedforward and back propagating networks, where a number of inputs are weighted in order to provide a predictive outcome. Neural networks are good for detecting non-linearities, and therefore preferred as a predictive tool in electrical load forecasting, yet also often criticized for low transparency and lack of interpretability because of the black box approach and using a large amount of data. Overfitting is still a challenging issue when applying neural networks to electrical demand prediction. It is known as the bias-variance trade-off. When a model is of very low complexity and yet scores well, it is highly biased, which signifies that the data fits the model accurately (the training set), and it will often perform poorly on new data (from the test set). The model should contain a complexity that is in coherence with the level of information embedded in the data. Somewhere in between is the optimal model, also referred to as the suitable model [45].

Urban area load is influenced by meteorological conditions; therefore, it is important to include impact of weather parameters on load prediction, yet this impact is governed by the prediction time, greater for long term, and decreases as the prediction time is narrowed. The electrical energy demand is influenced by the user behavior as well as weather conditions. Individual human behavior and weather are so random that a complex neural network would not predict the outcome better than a coin toss. Hence, if one is to analyze the load demand of larger area such as the urban area, systematic load behavior with correlation to weather parameters and continuous load profile should be investigated.

This work has uniqueness in electrical demand forecasting using regression tools through vertical approach, and it also considers the impact of meteorological parameters. This vertical approach uses less amount of data compared to continuous time series as well as neural network techniques.

The objectives of this work are to explore the use of regression tools for regional electrical load forecasting by correlating lower distinctive categorical levels (season,

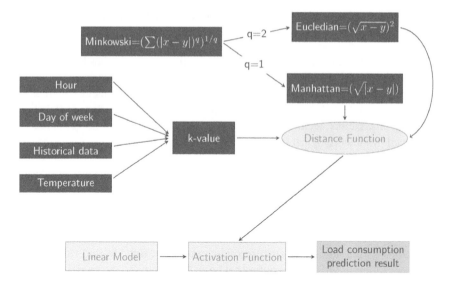

**FIGURE 4.9**   The regressor model for electric load demand forecasting.

day of the week) and weather parameters, see Figure 4.9. The vertical time approach is to consider a sample time period (e.g., seasonally and weekly) of data for four years, which will be tested for the same time period for the consecutive year. A vertical axis approach is shown to be competitive to ANN.

### 4.5.1   Feature Engineering for Electric Load Demand Forecasting

The following parameters are important for system electrical energy demand:

1. Time
2. Weather
3. Random effects

#### 4.5.1.1   Time

Apart from the seasonal effects, underlying patterns emerge in the system load demand. There are different peaks throughout the seasons, whether it is a winter peak or a summer peak. Emerging under this seasonal patterns are daily- and weekly-cycles. The daily routines of human behavior are manifested in systematic load patterns on a daily basis. Day of the week is also significant. Working day or non-working day (weekend or other calendar event) changes human activities, and whether it is a working day or not, influences load patterns. People might also during weekends shift their sleeping habits, as to wake up later, and thus change the diurnal load demand to delay the morning peak load demand. Sub-categorical levels such as working/non-working days are referred to in the literature as an indicator variable. Such an indicator variable composes a lower indicator level, with a binary switch of working days and non-working days/holidays (0 and 1). To give this property to our algorithms is very important as it makes prediction of forecast load more efficient. The use of such type of variables has been successfully employed in the forecasting of electric market [42, 46–49].

### 4.5.1.2 Weather

Weather variables play an important rule in changing load patterns. The effect of ambient temperature as well as past temperature on the load is necessary for prediction analysis; the indoor temperature on a hot summer day may reach its peak after sunset due to heat buildup in the construction materials of buildings. In addition to the daily heat buildup, a sequence of days with high temperature creates a new system peak. The time delay, from a shift in temperature until the change in electric load demand is observed, should be evaluated through the temporal similarity of cross-correlation between the load and different weather parameters: DryBulb, DewPnt, WetBulb, and Humidity. Dry bulb temperature (DBT) is the temperature measured from air, yet not exposed to solar radiation or moisture. Wet bulb temperature (WBT) measured from a thermometer where the bulb of the measurement device is soaked by a wet cloth. As long as the air is not saturated, evaporation from the moist cloth keeps the WBT lower than the DBT. From the DBT and WBT, one can then derive the relative humidity of the air and the dew point from a Mollier Chart by psychometric. In humid and hot conditions, it is likely that humidity will effect the load pattern in similar ways as temperature. Humidity explains the complex relation between temperature and load, and therefore mathematical models are not enough in a thorough analysis. Humidity is the amount of water vapor in the air, and might increase the gap between the actual and the apparent or felt temperature. When regulating temperature, the body utilizes evaporative cooling, and the rate of evaporation through the skin is correlated to humidity, and because of the conductive properties of water, we feel warmer at high humid conditions. Also, due to the seasonal changes of weather data, the correlation to the electrical load will vary during the year. Many electrical utilities are weather-sensitive such as heating and air conditioning. Electric loads are often classified into weather-sensitive load and non-weather-sensitive load. Temperature data is obviously a very important factor affecting the load. However, its value is often limited to the confidence level on weather forecasting. Therefore, unless the weather forecasting is very accurate, an underlying deterministic model is its premise. The complexity in the control system engineering of maintaining thermal comfort as well as optimizing for energy is important to know. At the same time, it is important to acknowledge that most houses are designed to resist the worst meteorological conditions. There are also limitations in the heating system itself that might cause load peaks, such as the inertia in the floor heating system, known as thermal lag. Therefore machine learning can help to use the weather parameters for load predictions in the built-environment [50–53].

### 4.5.1.3 Random Effects

Random disturbances may lead to a change in the number of electricity consumers due to many factors. Infrastructural changes in the urban area and maintenance work are random effects that are not detected by pattern recognition. Load patterns are consistent from year to year, and show reoccurring seasonal pattern. When the yearly load curves do not vary from year to year, it means that there are no economic trends. Load prediction analysis using machine learning can take care of random effects.

The effect of external parameters on load predictions can be considered through the machine learning approaches for different type of loads (e.g. rural area and urban area loads).

## 4.6   CASE STUDY 1: RURAL AREA ELECTRIC ENERGY LOAD

In this study, the dataset for rural area electric energy load is the data collected by a smart meter at a electric substation providing Nissedal Cabin Area in Bjønntjønn with power. It is a typical Norwegian rural power network with 125 cottages, and 478 kW peak demand. The dataset is hereby referred to as the Bjønntjønn dataset. The rural area load profile is illustrated in Figure 4.7. The smart meter collects data at every hour, as a point value, making it a dataset of hourly values. The weather information by Norwegian Institute of Bioeconomy Research (NIBIO) runs 52 weather stations with detailed information down to hourly resolution and freely downloadable on their web service (lmt.nibio.no). Among the 52 weather stations, three weather stations closest to Bjønntjønn Cabin Area are Bø, Gvarv, and Gjerpen. Based on the correlation analysis, the weather station with the strongest correlation of temperature to the load data from Bjønntjønn Cabin Area is identified, and used for the analysis.

## 4.7   CASE STUDY 2: URBAN AREA ELECTRIC LOAD

The dataset for urban area electric load contains 87648 collected datapoints from the urban area of Sydney in the region of New South Wales in Australia. These datapoints are collected at every 30 minutes, spanning from five years. Since it is the granularity of collected data observations that decides the lower limit of forecast window, this dataset gives the oppurtunity of 30 minutes predictions. The historical data is gathered by Australian Energy Market Operator (AEMO) and Bureau of Meteorology (BOM) from years 2006 to 2010, and hereafter referred to as the Sydney dataset. During the years 2006–2010, the maximum load was 14274.2 MW. In this study the purpose is to test the regression tools on the available real data of urban area.

## 4.8   RESULTS AND DISCUSSION

In this work, several regression tools have been analyzed and compared for different datasets. Based on the analysis of the data and regressors, a new vertical approach has been further developed and inferred to deal with the relatively low amount of data and load pattern; it has been in particular validated for the case studies (1) in the rural area and (2) in the urban area.

The vertical time approach also uses seasonal data for training and inference. The horizontal approach uses continuous datasets, i.e., it utilizes all data in a continuum from the start of the dataset until the time period used for inference. The illustration of horizontal and vertical approaches is presented in Figure 4.10.

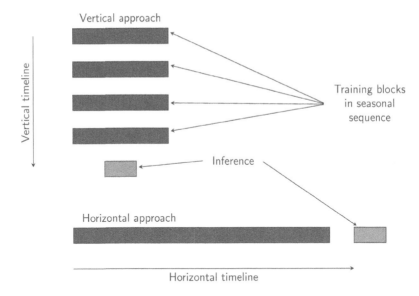

**FIGURE 4.10**   Illustration of vertical and horizontal approaches.

Vertical approach can be performed with minimum amount of data compared to continuous approach. Also, the vertical time approach predictive results are compared with prediction based on continuous time series data. In vertical approach, the training set, $D = \{x_i\}_{i=1}^{N}$, is partitioned into subsets by each season of the year, and then are merged together only containing seasonally information about the load pattern. In a dataset containing time observation for five years (e.g., 2016–2020), time is separately selected season-wise, and then merged to contain only the specific season for training, $D = \{x_{spring_i}\}_{i=2016}^{2019}$. In this study, the inferred test-set is for a week in the middle of the selected trained season for the following year $D = \{x_{week}\}_{i=monday}^{sunday}$. Seasons are divided into months, as shown in Table 4.1, where Season 1 is Winter, and Season 4 is Autumn.

### 4.8.1   CASE STUDY 1: RURAL AREA

In the case study of rural area load prediction, the regression analysis has been done on continuous time basis as well as using vertical time axis approach. The

**TABLE 4.1**
**Seasons**

| Season | Months | | |
|---|---|---|---|
| Season 1 | December | January | February |
| Season 2 | March | April | May |
| Season 3 | June | July | August |
| Season 4 | September | October | November |

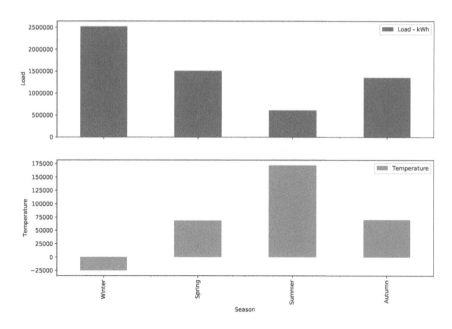

**FIGURE 4.11**   Load consumption and temperature profiles on seasonal basis.

correlation analysis of load and weather parameters has been analyzed to study the relation between meteorological parameters and electricity consumption. The hourly electrical loads of each season have been juxtaposed to the seasonal temperature, and negative correlation has been observed (Figure 4.11).

From this observation, it can be seen that vertical approach enables the algorithm to reveal complexity of load and temperature for better prediction results [54]. The relation between working days and non-working days affects the cycles of load consumption, and is noticeable in the latter part of of the holiday where load demand increases even more (Figure 4.12).

The load pattern shows autocorrelation (AC) to previous lags, as seen in Figure 4.13. The AC aids the feature extraction procedure in engineering for the optimal previous k-lag values to be selected for the predictive algorithm. The observed results from the autocorrelation function (ACF) plot (Figure 4.13) shows a steep linear decline in lags 0–5; after that the slope is almost horizontal (lags 6–15) before it makes a small bump at lag 17–20, for then again to increase its value for the 23rd lag (which is the 24th hour since unity lag is zero), and then a deep decrease. The ACF plot also shows strong dependencies on historical data values, which indicate that the time series is autoregressive. The further correlation analysis of the rural electrical load demand patterns reveals also a strong dependency on the day of the week. For the considered Norwegian rural load of holiday cabins, the Norwegian holidays are identified as Easter, labor day, national day, ascension day, Pentecost, and X-mas. The observed correlations between the load and temperature, load and working days/non-working days, and the intercorrelation of temperature and working days/non-working days for the rural area have been well within the good heuristic model for correlation-based

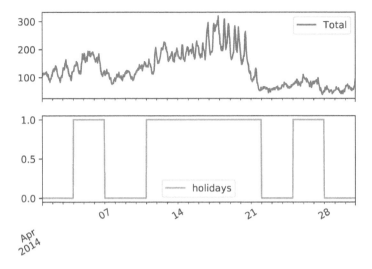

**FIGURE 4.12**   Load consumption related to working and non-working days.

feature selection. The heuristics of good correlation-based feature selection is based on the level of intercorrelation within the class and subset features. In the rural area, there is no correlation between the working days and temperature. A good feature set contains independent variables that have high positive or negative correlation to the dependent variable, and no correlation amongst the other dependent variables [55]. In further evaluation of the regressors, performance metrics are used and the results are presented in Table 4.2 and 4.3.

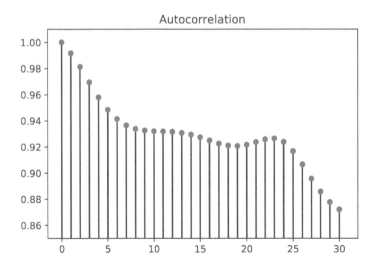

**FIGURE 4.13**   Autocorrelation of load consumption of the first 30 lags for Bjønntjønn Cabin Area 2014–2018.

**TABLE 4.2**

**Forecasting Results (24 Hours Prediction) for Season 1 (Winter) Trained with Time Feature Lags of 24-, 48- and 168-Hours**

| Features | Vertical Approach Winter | | | Continuous Approach Winter | | |
|---|---|---|---|---|---|---|
| | SMAPE | MAPE | MAE | SMAPE | MAPE | MAE |
| kNN AC | 9.88 | 10.06 | 26.07 | 9.72 | 9.74 | 25.60 |
| RF AC | 10.43 | 10.67 | 27.85 | 9.56 | 9.49 | 25.24 |
| kNN AC AR | 10.05 | 10.20 | 26.39 | 9.25 | 9.24 | 24.42 |
| RF AC AR | 10.87 | 11.03 | 28.67 | 10.34 | 10.34 | 26.91 |
| kNN AC T H | 9.48 | 9.66 | 25.09 | 9.05 | 9.09 | 23.89 |
| RF AC T H | 11.39 | 11.53 | 29.86 | 11.50 | 11.53 | 29.81 |
| kNN AC AR T H | 9.75 | 9.92 | 25.65 | **8.88** | **8.86** | 23.45 |
| RF AC AR T H | 12.03 | 12.18 | 31.56 | 10.88 | 10.96 | 28.06 |

In this work, different features in the regression tools (kNN and RF) have been studied to analyze how they perform. In Tables 4.2 and 4.3, the autocorrelation (AC), autoregression (AR), temperature (T) and holiday effects (H) have been studied separately and together (AC, AR, T, H) combined with the regressors. The performance metrics SMAPE, MAPE, and MAE have been chosen to make appropriate analysis of their performance (see section 4.5.4.2). MAE is the most straightforward error estimation, but is poor in order to understand the context it is given; therefore MAPE is more used, since it is normalized to the true value of the time series. Typially for the rural area, the load demand is low, opposite to the urban area, and occasionally

**TABLE 4.3**

**Forecasting Results (24 Hours Prediction) for Season 3 (Summer) Trained with Time Feature Lags of 24-, 48- and 168-Hours**

| Features | Vertical Approach Summer | | | Continuous Approach Summer | | |
|---|---|---|---|---|---|---|
| | SMAPE | MAPE | MAE | SMAPE | MAPE | MAE |
| kNN AC | 12.74 | 12.74 | **6.87** | 13.17 | 13.35 | 7.17 |
| RF AC | 14.70 | 14.78 | 8.07 | 15.27 | 15.47 | 8.49 |
| kNN AC AR | 13.17 | 13.24 | 7.11 | 13.28 | 13.43 | 7.23 |
| RF AC AR | 14.16 | 14.14 | 7.70 | 13.89 | 14.07 | 7.54 |
| kNN AC T H | 14.79 | 14.46 | 7.94 | 15.07 | 14.75 | 8.08 |
| RF AC T H | 16.53 | 16.10 | 8.80 | 17.05 | 16.48 | 9.14 |
| kNN AC AR T H | 14.27 | 14.07 | 7.68 | 14.41 | 14.14 | 7.71 |
| RF AC AR T H | 16.98 | 16.66 | 9.02 | 17.21 | 16.91 | 9.19 |

Note the big difference in MAE between the seasons; however, MAPE and SMAPE have more or less the same values. This is due to relatively higher load consumption in winter time that leads to a higher absolute error, but when compared in absolute percentage error, the error is not noticeable.

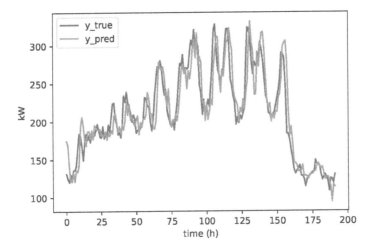

**FIGURE 4.14**  Prediction result without error estimation.

the rural area load reaches zero. At zero values the MAPE is obsolete and the performance is also measured by SMAPE.

Table 4.2 compares the vertical and continuous approach for winter season, whilst Table 4.3 compares the vertical and continuous approach for summer season.

The kNN regressor is compared to RF regressor, and it also uses autoregression. In the analysis, a visual inspection might aid to understand the predictive outcome. Prediction results are compared with and without error estimation (see Figures 4.14 and 4.15). The kNN and RF alone has no information about the finite gradient of the curvature. In Figure 4.14, the two graphs mostly appear to merely be shifted in time. To overcome this, the real value was compared to the error estimation (see Figure 4.15),

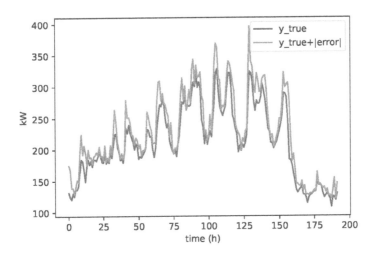

**FIGURE 4.15**  Prediction result with error estimation.

and increasingly peaking errors were shown. A simple form of autoregression is tried in order to mitigate the problem of peaking errors. It is a possible remedy, since the correlation analysis showed a strong autocorrelation to the first historical instances of the time series. Instead of a cumbersome ordinary least square-search (OLS) for the backshift operator parameter, only a backshift value is found based on Equation 4.24. The autoregressor is used to find the curvature and gives a finite gradient based on the latest update from the targeted vector (in this case, the load).

$$c = (L_{t-1} - L_{t-2})^{\frac{1}{p}} \qquad (4.24)$$

Autoregression is the simplest and most straightforward predictive model, based on the targeted vector itself, and at certain time window, it indicates the decline and incline of the time window, and gives a finite gradient for the curvature of load profiles. The joint learning of regression tools with autoregression predicts time series components of different characteristics. Other hybrid combinations can be done with MA, ARMA, and ARIMA models, to aid the regressor model in the predictions.

The load profile of the considered holiday resort (rural area) is categorized seasonally. In this work, regression tools are used for load predictive analysis. In the load predictive analysis, vertical time approach is used for a particular holiday time period. Vertical approach can be performed with minimum amount of data compared to continuous approach. Also, in vertical time approach, predictive results are compared with the prediction based on continuous time series (i.e., horizontal approach). The presented vertical approach methodology can also deal with the problems of irregularities and randomness in the dataset [56].

## 4.8.2 CASE STUDY 2: URBAN AREA

The dataset for urban area electric load contains 87648 collected datapoints from the urban area of Sydney in the region of New South Wales in Australia. The relative comparison of load prediction with MAPE for considered regression tools for 30 minutes and 24 hours is done using both horizontal and vertical approaches for all seasons. The results are shown in Table 4.4. It is found that the lowest MAPE is achieved with the use of previous load patterns together with indicator variables, and noticeably disregarding weather variables. This goes well with the previous analysis of correlation, which confirms that previous load patterns and indicator variables have higher correlation to the actual load than the weather parameters.

It has been observed from the test results, the lowest MAPE is found through RF regressor for 30 minutes prediction using vertical approach. For the 24-hour time period, kNN provides the lowest MAPE through vertical approach.

MAPE for 30 minutes prediction results using RF regressor varies between 1% and 2%, and provides very good results compared to other regressions techniques, which have been used in this work. The 24 hours prediction results using kNN regressor technique have MAPE of 2.61%, which is much better compared to other regressors. From the results, it has been observed that for short-term predictions (30 minutes), RF regressor should be used; and for long-term predictions (24 hours), kNN regressor should be considered [53].

## TABLE 4.4
## MAPE for Urban Area Load and Indicator Aggregated Version Test Results

| Time | Regressor | | |
|------|-----------|---|---|
| | Random Forest | k-Nearest Neighbour | Linear Regression |
| **Season One Horizontal Approach** | | | |
| 30 minutes | 1.11(9*) | 1.98(7**,1***) | 2.04 |
| 24 hours | 5.32(13*) | 6.53(4**,1***) | 5.15 |
| **Season One Vertical Approach** | | | |
| 30 minutes | 0.94(16*) | 1.85(8**,1***) | 1.76 |
| 24 hours | 5.88(13*) | 5.49(5**,2***) | 5.83 |
| **Season Three Horizontal Approach** | | | |
| 30 minutes | 1.12(17*) | 2.36(5**,1***) | 2.29 |
| 24 hours | 4.76(9*) | 5.41(19**,1***) | 5.27 |
| **Season Three Vertical Approach** | | | |
| 30 minutes | *0.86(17*)* | 1.19(6**,1***) | 2.15 |
| 24 hours | 2.71(17*) | *2.61(17**,1***)* | 4.26 |

*Note:* * n-estimator; ** k-value; ***q-value.

RF regressor, kNN regressor, and LR are used for analyzing the urban area electrical energy demand forecasting, using larger dataset of Sydney region. Data correlation over seasonal changes have been argued by means of improving MAPE. By examining the structure of various regressors, they are compared for the lowest MAPE. The regressors show good MAPE for short term (30 minutes) prediction, and RF regressor scores best in the range of 1–2% MAPE. kNN shows the best results for 24 hours prediction, with a MAPE of 2.61%. The prediction of the short-term 30 minutes electrical energy using vertical approach is relatively better through RF regression tool. For long-term prediction of 24 h, kNN regression tool can provide better results using vertical approach.

---

Urban area electrical energy demand forecasting is very important for generation scheduling and flexibility with consideration of renewable energy sources and possible demand side management. Urban area electrical energy demand predictions for short term (30 minutes) and long term (24 hours) are necessary for scheduling power generation units as well as for participating them in short term and day ahead energy market.

The seasonal patterns are repeating with the same upper and lower limits (e.g., repeating on annual basis), and can be further investigated for economic effects on the load behavior in the urban area of Sydney during the years 2006–2010. When investigating the Sydney dataset, we find that the load curves, yet containing cyclic and seasonal differences, do not contain significant changes on the system load due to changing economic trends [57]. When inspecting the daily and weekly load cycle, we can clearly see a load pattern emerging from a very low activity during the early hours of the day, into one peak at morning (between 8 and 10 hours), and another peak in the evening (between 19 and 21 hours). The same daily repeating patterns, with a low activity followed by two peaks, are also evident in the weekly cycle, except for the last two days of the week (Saturday and Sunday) when the peaks and general load are lower. It can be seen that urban area load predominantly reflects the domestic load, and it can be correlated to human behavior. The periodicity in the load patterns reveals a load demand that reflects a consumer lifestyle. When examining the features enlisted in the Sydney dataset, it has indicators "Date" and "Hour", four weather parameters, information about the electricity price, "ElecPrice" and information about the electricity load consumption, "SYSLoad". These features have been developed in the pre-processing to match the requirements of the prediction tool.

## 4.9  CONCLUSION

This work has explored the use of regression tools for electrical energy load forecasting through correlating weather parameters as well as the time period. Load prediction analysis using regression tools has been done on continuous time basis (horizontal) as well as using vertical time approach. The Pearson method and visual inspection of the vertical approach depict meaningful relation among pre-processing of data, test methods, and results for the examined regressors.

The application of regression tools has shown to be promising for predicting electric load within distributed network as well as for flexibility analysis. The distributed network are low-capacity networks with low amount of data that need flexible operation and analysis. RF regressor and kNN regressor are considered for analyzing the rural area and urban area electrical energy demand forecasting. In addition, LR is used for urban area due to the continous load patterns.

The methodology presented is developed to deal with the problems of irregularities and randomness in the time series. RF regressor yields good result on hourly time prediction in load forecasting. The kNN regressor has shown precise prediction in time series due to its capability to capture the nearest step in a time series based on the nearest neighbor principle.

Autocorrelation is a neat and practical approach to feature engineering that saves time for the appropriate actions to be made for feature extraction. The regression tools can handle the low amount of data, typical for the rural area, for day-ahead forecasting. In this work, the regression analysis for load prediction of rural area is done using vertical and continuous time approaches for day-ahead planning with 24 hours prediction. The vertical time approach uses seasonal data for training and inference, as opposed to continuous time approach that utilizes all data in a continuum from the start of the dataset until the time period used for inference. The regression tools can handle the low amount of data, and the prediction accuracy (through MAPE) matches with other techniques. It is observed that through load predictive analysis, the autocorrelation by vertical approach with kNN-regressor gives a low SMAPE. The kNN captures the lower boundaries of the load demand quite well. When analyzing the error, we find that the algorithms struggle for identifying and predicting the high peaks of the load demand. When the autoregression is given, it helps the algorithm to find the curvature of high peaks; even without capturing the overall trend of the load peak demand, MAPE can be improved by autoregression.

RF regressor, kNN regressor, and LR are used for analyzing the urban area electrical energy demand forecasting. The presented regression techniques can forecast electrical demand for short term (30 minutes) and long term (24 hours) using limited datasets. Vertical axis approach can have more competitiveness to ANN due to the use of low amount of data and considering the impact of meteorological parameters.

Load forecasting is the most fundamental application of smart grid, which provides essential input for flexibility such as demand response, topology optimization, and abnormally detection, facilitating the integration of intermittent clean energy sources.

## REFERENCES

[1] K. Kostkov, Omelina, P. Kyina, and P. Jamrich, "An introduction to load management," *Electric Power Systems Research*, vol. 95, pp. 184–191, 2013. [Online]. Available: http://www.sciencedirect.com/science/article/pii/S037877961200288X.

[2] P. Denholm and M. Hand, "Grid flexibility and storage required to achieve very high penetration of variable renewable electricity," *Energy Policy*, vol. 39, no. 3, pp. 1817–1830, 2011. [Online]. Available: http://www.sciencedirect.com/science/article/pii/S0301421511000292.

[3] M. G. Lauby and R. Villafranca, "Flexibility requirements and metrics for variable generation: Implications for system planning studies," *North American Electric Reliability Corporation*, 116–390 Village Blvd., 08540, Tech. Rep., Aug. 2010.

[4] C. Gorria, J. Jimeno, I. Laresgoiti, M. Lezaun, and N. Ruiz, "Forecasting flexibility in electricity demand with price/consumption volume signals," *Electric Power Systems Research*, vol. 95, pp. 200–205, 2013. [Online]. Available: http://www.sciencedirect.com/science/article/pii/S0378779612002933.

[5] E. Hirst, "Flexibility benefits of demand-side programsin electric utility planning," *The Energy Journal*, vol. 11, no. 1, 1990.

[6] C. B. Smith and K. E. Parmenter, "Chapter 1 - introduction," in *Energy Management Principles (Second Edition)*, second edition ed., C. B. Smith and K. E. Parmenter, Eds. Elsevier, 2016, pp. 1–12. [Online]. Available: http://www.sciencedirect.com/science/article/pii/B978012802506200001X.

[7] F. E. R. Commision. (2017) Federal energy regulatory commissin reports on demand response and advanced metering, [online]. [Online]. Available: https://www.ferc.gov/industries-data/electric/power-sales-and-markets/demand-response/reports-demand-response-and.

[8] L. I. Minchala-Avila, J. Armijos, D. Pesntez, and Y. Zhang, "Design and implementation of a smart meter with demand response capabilities," *Energy Procedia*, vol. 103, pp. 195–200, 2016, Renewable Energy Integration with Mini/Microgrid Proceedings of REM2016. [Online]. Available: http://www.sciencedirect.com/science/article/pii/S1876610216314825.

[9] NATO Advanced Study Institute on Demand-Side Management and Electricity End-Use Efficiency, and Almeida, Anibal T. de. and Rosenfeld, Arthur H., and North Atlantic Treaty Organization, *Demand-side management and electricity end-use efficiency/edited by Anibal T. De Almeida and Arthur H. Rosenfeld*. Kluwer Academic Publishers Dordrecht, 1988.

[10] A. Shivakumar, C. Taliotis, P. Deane, J. Gottschling, R. Pattupara, R. Kannan, D. Jak, K. Stupin, R. V. Hemert, B. Normark, and A. Faure-Schuyer, "Chapter 21 - need for flexibility and potential solutions," in *Europe's Energy Transition*, M. Welsch, S. Pye, D. Keles, A. Faure-Schuyer, A. Dobbins, A. Shivakumar, P. Deane, and M. Howells, Eds. Academic Press, 2017, pp. 149–172. [Online]. Available: http://www.sciencedirect.com/science/article/pii/B9780128098066000213.

[11] A. Benato, A. Stoppato, and S. Bracco, "Combined cycle power plants: A comparison between two different dynamic models to evaluate transient behaviour and residual life," *Energy Conversion and Management*, vol. 87, pp. 1269–1280, 2014. [Online]. Available: http://www.sciencedirect.com/science/article/pii/S0196890414005421.

[12] K. Van den Bergh and E. Delarue, "Cycling of conventional power plants: Technical limits and actual costs," *Energy Conversion and Management*, vol. 97, pp. 70–77, 2015. [Online]. Available: http://www.sciencedirect.com/science/article/pii/S0196890415002368.

[13] M. Uddin, M. Romlie, M. Abdullah, C. Tan, G. Shafiullah, and A. Bakar, "A novel peak shaving algorithm for islanded microgrid using battery energy storage system," *Energy*, vol. 196, p. 117084, 2020. [Online]. Available: http://www.sciencedirect.com/science/article/pii/S0360544220301912.

[14] G. Box and G. M. Jenkins, *Time Series Analysis: Forecasting and Control*. Holden-Day, 1976.

[15] S. Vemuri, W. L. Huang, and D. J. Nelson, "On-line algorithms for forecasting hourly loads of an electric utility," *IEEE Transactions on Power Apparatus and Systems*, vol. PAS-100, no. 8, pp. 3775–3784, 1981.

[16] N. D. Uri, "Forecasting peak system load using a combined time series and econometric model," *Applied Energy*, vol. 4, no. 3, pp. 219–227, 1978. [Online]. Available: http://www.sciencedirect.com/science/article/pii/0306261978900041.

[17] S. Rahman and R. Bhatnagar, "An expert system based algorithm for short term load forecast," *IEEE Transactions on Power Systems*, vol. 3, no. 2, pp. 392–399, 1988.

[18] M. J. Damborg, M. A. El-Sharkawi, M. E. Aggoune, and R. J. Marks, "Potential of artificial neural networks in power system operation," in *IEEE International Symposium on Circuits and Systems*, 1990, pp. 2933–2937, vol. 4.

[19] D. J. Sobajic and Y. Pao, "Artificial neural-net based dynamic security assessment for electric power systems," *IEEE Transactions on Power Systems*, vol. 4, no. 1, pp. 220–228, 1989.

[20] R. Fischl, M. Kam, J. Chow, and S. Ricciardi, "Screening power system contingencies using a back-propagation trained multiperceptron," in *1989 IEEE International Symposium on Circuits and Systems (ISCAS)*, 1989, pp. 486–489 vol. 1.

[21] H. Mori and S. Tsuzuki, "Power system topological observability analysis using a neural network model," in *Proc. of Second Symposium on Expert Systems Application to Power Systems*, 1989, pp. 385–391.

[22] A. Marthur, T. Samad, and K. Anderson, "Neural networks and how the utility industry can benefit from them," in EPRI Conference on Expert Systems Applications for the Electric Power Industry, Scottsdale, Arizona, 1989.

[23] S. Matsuda and Y. Akimoto, "The representation of large numbers in neural networks and its application to economical load dispatching of electric power," in International Joint Conference on Neural Networks. IEEE, 1989, pp. 587–592.

[24] D. C. Park, M. A. El-Sharkawi, R. J. Marks, L. E. Atlas, and M. J. Damborg, "Electric load forecasting using an artificial neural network," *IEEE Transactions on Power Systems*, vol. 6, no. 2, pp. 442–449, 1991.

[25] A. G. Bakirtzis, V. Petridis, S. J. Kiartzis, M. C. Alexiadis, and A. H. Maissis, "A neural network short term load forecasting model for the greek power system," *IEEE Transactions on Power Systems*, vol. 11, no. 2, pp. 858–863, 1996.

[26] C.-C. Hsu and C.-Y. Chen, "Regional load forecasting in taiwanapplications of artificial neural networks," *Energy Conversion and Management*, vol. 44, no. 12, pp. 1941–1949, 2003. [Online]. Available: http://www.sciencedirect.com/science/article/pii/S019689040200225X

[27] C. Lu, H. Wu, and S. Vemuri, "Neural network based short term load forecasting," *IEEE Transactions on Power Systems*, vol. 8, no. 1, pp. 336–342, 1993.

[28] T. Hastie, R. Tibshirani, and J. Friedman, *The Elements of Statistical Learning, Series*, Springer Series in Statistics. Springer New York Inc., 2001.

[29] A. Nielsen, *Practical Time Series Analysis: Prediction with Statistics and Machine Learning*. O'Reilly Media, Inc, 2019.

[30] N. K. Ahmed, A. F. Atiya, N. E. Gayar, and H. El-Shishiny, "An empirical comparison of machine learning models for time series forecasting," *Econometric Reviews*, vol. 29, no. 5–6, pp. 594–621, 2010. [Online]. Available: https://doi.org/10.1080/07474938.2010.481556.

[31] L. Wen, K. Zhou, S. Yang, and X. Lu, "Optimal load dispatch of community microgrid with deep learning based solar power and load forecasting," *Energy*, vol. 171, pp. 1053–1065, 2019. [Online]. Available: http://www.sciencedirect.com/science/article/pii/S0360544219300775.

[32] H. J. Sadaei, P. C. de Lima e Silva, F. G. Guimares, and M. H. Lee, "Short-term load forecasting by using a combined method of convolutional neural networks and fuzzy time series," *Energy*, vol. 175, pp. 365–377, 2019. [Online]. Available: http://www.sciencedirect.com/science/article/pii/S0360544219304852.

[33] K. P. Murphy, *Machine Learning: A Probabilistic Perspective*. The MIT Press, 2012.

[34] S. M. Weiss and N. Indurkhya, "Rule-based machine learning methods for functional prediction," *Journal of Artificial Intelligence Research*, vol. 3, no. 1, p. 383403, Dec. 1995.

[35] J. R. Quinlan, "Combining instance-based and model-based learning," in *Proceedings of the tenth international conference on machine learning*, 1993, pp. 236–243.

[36] P. Virtanen, R. Gommers, T. E. Oliphant, M. Haberland, T. Reddy, D. Cournapeau, E. Burovski, P. Peterson, W. Weckesser, J. Bright, S. J. van der Walt, M. Brett, J. Wilson, K. J. Millman, N. Mayorov, A. R. J. Nelson, E. Jones, R. Kern, E. Larson, C. J. Carey, I. Polat, Y. Feng, E. W. Moore, J. VanderPlas, D. Laxalde, J. Perktold, R. Cimrman, I. Henriksen, E. A. Quintero, C. R. Harris, A. M. Archibald, A. H. Ribeiro, F. Pedregosa, P. van Mulbregt, and SciPy 1.0 Contributors, "SciPy 1.0: Fundamental algorithms for scientific computing in Python," *Nature Methods*, vol. 17, pp. 261–272, 2020.

[37] M. N. Nounou and B. R. Bakshi, "Chapter 5 - multiscale methods for denoising and compression," in *Wavelets in Chemistry*, ser. Data Handling in Science and Technology, B. Walczak, Ed. Elsevier, 2000, vol. 22, pp. 119 – 150. [Online]. Available: http://www.sciencedirect.com/science/article/pii/S0922348700800301.

[38] T. R. Derrick and J. M. Thomas, *Time Series Analysis: The Cross-Correlation Function*. Kinesiology Publications, 2004. [Online]. Available: https://lib.dr.iastate.edu/kin\.pubs/46.

[39] D. Patidar, B. C. Shah, and M. Mishra, "Performance analysis of k nearest neighbors image classifier with different wavelet features," *2014 International Conference on Green Computing Communication and Electrical Engineering (ICGCCEE)*, pp. 1–6, 2014.

[40] L. Breiman, "Random forests," *Springer, Machine Learning*, 2001.

[41] S. Jukna, *The Entropy Function*. Springer Berlin Heidelberg, 2011, pp. 313–326. [Online]. Available: https://doi.org/10.1007/978-3-642-17364-6\.22.

[42] M. Q. Raza and A. Khosravi, "A review on artificial intelligence based load demand forecasting techniques for smart grid and buildings," *Renewable and Sustainable Energy Reviews*, vol. 50, pp. 1352–1372, 2015. [Online]. Available: http://www.sciencedirect.com/science/article/pii/S1364032115003354.

[43] N. J. Johannesen, M. Kolhe, and M. Goodwin, "Deregulated electric energy price forecasting in nordpool market using regression techniques," in *2019 IEEE Sustainable Power and Energy Conference (iSPEC)*, 2019, pp. 1932–1938.

[44] S. Fallah, R. Deo, M. Shojafar, M. Conti, and S. Shamshirband, "Computational intelligence approaches for energy load forecasting in smart energy management grids: State of the art, future challenges, and research directions," *Energies*, vol. 11, no. 3, p. 596, Mar. 2018. [Online]. Available: http://dx.doi.org/10.3390/en11030596.

[45] P. Bacher and H. Madsen, "Identifying suitable models for the heat dynamics of buildings," *Energy and Buildings*, vol. 43, no. 7, pp. 1511–1522, 2011. [Online]. Available: http://www.sciencedirect.com/science/article/pii/S0378778811000491.

[46] A. Lahouar and J. B. H. Slama, "Random forests model for one day ahead load forecasting," *IREC2015 The Sixth International Renewable Energy Congress*, pp. 1–6, 2015.

[47] F. H. Al-Qahtani and S. F. Crone, "knn," *The 2013 International Joint Conference on Neural Networks (IJCNN)*, pp. 1–8, 2013.

[48] M. Usman and N. Arbab, "Factor affecting short term load forecasting," *Journal of Clean Energy Technologies*, vol. 2, 2014.

[49] N. J. Johannesen, M. Lal Kolhe, and M. Goodwin, "Load demand analysis of nordic rural area with holiday resorts for network capacity planning," in *2019 4th International Conference on Smart and Sustainable Technologies (SpliTech)*, 2019, pp. 1–7.

[50] M. Mastouri and N. Bouguila, "A methodology for thermal modelling and predictive control for building heating systems," *2017 18th International Conference on Sciences and Techniques of Automatic Control and Computer Engineering (STA)*, pp. 568–573, 2017.

[51] T. Salque, D. Marchio, and P. Riederer, "Neural predictive control for single-speed ground source heat pumps connected to a floor heating system for typical french dwelling," *Building Services Engineering Research and Technology*, vol. 35, no. 2, pp. 182–197, 2014. [Online]. Available: https://doi.org/10.1177/0143624413480370.

[52] K. Y. Lee, T. I. Choi, C. C. Ku, and J. H. Park, "Short-term load forecasting using diagonal recurrent neural network," in *[1993] Proceedings of the Second International Forum on Applications of Neural Networks to Power Systems*, 1993, pp. 227–232.

[53] N. J. Johannesen, M. Kolhe, and M. Goodwin, "Relative evaluation of regression tools for urban area electrical energy demand forecasting," *Journal of Cleaner Production*, vol. 218, pp. 555–564, 2019. [Online]. Available: http://www.sciencedirect.com/science/article/pii/S0959652619301192.

[54] N. J. Johannesen, M. Kolhe, and M. Goodwin, "Comparison of regression tools for regional electric load forecasting," in IEEE 3rd International Conference on Smart and Sustainable Technologies (SpliTech), 2018, pp. 1–6.

[55] M. A. Hall, "Correlation-based feature selection for discrete and numeric class machine learning," in *Proceedings of the Seventeenth International Conference on Machine Learning*, ser. ICML '00. San Francisco, CA, USA: Morgan Kaufmann Publishers Inc., 2000, pp. 359–366. [Online]. Available: http://dl.acm.org/citation.cfm?id=645529.657793.

[56] N. J. Johannesen, M. L. Kolhe, and M. Goodwin, "Smart load prediction analysis for distributed power network of holiday cabins in norwegian rural area," *Journal of Cleaner Production*, vol. 266, p. 121423, 2020. [Online]. Available: http://www.sciencedirect.com/science/article/pii/S0959652620314700.

[57] T. Hong, M. Gui, M. E. Baran, and H. L. Willis, "Modeling and forecasting hourly electric load by multiple linear regression with interactions," in *IEEE PES General Meeting*, 2010, pp. 1–8.

# 5 Effective Strategies of Flexibility in Modern Distribution Systems

## *Reconfiguration, Renewable Sources and Plug-in Electric Vehicles*

*Abdollah Kavousi-Fard[1], Mojtaba Mohammadi[1], and A.S. Al-Sumaiti[2]*
[1]Department of Electrical and Electronics Engineering, Shiraz University of Technology, Shiraz, Iran
[2]Electrical Engineering and Computer Science Department, Advanced Power and Energy Center, Khalifa University, Abu Dhabi, United Arab Emirates

## CONTENTS

## NOMENCLATURE

| | |
|---|---|
| $a, b$ | Wöhler diagram constants |
| $Cost_{sub}$ | Cost of power purchased from the upstream grid |
| $Cost_{P_{loss}}$ | Cost of active power losses |
| $Cost_{PEV}$ | Cost of total plug-in vehicles |
| $Cost_{Rel}$ | Cost of reliability |
| $Cost_{OP}$ | Cost of operation of plug-in vehicles |
| $Cost_{deg}$ | Cost of BDC in plug-in vehicles |
| Cost | Overall cost of the grid |
| $C_{sub}t/C_{loss}t/C_{PEV}t/$ | market price/loss cost/vehicle to grid cost |
| $Ci$ | Interruption price |
| $C_{bat}$ | Battery investment cost ($). |
| $DoD_i/DoD_f$ | Initial/final depth of the charge during a discharge cycle. |
| $E_{bat}$ | Available battery charge |
| $ED_{,v}{}^t$ | The total energy required by plug-in electric vehicles in a fleet |
| $E_v{}^t$ | Usable energy of energy storages in a fleet |
| $E_v{}^{ini}/E_v{}^{fin}$ | Initial/final energy in plug-in electric vehicle fleet |
| $E_v{}^{min}/E_v{}^{max}$ | Minimum/maximum energy in energy storages of a fleet |
| $L_{a(i)}$ | Average demand on bus $i$ |
| $N_L/N_{br}/N_{bus}$ | Number of loops/branches/buses of grid |
| $n$ | Number of probabilistic variables |
| $N_v$ | Total number of plug-in electric vehicles |
| $N_{dis}$ | Number of discharge cycles |
| $N_c$ | Number of life cycles |
| $N_{tie}/N_{sw}$ | Number of remote control switches (tie/sectionalizing) |
| $P_{sub}{}^t/P_{sub}{}^{max}$ | Hourly/max injected power from substation grid |
| $P_{loss}{}^t$ | Hourly resistive power loss of the grid |
| $P_{c,v}{}^t/P_{d,v}{}^t$ | Charging/discharging capacity of a plug-in electric vehicle fleet |
| $P_{c,v}{}^{min}/P_{c,v}{}^{max}$ | Minimum/maximum charging capacity of a plug-in electric vehicle fleet |
| $P_{d,v}{}^{min}/P_{d,v}{}^{max}$ | Min/max discharging capacity of a plug-in electric vehicle fleet |
| $P_v{}^t$ | Charge/discharge energy rate of a plug-in electric vehicle fleet |
| $P_i{}^t/Q_i{}^t$ | Hourly injected active/reactive power at bus $i$ |
| $P_x{}^x/P_y{}^y$ | Covariance of input variable $X$/output variable $Y$ |
| $ri$ | fix rate of the $i$th part |
| $rs$ | Average blackout duration of the grid |
| $S_j{}^{it}/S_{ij}{}^{max}$ | Hourly/max apparent power flow between bus $i$ and $j$ |
| $T$ | Planning time |
| $Tie_k{}^t/S_{wk}{}^t$ | Status of tie/sectionalizing remote control switches |
| $t'$ | Time in which SOC is adjusted to a certain value |
| $U_v{}^t$ | Status of the connection of a fleet |
| $U_{c,v}{}^t/U_{d,v}{}^t/U_{i,v}{}^t$ | Status of a fleet (i.e., charge/discharge/idle mode) |
| $US$ | Annual system blackout duration |

| $V_i^t/\delta_i^t$ | Voltage value/voltage angle of bus |
| $V_i^{min}/V_i^{max}$ | Min/max voltage of bus |
| $W^k$ | weighting factor of the sample point |
| $Y_{ij}/\theta_{ij}$ | Magnitude/phase of impedance between bus $i$ and $j$ |
| $\lambda S$ | Average failure rate of the grid |
| $\lambda i$ | Failure rate |
| $\mu i$ | Mean input variables |

## 5.1 INTRODUCTION

*Flexibility* in power grids is defined as the ability of systems to respond fast and efficiently to disturbance or sudden changes in demand and generation. Flexibility is especially prized in modern power system with high penetration of grid-connected technologies such as plug-in electric vehicles (PEVs), renewable energy sources, etc. [1]. As one of the key subjects of the modern electrical grids, this chapter focuses on the operation and flexibility of the electric distribution systems (EDSs) from the technical and formulation aspects. In this chapter, we investigate the impact of some popular new technologies, including PEVs, feeder reconfiguration process (FRP), and wind turbines (WT), on the operation, management, flexibility of EDSs. EDSs were initially designed as passive systems, which provided one-way interaction between the upstream grid and consumers. These networks are the last level in the delivery of electrical energy to consumers, and thus act as an interface between the transmission system and end-users. In recent years, the emergence of new technologies has made traditional EDSs an active system with bidirectional power flow. One of the most penetrated and popular technologies in the last years is the plug-in electric vehicles. PEVs, which can be considered as mobile energy storage or mobile loads, are hourly distributed in the grid and can provide many benefits to the grid. For instance, optimal coordination of these devices can highly increase the flexibility and reliability of the system, reduce power loss, decrease operation cost and, etc. [2–4]. Unlike the old fossil fuel-based vehicles, the clean and nature-friendly operation of PEVs can reduce greenhouse gas emissions. To this end, this technology has attracted the attention of many engineers, researchers, and companies around the globe. By the use of the vehicle to grid (V2G) technology, PEVs can also play a significant role in the scheduling of the grid by participating in the electricity market [5]. Despite the benefits of PEVs for the grid, integration of these devices to the system can create many challenges. For instance, not concerning the charging demand of the PEVs can cause feeder congestion, transformers overloading, and increase power loss [4]. Therefore, an efficient management strategy concerning these systems is required. In refs. [6, 7], some smart charging strategies are presented to manage the energy demand of the PEVs and improve the flexibility within the EDSs. The impact of different kinds of charging strategies of PEVs on the operation of smart microgrids is investigated in ref. [8]. V2G is devised to change the passive role of PEVs to an active role by providing an infrastructure for PEVs to exchange energy with the grid in different time intervals and participate in the energy market. Although the deployment of V2G can improve the flexibility and reliability of the system, it increases its complexity at the same time [9].

Another useful strategy for the optimal operation of EDSs is the FRP. This strategy, which is strongly affected by the presence of the PEVs, can provide many benefits for the electric EDS including increasing flexibility and reliability, preventing feeder congestion, reducing power loss, etc. [3, 10]. The FRP can be defined as an operational process that changes the feeders' topology structure using sectionalizing switches and tie switches where the radial structure of the system is preserved simultaneously. Sectionalizing switches are normally closed switches, and tie switches are normally open switches. The main goal of the FRP is to optimize certain objectives considering some technical constraints [11, 12]. In ref. [13], a neural network-based method is employed to investigate the impact of FRP on the power loss within EDSs. Reference [14] examines the effects of FRP on the power loss within EDSs using graph theory. Similar works concerning the impacts of FRP on the power loss within EDSs can be found in refs. [15–17]. Besides the useful effects of FRP on the power loss, this technology can also improve the profile of the voltage [18], decrease the operation cost of the system [19], increase the flexibility and reliability of the system [20], and improve the load balance [21]. While each of the above works has investigated the effects of FRP on the operation of EDSs from different points of views, none of them has analyzed its impact on the operation and flexibility of EDSs with high penetration of the V2G PEVs. This issue is investigated in this chapter. In the following, we demonstrate that FRP can facilitate the performance of the next-generation transportation fleets.

Since renewable energy sources (e.g., WT, photovoltaic) are growing rapidly in the EDSs, it is essential to examine their performance and impact on the grid. In this chapter, we focus on the effects of WTs, FRP, and PEVs on the operation and flexibility of EDSs. The implementation of WT units in EDSs has two main challenges: (1) these energy sources are strongly affected by nature and have a volatile output power, and (2) there is a correlation between the output power of WT units in a WT farm. It is declared that the WTs' output power forecast error is unavoidable [22]. Therefore, there is always an uncertainty associated with the output power of WTs in the one day-ahead operation scheduling of EDSs. In order to address these issues, a novel stochastic framework based on the unscented transform (UT) is presented to model the uncertainties associated with WTs' output power within EDSs. UT is a novel approach that can be used to model the uncertainty in correlated problems [23]. UT is also well-suited for state estimation problems and has shown promising results in nonlinear transformations [24]. In this chapter, UT is also employed to model the uncertainties associated with loads' power demand, charge/discharge time, the number of PEV fleets in system, market price, charging/discharging state of energy storages, and the number of PEVs.

## 5.2 EFFECTIVE STRATEGIES OF FLEXIBILITY IN ELECTRIC DISTRIBUTION SYSTEMS

As the next generation of power systems, EDSs should be equipped with many technologies and devices to be able to respond to the appearing demands. In the section, we investigate the impact of some of these technologies on the operation and flexibility of modern EDSs.

### 5.2.1  PLUG-IN ELECTRIC VEHICLES FLEETS

Decreasing the dependence on fossil fuels and reducing greenhouse emissions are the main reasons that have led humans to choose electric-based vehicles as the next generation of transportation systems [25]. Generally, electric vehicles can be organized into three different categories: hybrid electric vehicles, plug-in electric vehicles, and plug-in hybrid electric vehicles. A PEV fleet is defined as a random subset of all PEVs within the transportation system. As the main part of the future transportation systems, PEVs play a significant role in the energy infrastructure. From electric companies' point of view, PEVs are considered as the distributed non-stationary loads, which can cause major problems for the grid if not managed properly. For instance, charging/discharging a large number of PEVs at different hours of the day can impose sudden load or sudden additional power to the system and upset the balance between generation and consumption. To overcome the problems associated with the penetration of PEV fleets in EDSs, characteristics, and travel patterns of PEVs must be considered in the planning and scheduling of the EDSs. On the other hand, the presence of the PEVs in the system can increase the maneuverability of the system operators and improve the flexibility and reliability of the system. For instance, PEVSs' available battery capacities can be employed to eliminate peak-consumption points and prevent blackouts in post-contingency times. The model of PEV fleets in the EDSs can be determined by specifying their departure and arrival time and location. To this end, from now on we consider some assumptions regarding the characteristics of PEV fleets. Characteristics of a fleet are determined based on the number of PEVs in the fleet, PEVs' charging state ratio (CSR) (the ratio of existing energy to the capacity of the battery), and battery capacity of the PEVs. It has been shown that on average, during a day, each driver makes three long trips along with several short trips of less than ten minutes [26]. Due to the small impact of short trips and also in order to maintain scheduling on an hourly basis, short trips are neglected in this work. It is also assumed that PEVs' departure location at the beginning of the day and their final destination at the end of the day are the same. According to ref. [27], the amount of energy required for the departure path and the energy required for the returning direction are considered equal. Statistics show that a PEV fleet travels about 12,000 (mi) a year [28]. In this regard, on average, each PEV has 3.65 (kW/mile) and 9 (kW/day) energy consumption rates [29]. In order to have more realistic results, it is assumed that PEVs' batteries are fully charged (i.e., 100% CSR) at the beginning of the day. During a day, PEVs can charge/discharge in charging stations and exchange energy with the grid regarding the cost and technical constraints such as battery max/min capacity, the depth of the discharge (DoD), charging/discharging rates, etc. According to ref. [30], in order to decelerate the aging process of the batteries, the maximum depth of the discharge should not exceed 20% CSR. The total cost of a V2G-PEV fleet includes the cost of energy received/injected by PEVs within the fleet.

### 5.2.2  WIND TURBINES

WTs are widespread renewable energy sources that usually being implemented at the distribution level. Due to their clean nature, cost-effective operation, and wide availability, the idea of wind farms has attracted both electric utilities and end-users in the last decade. These devices are introduced as a solution to solve the fossil-fuel-based energies and green-house emission problems in power systems [30]. However, due to the uncertainties associated with the wind speed and wind strength during the day, WTs have unstable output power generation, and therefore are not considered as completely reliable energy sources [31]. In this way, a suitable probabilistic model is required to capture these probabilities. To this end, in this work, UT is presented to capture the uncertainties of WTs output power and the correlation between WTs in a wind farm.

### 5.2.3 FEEDER RECONFIGURATION PROCESS

One of the most important topics in EDSs, which can highly affect the operation of the system, is FRP. As mentioned before FRP is the process of optimal adjusting of remote control switches (RCS) (i.e., tie switches and sectionalizing switches) in a way that the cost objective function is minimized and technical limits are satisfied. The main benefit associated with the FRP is that it can improve the system from different aspects (e.g., operation, flexibility, reliability, cost, etc.) with a little cost. For instance, in peak load hours, FRP can avoid feeder congestion by transferring a portion of loads on feeders with the available capacity. It is also demonstrated that FRP can improve the average annual outage of the grid up to 30% [32]. In order to achieve the optimal FRP, four infrastructures are required: (1) the metering infrastructure, which includes several metering devices that measure the vale of voltages and currents within the grid, (2) actuators (i.e., remote control switches), (3) communication platform, and (4) central control. The communication platform transmits information (e.g., data of metering devices, central control commands, etc.) between different components of the system, and the central control is responsible for analyzing the collected data and sending suitable commands to the actuators. A graphical illustration of components of reconfigurable EDSs is presented in Figure 5.1. Although FRP can be highly beneficial for the grid, utilizing this technology is associated with several challenges. The biggest challenges are the high influence of uncertainties on the process and the optimality of process. These issues are addressed in this work.

As mentioned before, there is a high level of uncertainty associated with the operation of EDSs. Therefore, a powerful uncertainty modeling tool is required to capture uncertainty impacts. Note that the stochastic parameters in this problem are the number of PEVs in a fleet, active and reactive load demand of the buses, the market price at different time intervals during a day, output power of the WTs, correlation between WTs, and PEVs' charging/discharging patterns. In the next section, the mathematical model of the UT method is explained in detail.

## 5.3 PROBABILISTIC LOAD FLOW BASED ON UNSCENTED TRANSFORM METHOD

The operation of EDSs incorporates several stochastic parameters which can affect the practicability aspect of the analysis. To model these uncertain parameters, in this work, a probabilistic framework based on the UT is utilized. As a powerful stochastic modeling tool, UT can be utilized for modeling uncertainties in correlated and uncorrelated problems. Regarding technical classifications, there are three main methods

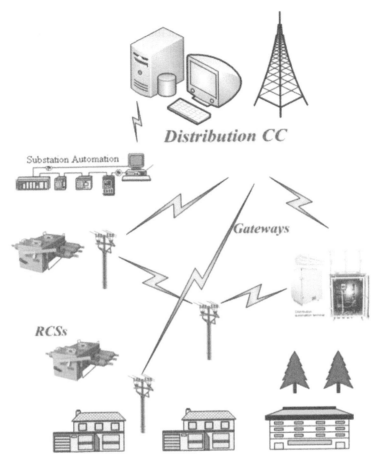

**FIGURE 5.1**   A graphical illustration of different components of a typical reconfigurable EDS [32].

for modeling uncertainties in problems: (1) Monte Carlo simulation method (MCSM), (2) analytical models (AMs), and (3) approximate methods [23]. The biggest drawback associated with the MCSM is its high computational process for converging. On the other hand, AMs have solved the high computational problem, but the performance of these methods depend on some mathematical assumption. Both the above deficiencies are solved in the third group, and therefore, approximate methods can be more useful for modeling uncertainties in problems. As one of the most popular approximate methods, UT technique is employed as a powerful modeling tool for both correlated and non-correlated problems. Due to its benefits such as easy implementation, low computation, etc., UT transform has attracted the attention of many researchers recently. A comprehensive comparison between UT and other methods is described in ref. [24]. In order to capture the uncertainties in a problem with $v$ parameters, UT solves the problem $2v+1$ times. To explain the UT model, consider $y = f(X)$ as the nonlinear problem that we tend to model its uncertainties. Note that $y$

is the output vector, $f$ is the non-linear function and $x$ is the input vector of size $n$ (i.e., it is the number of uncertain parameters) with the mean value of $\mu_x$. The covariance matrix $P_{xx}$ is also considered to model the standard deviation of the parameters. The behavior of parameters from correlated and non-correlated standpoints is determined based on the elements of $P_{xx}$. Diagonal elements of the $P_{xx}$ present the standard deviation of uncertain parameters and the non-symmetric elements present the covariance between different uncertain variables. The output's mean value ($\mu_y$) and covariance ($P_{yy}$) are calculated by the use of $x$, $P_{xx}$, and the following stages:

Stage 1: Generate $2n+1$ points from the input data [24]:

$$x^0 = \mu \tag{5.1}$$

$$x^k = \mu + \left(\sqrt{\frac{n}{1-W^0}}\, P_{xx}\right)_k \;; k = 1,2,...,n$$

$$x^k = \mu - \left(\sqrt{\frac{n}{1-W^0}}\, P_{xx}\right)_k \;; k = 1,2,...,n \tag{5.2}$$

where, $k$ presents the $k$th row or column of the matrix, and $W^0$ is the weight of $\mu$.

Stage 2: Compute the weighting factors as follows:

$$W^0 = W^0 \tag{5.3}$$

$$W^k = \frac{1-W^0}{2n}; k = 1,2,...,n \tag{5.4}$$

$$W^{k+n} = \frac{1-W^0}{2n}; k+n = n+1,...,2n \tag{5.5}$$

Note that the summation of the weighting factors should be equal to one:

$$\sum_{k=0}^{2n} W^k = 1 \tag{5.6}$$

Stage 3: Compute $2n+1$ output points using $2n+1$ inputs and the function $f$:

$$y^k = f(X^k) \tag{5.7}$$

Stage 4: Compute $\mu_y$ and $P_{yy}$ for the output points [24]:

$$\mu_y = \sum_{k=0}^{2n} W^k Y^k \tag{5.8}$$

$$P_{yy} = \sum_{k=0}^{2n} W^k (Y^k - \mu_y)(Y^k - \mu_y)^T \tag{5.9}$$

Figure 5.2 shows the sequential stages of the UT technique as a flowchart diagram.

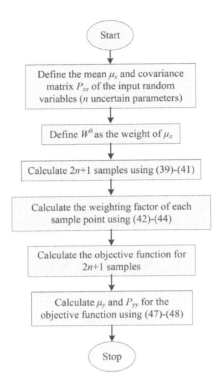

FIGURE 5.2   Flowchart diagram of the UT.

## 5.4   MATHEMATICAL MODELING OF SOME FLEXIBILITY STRATEGIES IN EDS: CASE OF PEVS, WTS, AND FRP

In this section, we present the mathematical model of EDSs considering the proposed flexibility improvement strategies (i.e., PEVs, WTs, FRP). The proposed model is presented as a constraint optimization problem. This model is utilized in the next section to simulate the operation of EDSs in presence of the proposed technologies. The mathematical model of the problem is presented as follows:

### 5.4.1   COST OBJECTIVE FUNCTION

The total cost objective function incorporates the operation and reliability costs. The operation cost includes the cost of energy purchased/sold from/to the main grid, the cost of power loss, and the cost of power drawn/injected from/to grid by PEVs.

$$\text{Min} \quad \text{Cost} = \text{Cost}_{\text{Rel}} + \text{Cost}_{\text{Opr}} \qquad (5.10)$$

$$\text{Cost}_{\text{Opr}} = \text{Cost}_{\text{sub}} + \text{Cost}_{\text{Ploss}} + \text{Cost}_{\text{PEV}} \qquad (5.11)$$

$$\text{Cost}_{\text{sub}} = \sum_{t=1}^{T} C_{\text{sub}}^t P_{\text{sub}}^t \tag{5.12}$$

$$\text{Cost}_{\text{Ploss}} = \sum_{t=1}^{T} C_{\text{loss}}^t P_{\text{loss}}^t \tag{5.13}$$

The first-term in Equation (5.10), which is called the reliability cost (RC) of the system, presents the energy interruption cost (EIC) of the consumers. RC can be considered as an index to evaluate the economic loss to the consumers due to disruption in the power supply process. RC can also help system planners and operators to make optimum planning or operation decisions, evaluate the economic and technical justifications for system improvements, find system strengths and weaknesses, etc. [10]. EIC is calculated based on consumer side reliability indicators such as the annual blackout time, average blackout duration, and outage rates as follows:

$$\lambda_S = \sum_i \lambda_i \quad ; \quad U_S = \sum_i \lambda_i r_i \tag{5.14}$$

$$r_S = \frac{U_S}{\lambda_S} = \frac{\sum_i \lambda_i r_i}{\sum_i \lambda_i} \tag{5.15}$$

$$\text{Cost}_{\text{Rel}} = \sum_{i=1}^{N_{\text{br}}} \text{ECOST}_i = \sum_{i=1}^{N_{\text{br}}} L_{a(i)} C_i \lambda_i \tag{5.16}$$

$C_i$, which presents the EIC, is computed using the consumer economic loss function (CELF). Illustration of a common CELF can be found in Figure 5.3.

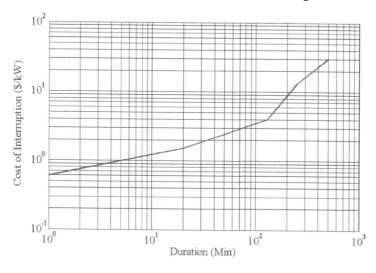

**FIGURE 5.3**   Graphical illustration of CELF [10].

The third term in Equation (5.11) presents the costs of PEVs. This term incorporates both PEV operation cost and battery degradation costs (BDC), and is calculated by the following equations:

$$\text{Cost}_{PEV} = \text{Cost}_{Op}^{PEV} + \text{Cost}_{deg} \tag{5.17}$$

$$\text{Cost}_{Op}^{PEV} = \sum_{t=1}^{T}\sum_{n=1}^{N_v} U_v^t C_v^t P_v^t \tag{5.18}$$

The PEV operation cost depends on the number of PEVs in the fleet and their charging/discharging power. BDC is a detrimental occurrence resulted from V2G technology. BDC is caused due to the additional cycling operation of batteries. The literature shows that the battery depth of discharge has a high influence on the BDC through Wöhler curve [33]. Wöhler diagram demonstrates the relation between the battery depth of the discharge and the average number of cycles before battery failure. Figure 5.4 shows the diagram for a typical Wöhler curve. According to Figure 5.4, the average number of cycles before battery failure is inversely related to the battery depth of the discharge meaning that increasing the depth of the discharge decreases the number of cycles. The Wöhler curve is obtained using the following equation:

$$N_c(\text{DoD}) = a.\text{DoD}^b \tag{5.19}$$

The type of the battery is determined based on $a$ and $b$, and the maximum cycles of the battery before its failure is calculated using Equation (5.19).

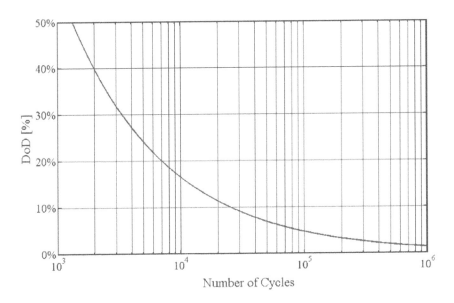

**FIGURE 5.4** Graphical illustration of the Wöhler curve for a typical lithium ion energy storage [3].

It is shown that BDC can be computed as a function of the depth of discharge. In this regard, BDC for one discharge from a completely charged state (DoD = 0) to a certain level of charge (DoD = DoD$_s$) can be calculated based on the following equation:

$$C_d(0,DoD_s) = \frac{C_{Bat} \times DoD_s \times E_{bat}}{N_c(DoD_s)} \tag{5.20}$$

Based on Equation (5.20), Equation (5.21) is devised which presents BDC for a discharge cycle between two charge levels (i.e., $DoD_i$, $DoD_f$).

$$C_d(DoD_i,DoD_f) = C_d(0,DoD_f) - C_d(0,DoD_i) \tag{5.21}$$

Therefore, the total BDC in a definite time can be calculated as follows:

$$Cost_{deg} = \sum_{k=1}^{N_{dis}} C_d^k(DoD_i,DoD_f) \tag{5.22}$$

### 5.4.2 Control Variables

There are several control variables in the optimization problem, including (1) hourly charge/discharge/idle status of PEV fleets, (2) ON/OFF status of remote control switches, and (3) the amount of hourly drawn/injected power from/to the grid by PEV fleets in each time interval. The vector-shaped control variables are presented as follows:

$$X = [X_{Tie}, X_{Sw}, X_{P_v}, X_{U_c}, X_{U_d}, X_{U_i}] \tag{5.23}$$

$$X_{Tie} = [\overline{Tie}_1, \overline{Tie}_2..., \overline{Tie}_{N_{Tie}}] \tag{5.24}$$

$$X_{Sw} = [\overline{Sw}_1, \overline{Sw}_2,..., \overline{Sw}_{N_{Sw}}] \tag{5.25}$$

$$\overline{Tie}_k = [Tie_k^1, Tie_k^2,..., Tie_k^T] \quad k = 1: N_{Tie} \tag{5.26}$$

$$\overline{Sw}_k = [Sw_k^1, Sw_k^2,..., Sw_k^T] \quad k = 1:N_{Sw} \tag{5.27}$$

$$X_{P_v} = [\overline{P}_1, \overline{P}_2,..., \overline{P}_{N_v}] \tag{5.28}$$

$$X_{U_c} = [\overline{U}_{c,1}, \overline{U}_{c,2},..., \overline{U}_{c,N_v}] \tag{5.29}$$

$$X_{U_d} = [\overline{U}_{d,1}, \overline{U}_{d,2},..., \overline{U}_{d,N_v}] \tag{5.30}$$

$$X_{U_i} = [\overline{U}_{i,1}, \overline{U}_{i,2},..., \overline{U}_{i,N_v}] \tag{5.31}$$

$$\overline{P}_k = [P_k^1, P_k^2,..., P_k^T] \quad k = 1: N_v \tag{5.32}$$

$$\overline{U}_{c,k} = [U_{c,k}^1, U_{c,k}^2,..., U_{c,k}^T] \quad k = 1: N_v \tag{5.33}$$

$$\overline{U}_{d,k} = [U_{d,k}^1, U_{d,k}^2,..., U_{d,k}^T] \quad k = 1: N_v \tag{5.34}$$

$$\overline{U}_{i,k} = [U_{i,k}^1, U_{i,k}^2,..., U_{i,k}^T] \quad k = 1: N_v \tag{5.35}$$

### 5.4.3 Technical Constraints

In order to obtain a feasible solution, several technical constraints must be taken into consideration during the process of solving the optimization problem. The first constraints are the AC power flow limitations. These constraints indicate the balance between the power demand and generation in the system as follows:

$$
\begin{cases}
P_i^t = \sum_j \left|V_i^t\right|\left|V_j^t\right|\left|Y_{ij}\right|\cos\left(\theta_{ij} + \delta_i^t - \delta_j^t\right) \\
Q_i^t = \sum_j \left|V_i^t\right|\left|V_j^t\right|\left|Y_{ij}\right|\sin\left(\theta_{ij} + \delta_i^t - \delta_j^t\right)
\end{cases}
\tag{5.36}
$$

Limitations on the max/min voltage of buses, exchanged power with the main grid, and VAR flow capacity of feeders are presented as follows:

$$
V_i^{\min} \le V_i^t \le V_i^{\max}
\tag{5.37}
$$

$$
\left|P_{sub}^t\right| \le P_{sub}^{\max}
\tag{5.38}
$$

$$
S_{ij}^t \le S_{ij}^{\max}
\tag{5.39}
$$

To avoid loop formation and islanded buses during FRP, after each switching process, the radiality of the grid is checked. If a loop is detected, one of the tie-switches within the loop must be opened. In this regard, $N_L$ presents the number of loops in the grid as follows:

$$
N_L = N_{br} - N_{bus} + 1
\tag{5.40}
$$

Since PEVs operate in V2G mode, they can exchange energy with the grid at non-traveling hours. The following constraints present the operation mode of PEV fleets (i.e., charge, discharge, idle) as well as min/max power exchange limits and energy capacity of the PEV fleets [34].

$$
U_{c,v}^t + U_{d,v}^t + U_{i,v}^t = U_v^t
\tag{5.41}
$$

$$
U_{c,v}^t P_{c,v}^{\min} \le P_{c,v}^t \le U_{c,v}^t P_{c,v}^{\max}
\tag{5.42}
$$

$$
U_{d,v}^t P_{d,v}^{\min} \le P_{d,v}^t \le U_{d,v}^t P_{d,v}^{\max}
\tag{5.43}
$$

$$
E_v^t = E_v^{ini} + \sum_{m=1}^{t}\left(U_{c,v}^m P_{c,v}^m \eta_{c,v} - U_{d,v}^m P_{d,v}^m \eta_{d,v}\right)
$$

$$
- \sum_{m=1}^{t}(1 - U_v^m)E_{D,v}^m
\tag{5.44}
$$

$$
P_v^t = E_v^t - E_v^{t-1}
\tag{5.45}
$$

$$
E_v^{\min} \le E_v^t \le E_v^{\max}
\tag{5.46}
$$

$$
E_v^{fin} = E_v^{ini}
\tag{5.47}
$$

where, $U_v^t$ is a binary variable that indicates the grid connection status of the PEV fleets.

As mentioned before, the PEV batteries are fully charged at the beginning of the day meaning that in the first trip, PEVs leave the station with 100% charged batteries.

$$E_v^{t'} = E_v^{max} \tag{5.48}$$

## 5.5   AN IEEE EXAMPLE AND NUMERICAL SIMULATIONS

In this section, the IEEE 69-bus test system is employed as an example to show how much of the flexibility strategies can affect the entire system. In order to analyze the impact of each strategy, four cases are simulated here:

Case 1: This case demonstrates the EDS cost function (CF) value neglecting all flexibility strategies (i.e., FRP, PEVs, WTs).
Case 2: This case demonstrates the EDS CF value considering only FRP.
Case 3: This case demonstrates the EDS CF value considering FRP and WTs.
Case 4: This case demonstrates the EDS CF value considering all three flexibility strategies.

### 5.5.1   SYSTEM MODEL AND ASSUMPTIONS

In order to investigate the operation of EDSs in the presence of the above technologies (i.e., PEVs, WTs, and FRP), the proposed problem is tested on the IEEE 69-bus test system. Figure 5.5 shows the schematic diagram of the test system considering

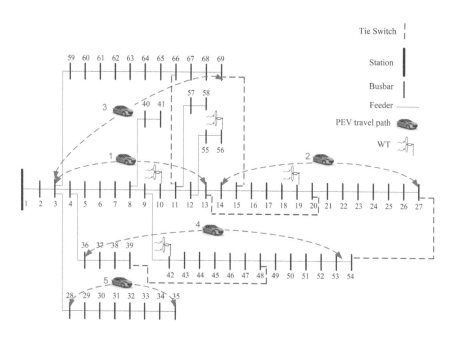

**FIGURE 5.5**   IEEE 69-bus electric distribution system.

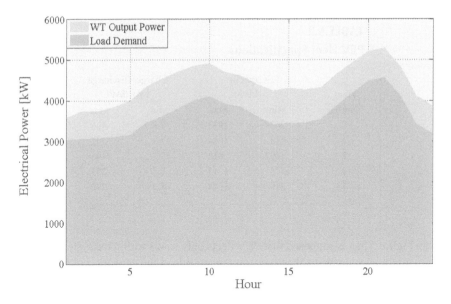

**FIGURE 5.6**   Wind turbine output power/total load demand of the grid.

PEVs, WTs, and remote control switches. As can be seen from Figure 5.5, the test system includes five tie switches and 68 sectionalizing switches in the grid. Also, each feeder is equipped with a protective system (i.e., sectionalizer), and there is a circuit breaker in the grid's point of common coupling (i.e., bus 1). To highlight the effects of WTs on the EDSs, four WTs are considered in the system. It is assumed that all WTs follow a power generation pattern similar to Figure 5.6 and also there is a correlation between WTs on buses 10 and 19 as well as WTs on buses 42 and 56. The correlation coefficient (CC) among WTs is considered as 0.5. In the case of PEVs, five fleets with five different driving patterns are considered in the system. The complete data related to the PEV fleets such as the number of PEVs in fleets and their departure and arrival time/location are specified in Table 5.1. As can be seen

**TABLE 5.1**
**PEV Fleet Movement Specifications**

| Fleet Number | Number of PEVs | First Trip | | | | Second Trip | | | |
|---|---|---|---|---|---|---|---|---|---|
| | | Departure | | Arrival | | Departure | | Arrival | |
| | | Time | Bus | Time | Bus | Time | Bus | Time | Bus |
| 1 | 68 | 6:00 | 3 | 8:00 | 13 | 17:00 | 13 | 19:00 | 3 |
| 2 | 40 | 7:00 | 14 | 8:00 | 27 | 16:00 | 27 | 17:00 | 14 |
| 3 | 20 | 5:00 | 3 | 7:00 | 69 | 16:00 | 69 | 18:00 | 3 |
| 4 | 32 | 5:00 | 36 | 6:00 | 54 | 17:00 | 54 | 18:00 | 36 |
| 5 | 40 | 7:00 | 28 | 9:00 | 35 | 18:00 | 35 | 20:00 | 28 |

**TABLE 5.2**

**PEV Fleet Specifications**

| Fleet | Capacity (kWh) | | Charge/Discharge rate (kW) | |
|---|---|---|---|---|
| Number | Min | Max | Min | Max |
| 1 | 263 | 1973 | 7.3 | 496 |
| 2 | 219 | 1644 | 7.3 | 292 |
| 3 | 109 | 822 | 7.3 | 146 |
| 4 | 175 | 1315 | 7.3 | 233 |
| 5 | 219 | 1644 | 7.3 | 292 |

from Figure 5.5, it is supposed that PEV fleets only travel within the system although there might be some fleets that their travel is outside of the network. For these cases, the same procedure as described before can be utilized for the times that fleets travel inside the system area. In this work, 315 ($) investment cost is considered for each PEV battery, and also, PEVs use lithium ion batteries with Wöhler constants of $a = 1331$ and $b = -1.825$ [35]. The complete data related to the technical characteristics of PEV fleets can be found in Table 5.2.

The predicted energy price [31] and forecasted load demand of the grid are presented in Figure 5.7 and Figure 5.6, respectively. Finally, in order to model the costs related to the reliability index, the EIC is calculated based on Figure 5.3. Other data related to failure rate, repair rate, etc. are obtained from ref. [10].

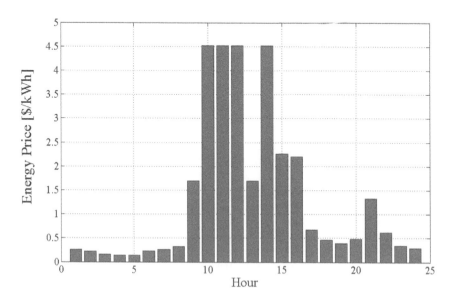

**FIGURE 5.7**   Predicted hourly energy price [31].

## TABLE 5.3
## Optimization Results Corresponding to Active Power Losses in the Grid

| Method | Power Loss [kW] | Minimum Voltage (pu) | Open Switches |
|---|---|---|---|
| Initial network | 225.0 | 0.90921 | s11-66, s13-20, s15-69, s27-54, s39-48 |
| Fuzzy genetic algorithms [36] | 102.6 | 0.93211 | s11-66, s13-20, s14-15, s50-51, s44-45 |
| Refined genetic [37] | 102.1 | 0.93211 | s11-66, s13-20, s14-15, s50-51, s47-48 |
| Near-optimum solution [38] | 106.632 | 0.93211 | s11-66, s17-18, s67-68, s21-22, s47-48 |
| Hybrid particle swarm optimization [39] | 99.62 | 0.94282 | s11-66, s13-20, s14-15, s50-51, s47-48 |
| Clonal selection algorithm [10] | 99.62 | 0.94282 | s11-66, s13-20, s14-15, s50-51, s47-48 |

### 5.5.2 RESULTS AND ANALYSIS

In order to observe the operation of the system from different aspects (e.g., FRP, charge/discharge of PEV fleets, etc.) during a day, the simulation is done for 24 hours. In this section, first, the stochastic framework is neglected, and the analysis is done for peak load data. In order to show the effects of FRP on the system's power loss, Table 5.3 compares the deterministic results obtained from several methods in terms of power loss, voltage, and open switches. Note that in this case, WTs and PEVs are neglected. Table 5.3 demonstrates that FRP can significantly reduce the power loss within the grid such that using this technology almost halved the power loss.

To see the positive effect of FRP on the system, the results of the optimization of the power loss, EIC, grid average outage frequency index (GAOFI), grid average outage duration (GAOD), energy not supplied, and maximum voltage deviation (MVD) are presented in Table 5.4. As can be seen from Table 5.4, the optimal FRP can highly improve both the reliability and operational indices. Also, in this table, the number of open switches in each optimization case is demonstrated.

While the above results provide useful information about the operation of the grid, in order to obtain more precise results, the high level of uncertainties in the system should be

## TABLE 5.4
## Impact of FRP on Several Objective Functions in the Grid

| Case | Before FRP | After FRP | Open Switches |
|---|---|---|---|
| Loss (kW) | 225 | 99.62 | s11-66, s13-20, s14-15, s50-51, s47-48 |
| MVD (pu) | 0.090559 | 0.0513620 | s11-66, s13-20, s14-15, s50-51, s39-48 |
| Cost of losses and EIC ($) | 189,959.90 | 127,721.13 | s11-66, s60-61, s16-17, s45-46, s39-48 |
| GAOFI (failure/Cust.yr) | 1.719368 | 1.419796 | s11-66, s2-17, s14-15, s58-59, s47-48 |
| GAOD (hr/Cust.yr) | 14.274709 | 12.59105 | s40-41, s2-17, s14-15, s55-57, s47-48 |
| Energy not supplied cost (kWh/Cust.yr) | 3.168866 | 2.797363 | s40-41, s14-15, s14-15, s56-57, s47-48 |

considered in the simulation. To this end, UT technique is employed to model the uncertainties associated with the number of PEVs in a fleet, output power of WTs, correlation between WTs load demand, CSR, PEVs charging patterns, and energy market price.

One of the main challenges regarding the deployment of WTs in EDs is their unreliable power generation such that in some hours during a day, it is possible that WTs output power suddenly goes to zero. In these cases, PEV batteries can be employed as backup energy sources. In fact, by the use of V2G technology, a high number of PEV batteries in charging stations or parking lots can compensate for the sudden power shortages caused by WTs. Therefore, the presence of PEVs along with V2G technology can highly improve the grid's flexibility. The results related to the CF of first and second cases as well as the open switches are demonstrated in Table 5.5. As can be seen from this table, the FRP has highly improved the performance of the system from the expected cost point of view. It is worth noting that this cost reduction is done without any additional investment cost and only by readjusting the structure of the grid during the day. Table 5.5 shows that FRP has reduced the total cost of the system about 10,000 ($), which is considerable.

**TABLE 5.5**

**Expected CF Value in First and Second Cases**

| Hour | Cost ($)×10⁵ | | Open Switches After FRP |
|---|---|---|---|
| | Case 1 | Case 2 | |
| 1 | 1.2285080 | 0.9382242 | s69, s17, s11, s53, s73 |
| 2 | 1.2347423 | 1.1584340 | s41, s17, s14, s72, s73 |
| 3 | 1.2405938 | 1.2405511 | s69, s70, s71, s72, s73 |
| 4 | 1.2521136 | 0.9824330 | s35, s70, s14, s57, s73 |
| 5 | 1.2704854 | 1.1972670 | s69, s20, s71, s72, s64 |
| 6 | 1.3962585 | 1.2912419 | s41, s16, s45, s49, s21 |
| 7 | 1.4589467 | 1.3648178 | s40, s70, s71, s72, s73 |
| 8 | 1.5380414 | 1.4371504 | s69, s16, s11, s49, s22 |
| 9 | 1.6734438 | 1.0258414 | s69, s70, s45, s58, s73 |
| 10 | 1.8448627 | 1.7346375 | s69, s16, s14, s72, s64 |
| 11 | 1.7588239 | 1.0396189 | s39, s16, s45, s52, s73 |
| 12 | 1.7244288 | 1.1241017 | s37, s70, s71, s54, s73 |
| 13 | 1.5136427 | 0.9536690 | s69, s70, s11, s57, s73 |
| 14 | 1.5354580 | 0.8983285 | s42, s18, s71, s58, s22 |
| 15 | 1.4622213 | 0.8691497 | s69, s70, s71, s58, s23 |
| 16 | 1.4682687 | 0.9421740 | s35, s70, s71, s53, s26 |
| 17 | 1.4437991 | 1.3006158 | s69, s16, s71, s49, s73 |
| 18 | 1.5748583 | 0.9267403 | s69, s70, s71, s55, s26 |
| 19 | 1.6954650 | 0.9701970 | s69, s70, s14, s57, s24 |
| 20 | 1.8232220 | 1.0836064 | s40, s70, s14, s54, s73 |
| 21 | 1.8950979 | 1.0414487 | s69, s15, s71, s54, s73 |
| 22 | 1.6739977 | 0.9976958 | s69, s19, s11, s53, s73 |
| 23 | 1.3849769 | 0.8393500 | s38, s70, s71, s55, s73 |
| 24 | 1.2910894 | 0.7627572 | s42, s70, s71, s52, s26 |
| Total | 36.3833471 | 26.1200513 | – |

**TABLE 5.6**

**Expected CF Value in Case 3 and Case 4**

| Hour | Cost ($)×$10^5$ | | |
|------|--------|--------|--------|
| | Case 1 | Case 3 | Case 4 |
| 1 | 1.2285080 | 0.2795030 | 0.3929130 |
| 2 | 1.2347423 | 0.2077981 | 0.2193076 |
| 3 | 1.2405938 | 0.2998762 | 0.1462423 |
| 4 | 1.2521136 | 0.1844929 | 0.1914981 |
| 5 | 1.2704854 | 0.1649648 | 0.1417301 |
| 6 | 1.3962585 | 0.6095251 | 0.1878506 |
| 7 | 1.4589467 | 0.4476875 | 0.2363975 |
| 8 | 1.5380414 | 0.4965885 | 0.1492854 |
| 9 | 1.6734438 | 0.6848564 | 0.5436086 |
| 10 | 1.8448627 | 1.5233252 | 1.4131962 |
| 11 | 1.7588239 | 1.0246619 | 1.1735034 |
| 12 | 1.7244288 | 1.1104679 | 1.4479373 |
| 13 | 1.5136427 | 0.8321381 | 0.7739744 |
| 14 | 1.5354580 | 0.5966907 | 0.6199722 |
| 15 | 1.4622213 | 0.8520885 | 0.6584877 |
| 16 | 1.4682687 | 0.9382028 | 0.4949853 |
| 17 | 1.4437991 | 0.8617956 | 0.6344768 |
| 18 | 1.5748583 | 0.6522869 | 0.4528320 |
| 19 | 1.6954650 | 0.4166146 | 0.3078979 |
| 20 | 1.8232220 | 0.5133001 | 0.3048827 |
| 21 | 1.8950979 | 0.8346384 | 0.7427245 |
| 22 | 1.6739977 | 0.4709634 | 0.4480455 |
| 23 | 1.3849769 | 0.3794920 | 0.2496771 |
| 24 | 1.2910894 | 0.6695592 | 0.1837946 |
| Total | 36.3833471 | 15.051517 | 12.1152207 |

Similar to Table 5.5, Table 5.6 provides a comparison among cases 1, 3, and 4. According to the third column of Table 5.6, which is corresponding to the third case, the existence of the correlated WTs along with FRP can reduce the cost of the grid in all hours. Comparing the results of case 2 and case 3 shows that although WTs are highly useful to the grid, FRP plays a greater role in reducing costs. Besides the impact of WTs in reducing the total cost of the system, these units can be employed as backup energy sources and improve the flexibility of the system. While it might be expected that the presence of PEVs in the grid as additional non-stationary loads should enhance the cost of the system, it can be seen from Table 5.6 that in the last case, even though the cost has increased in some hours, the overall cost of the grid has decreased. Additionally, immobile PEVs can be employed as emergency energy sources such that in face of disturbance, anomaly, or sudden changes in the power generation or load in the system, their battery storages can act as a fast response energy source and maintain the system's normal operation for a while. This ability of PEVs can highly increase the maneuverability of the system operators and improve the flexibility and reliability indices. The above benefits are the results of the bidirectional power flow between PEVs and the grid caused by V2G technology.

**TABLE 5.7**

**Fleets Aging Cost**

| Fleet | 1 | 2 | 3 | 4 | 5 |
|---|---|---|---|---|---|
| Cost(\$)×10³ | 2.2893 | 2.2894 | 0.5579 | 2.6553 | 7.2820 |

Table 5.7 shows the BDC of the PEV fleets in the grid. According to this table, total BDC of the fleets is about 15,000 (\$) without which the overall cost of the grid in the fourth case will be reduced. The BDC, which shows the battery aging process, is the undesired result of V2G. However; according to the results in Table 5.5, Table 5.6, and Table 5.7, it can be seen that the use of V2G PEVs in the system is economically justified.

The complete data related to charging/discharging/idle state of PEVs and the amount of power exchanged between PEVs and the grid are shown in Table 5.8. It is

**TABLE 5.8**

**Power Exchanged between PEVs and the Grid Considering FRP (kW)**

| | Fleet Number | | | | |
|---|---|---|---|---|---|
| Hour | 1 | 2 | 3 | 4 | 5 |
| 1 | – 185.1760 | –161.5040 | 77.4342 | 233.0000 | 85.8248 |
| 2 | 0 | – 11.3354 | 0 | 353.0000 | 292.0000 |
| 3 | 178.1007 | – 186.0845 | 69.2791 | 233.0000 | – 292.0000 |
| 4 | 18.6995 | 292.0000 | 132.4660 | 233.0000 | 201.5192 |
| 5 | 417.0954 | 177.6639 | 96.1760 | 233.0000 | 105.8946 |
| 6 | 496.0000 | 233.0566 | 0 | 0 | 292.0000 |
| 7 | 0 | 287.6279 | 0 | – 15.7700 | 292.0000 |
| 8 | 0 | 0 | – 88.3243 | – 138.7809 | 0 |
| 9 | – 203.0000 | – 258.0808 | 0 | 162.4394 | 0 |
| 10 | – 265.1456 | 66.7982 | – 38.6223 | – 206.9205 | – 64.4983 |
| 11 | 265.2811 | – 32.1854 | 103.3094 | 233.0000 | 92.8952 |
| 12 | 0 | 86.7178 | – 108.5810 | – 166.4908 | – 93.9245 |
| 13 | –207.0920 | 73.4535 | 75.1110 | 223.9765 | 33.9316 |
| 14 | – 164.9914 | – 70.4516 | – 96.9980 | – 206.2808 | – 266.6978 |
| 15 | – 66.6270 | 97.1193 | – 22.7919 | 126.0713 | – 113.4290 |
| 16 | – 112.6220 | – 170.0440 | – 99.1999 | –39.8829 | 0 |
| 17 | 496.0000 | 0 | 0 | –127.1253 | – 35.6369 |
| 18 | 0 | – 22.2292 | 0 | 0 | – 45.7059 |
| 19 | 0 | 126.9064 | 105.9714 | 131.4604 | 0 |
| 20 | – 74.4737 | – 161.3354 | – 80.5058 | – 72.0184 | 0 |
| 21 | – 433.5433 | – 108.9425 | – 107.8253 | – 113.5519 | 40.9461 |
| 22 | 428.5090 | 183.9245 | 73.3672 | 25.0452 | 39.0776 |
| 23 | – 210.8925 | – 194.6081 | – 77.2715 | 42.4009 | 0 |
| 24 | 239.1899 | 111.5328 | 61.6693 | – 233.0000 | – 191.7853 |

worth noting that in this table, the positive/negative values present the charging/discharging status, and the zero values show the idle operation mode. A PEV is in idle operation mode if it is not connected to the grid (e.g., while it is being driven). The results in Table 5.8 show that most PEVs tend to charge at the beginning of the day and inject their energy to the grid at the peak load hours (i.e., end of the day). This behavior of PEVs is beneficial in several aspects. For instance, by supplying some loads in peak hours, they prevent power plants from operating at points close to their maximum power generation capacity or they can decrease the cost of the system.

### 5.5.3 SENSITIVITY ANALYSIS

In order to examine the effects of the stochastic parameters' standard deviation and CC of WTs on the CF, a sensitivity assessment is carried out in this section. In the following, two simulations regarding the relation between CC of WTs and standard deviation of parameters are carried out. The simulations are performed on the fourth case (i.e., where all three elements are considered in the CF) of the last subsection. Initially, we investigate the effect of different CC on the CF. In this way, the CF-CC diagram for various WT CCs in the range of [0.1,1] with a step size of 0.02 is presented in Figure 5.8. As can be seen from this figure, by increasing CC from 0.1 to 0.7, the value of CF has increased and after 0.7, increasing CC has an inverse effect on the CF. This result shows that WTs CC has a high impact on the CF through WT output power.

In the second part, an investigation regarding the impact of parameters' standard deviation on the CF is performed. Since in this work, the normal distribution function with the mean value of the base case is employed to model uncertainties in the

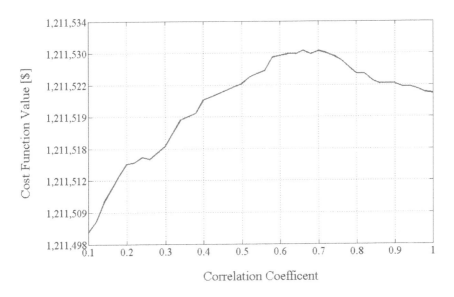

**FIGURE 5.8** Cost function-correlation coefficient diagram.

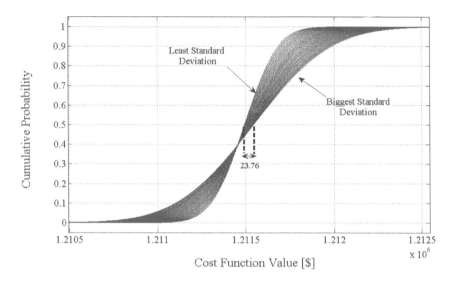

**FIGURE 5.9**   Cumulative probability-cost function diagram.

system, for investigating the impact of different standard deviation values on the CF, in each scenario all of the standard deviation values of all parameters are changed and the problem is run. The standard deviation value of each parameter is changed from 0.5 to 4 times its base value with a step size of 0.1. Figure 5.9 shows the results. As can be seen from Figure 5.9, the standard deviation of parameters and the standard deviation of the CF has a direct relation (which is what we expected), and the value of the CF has increased 23.7640 ($).

## 5.6   CONCLUSION

This chapter investigated the significance of some flexibility strategies in EDSs including FRP, V2G PEVs, and WTs. UT technique was introduced as a popular and successful approach to capture the uncertainties associated with the number of PEVs in a fleet, WTs' output power, and correlation of WTs, load demand, market price, and PEVs' charge/discharge pattern. A problem formulation was also developed to show how each strategy may be modeled in an EDS, and its performance was tested on the IEEE 69-bus test system. The results demonstrated that V2G technology, by providing a bidirectional power flow among PEVs and the grid, can highly decrease the cost of the system. Therefore, V2G can be assumed as a key flexibility strategy for the operator to minimize the costs and improve the technical aspects of the grid. In the case of WTs, results showed that the deployment of renewable energy sources such as WTs in the EDSs can improve the system from both the operation and reliability standpoints. Additionally, it was seen that FRP can facilitate the operation of the system in the presence of PEVs, and thus may be assumed as a strategic approach for reinforcing the system flexibility. Moreover, a sensitivity analysis was provided, and its results show that there is a direct relationship between the correlation of

WTs and CF in low CC values. This relation is inverse with high CCs. Also, it was concluded that changing the standard deviation of the stochastic parameters in the system does not have a high influence on the CF.

## REFERENCES

[1] Cochran, J., Miller, M., Zinaman, O., Milligan, M., Arent, D., Palmintier, B., O'Malley, M., Mueller, S., Lannoye, E., Tuohy, A. and Kujala, B., *Flexibility in 21st century power systems* (No. NREL/TP-6A20-61721). National Renewable Energy Lab. (NREL), Golden, CO (United States), 2014.

[2] Kavousi-Fard, Abdollah, Taher Niknam, and Mahmud Fotuhi-Firuzabad. "Stochastic reconfiguration and optimal coordination of V2G plug-in electric vehicles considering correlated wind power generation." *IEEE Transactions on Sustainable Energy* 6, no. 3 (2015): 822–830.

[3] Rostami M. A., M. Raoofat, A. Abunasri, and A. Kavousi-Fard. "Smart automated distribution system considering plug-in hybrid electric vehicle." *Journal of Intelligent & Fuzzy Systems* 28, no. 6 (2015): 2481–2492.

[4] Lei, Ming, and Mojtaba Mohammadi. "Hybrid machine learning based energy policy and management in the renewable-based microgrids considering hybrid electric vehicle charging demand." *International Journal of Electrical Power & Energy Systems* 128 (2021): 106702.

[5] Tookanlou M. B., M. Marzband M., A. Al Sumaiti A., and A. Mazza. Cost-benefit analysis for multiple agents considering an electric vehicle charging/discharging strategy and grid integration. In 2020 IEEE 20th Mediterranean Electrotechnical Conference (MELECON) 2020 Jun 16 (pp. 19–24). IEEE.

[6] Darabi Z. and M. Ferdowsi. "An event-based simulation framework to examine the response of power grid to the charging demand of plug-in hybrid electric vehicles." *IEEE Transactions on Industrial Informatics.* 10, no. 1 (2013): 313–322.

[7] Kennel F., D. Görges, and S. Liu. "Energy management for smart grids with electric vehicles based on hierarchical MPC." *IEEE Transactions on Industrial Informatics.* 9, no. 3 (2012):1528–1537.

[8] M. Dabbaghjamanesh, A. Kavousi-Fard and J. Zhang, "Stochastic Modeling and Integration of Plug-In Hybrid Electric Vehicles in Reconfigurable Microgrids With Deep Learning-Based Forecasting," in IEEE Transactions on Intelligent Transportation Systems, doi: 10.1109/TITS.2020.2973532.

[9] Sortomme, Eric and Mohamed A. El-Sharkawi. "Optimal charging strategies for unidirectional vehicle-to-grid." *IEEE Transactions on Smart Grid* 2, no. 1 (2010): 131–138.

[10] Kavousi-Fard A. and T. Niknam. "Optimal distribution feeder reconfiguration for reliability improvement considering uncertainty." *IEEE Transactions on Power Delivery* 29, no. 3 (2013):1344–1353.

[11] Abbasi R.S. and Mehmood T. Feeder reconfiguration techniques: A review. In 2014 International Conference on Energy Systems and Policies (ICESP) 2014 Nov 24 (pp. 1–5). IEEE.

[12] Kavousi-Fard A. and M. R. Akbari-Zadeh. "Reliability enhancement using optimal distribution feeder reconfiguration." *Neurocomputing* 106 (2013): 1–11.

[13] Kim H., Y. Ko, and K. H. Jung. "Artificial neural-network based feeder reconfiguration for loss reduction in distribution systems." *IEEE Transactions on Power Delivery* 8, no. 3 (1993): 1356–1366.

[14] Lopez E., H. Opazo, L. Garcia, and P. Bastard. "Online reconfiguration considering variability demand: Applications to real networks." *IEEE Transactions on Power Systems* 19, no. 1 (2004): 549–553.

[15] Goswami S. K. and S.K. Basu. "A new algorithm for the reconfiguration of distribution feeders for loss minimization." *IEEE Transactions on Power Delivery* 7, no. 3 (1992):1484–1491.

[16] Morton A. B. and I. M. Mareels. "An efficient brute-force solution to the network reconfiguration problem." *IEEE Transactions on Power Delivery*15, no. 3 (2000): 996–1000.

[17] Taylor T. and D. Lubkeman. "Implementation of heuristic search strategies for distribution feeder reconfiguration." *IEEE Transactions on Power Delivery*5, no. 1 (1990): 239–246.

[18] Kavousi-Fard, Abdollah, Taher Niknam, Hoda Taherpoor, and Alireza Abbasi. "Multi-objective probabilistic reconfiguration considering uncertainty and multi-level load model." *IET Science, Measurement & Technology* 9, no. 1 (2014): 44–55.

[19] Niknam T., A. Kavousifard, S. Tabatabaei, J. Aghaei. "Optimal operation management of fuel cell/wind/photovoltaic power sources connected to distribution networks." *Journal of Power Sources* 196, no. 20 (2011): 8881–8896.

[20] Kavousi-Fard, Abdollah and Taher Niknam. "Multi-objective stochastic distribution feeder reconfiguration from the reliability point of view." *Energy* 64 (2014): 342–354.

[21] Kashem, M. A., Velappa Ganapathy, and G. B. Jasmon. "Network reconfiguration for load balancing in distribution networks." *IEE Proceedings-Generation, Transmission and Distribution* 146, no. 6 (1999): 563–567.

[22] Khosravi, Abbas, Saeid Nahavandi, and Doug Creighton. "Prediction intervals for short-term wind farm power generation forecasts." *IEEE Transactions on sustainable energy* 4, no. 3 (2013): 602–610.

[23] Papari, Behnaz, Chris S. Edrington, Indranil Bhattacharya, and Ghadir Radman. "Effective energy management of hybrid AC–DC microgrids with storage devices." *IEEE transactions on smart grid* 10, no. 1 (2017): 193–203.

[24] Aien, Morteza, Mahmud Fotuhi-Firuzabad, and Farrokh Aminifar. "Probabilistic load flow in correlated uncertain environment using unscented transformation." *IEEE Transactions on Power systems* 27, no. 4 (2012): 2233–2241.

[25] Coria, Gustavo E., Angel M. Sanchez, Ameena S. Al-Sumaiti, Guiseppe A. Rattá, Sergio R. Rivera, and Andrés A. Romero. "A Framework for Determining a Prediction-Of-Use Tariff Aimed at Coordinating Aggregators of Plug-In Electric Vehicles." *Energies* 12, no. 23 (2019): 4487.

[26] Shahidinejad, Soheil, Shaahin Filizadeh, and Eric Bibeau. "Profile of charging load on the grid due to plug-in vehicles." *IEEE Transactions on Smart Grid* 3, no. 1 (2011): 135–141.

[27] Santos, Adella, Nancy McGuckin, H. Y. Nakamoto, D. Gay, and Susan Liss. *Summary of Travel Trends: 2009 National Household Travel. Federal Highway Administration, Washington, DC* (2011).

[28] Roe, Curtis, A. P. Meliopoulos, Jerome Meisel, and Thomas Overbye. "Power system level impacts of plug-in hybrid electric vehicles using simulation data." In *2008 IEEE Energy 2030 Conference*, pp. 1–6. IEEE, 2008.

[29] Tomić, Jasna, and Willett Kempton. "Using fleets of electric-drive vehicles for grid support." *Journal of power sources* 168, no. 2 (2007): 459–468.

[30] Kavousi-Fard A., A. Khosravi, S. Nahavandi. "A new fuzzy-based combined prediction interval for wind power forecasting." *IEEE Transactions on Power Systems*31, no. 1 (2015): 18–26.

[31] Moghaddam A. A., A. Seifi, T. Niknam, and M. R. Pahlavani. "Multi-objective operation management of a renewable MG (micro-grid) with back-up micro-turbine/fuel cell/battery hybrid power source." *Energy*36, no. 11 (2011):6490–64507.

[32] Kavousi-Fard, Abdollah, Taher Niknam, and Mahmud Fotuhi-Firuzabad. "A Novel stochastic framework based on cloud theory and $\theta$-modified Bat Algorithm to solve the distribution feeder reconfiguration." *IEEE Transactions on Smart Grid* 7.2 (2015): 740–750.

[33] Dallinger, David, Jochen Link, and Markus Büttner. "Smart grid agent: Plug-in electric vehicle." *IEEE Transactions on Sustainable Energy* 5, no. 3 (2014): 710–717.

[34] Khodayar, Mohammad E., Lei Wu, and Mohammad Shahidehpour. "Hourly coordination of electric vehicle operation and volatile wind power generation in SCUC." *IEEE Transactions on Smart Grid* 3, no. 3 (2012): 1271–1279.

[35] Soylu, Seref, ed. Electric vehicles: the benefits and barriers. BoD–Books on Demand, 2011.

[36] Li, Liu, and Chen Xue-Yun. "Reconfiguration of distribution networks based on fuzzy genetic algorithms." *Proceedings-Chinese Society of Electrical Engineering* 20, no. 2 (2000): 66–69.

[37] Bi, Pengxiang, Jian Liu, Chunxin Liu, and W. Y. Zhang. "A refined genetic algorithm for power distribution network reconfiguration." *Automation of Electric Power Systems* 26, no. 2 (2002): 57–61.

[38] Shirmohammadi, Dariush, and H. Wayne Hong. "Reconfiguration of electric distribution networks for resistive line losses reduction." *IEEE Transactions on Power Delivery* 4, no. 2 (1989): 1492–1498.

[39] Naka, Shigenori, Takamu Genji, Toshiki Yura, and Yoshikazu Fukuyama. "A hybrid particle swarm optimization for distribution state estimation." *IEEE Transactions on Power systems* 18, no. 1 (2003): 60–68.

# 6 Resource-Driven Flexibility Planning of Active Distribution Network

*Mahnaz Moradijoz*
Tarbiat Modares University, Tehran, Iran

## CONTENTS

## 6.1   INTRODUCTION

Distribution network flexibility needed to cope with the distributed energy resources (DERs) power output variations or network topology variations in normal or contingency conditions can be divided into two categories:

- Behavioral flexibility.
- Structural flexibility.

Behavioral flexibility is the network ability to economically satisfy power balance constraints by changing the power generation level. Therefore, this type of flexibility mainly affects the network operational cost. Structural flexibility is the network ability to change its topological layout to improve technical indices. Network reconfiguration is the traditional form of this type of flexibility which is generally carried out to improve losses. The advent of DER in distribution networks, referred to as active distribution networks (ADNs), introduces novel form of structural flexibility which is mainly affects network reliability indices. Structural flexibility enables ADNs to be operated reliably by islanding the faulted part of the system in the form of flexible micro-grids. In other words, the more this type of flexibility is, the more restored loads in flexible micro-grids will be.

Regarding the above mentioned definitions for the network flexibility, flexibility providers in the network can be classified into two groups:

- Flexibility providers in normal conditions.
- Flexibility providers in contingency conditions.

In normal conditions, flexibility required for covering load variations of passive distribution networks mainly is provided by the upstream grid. However, in ADNs, DERs can contribute in flexibility enhancement of the network to cope with the net load variations. It should be noted that, different types of DERs have different capability in providing this type of flexibility. So that, some type of DERs such as renewable-based ones increases the need for this type of flexibility. Conversely, due to the flexibility in power generation and consumption along with temporal flexibility, energy storage systems can make a significant contribution to increasing the network flexibility to respond to net load variations under normal conditions. Different flexibility enhancement capability of DERs highlights the necessities of developing integrated distribution system planning models considering these resource-driven flexibility providers. Similarly, in contingency conditions, passive distribution networks rely exclusively on the upstream grid for load restoration, whereas in ADNs, DERs may restore interrupted load points in the form of intentional islands. In order to enable DER participation in flexibility enhancement, different infrastructures should be provided. One of the most important infrastructures is the existence of the network-driven flexibility providers such as sectionalizing switches. Optimal placement of the network-driven flexibility providers has a great impact on the DERs' capability in flexibility enhancement. This fact emphasizes the need for appropriate network-driven flexibility providers' allocation model along with applicable resource-driven flexibility providers' expansion planning model.

Given the above, it is clear that utilizing the embedded flexibility in DERs is tied to the planning of both resource-driven flexibility providers and network-driven flexibility providers in ADNs.

In this chapter, an ADN expansion planning framework is presented considering the flexibility provided by resource-driven ones, i.e., DERs. To this end, first the investment environment governing DER deployment is determined. In bundled environment, in which distribution network operators (DNOs) is allowed to perform investment in DER deployment, DERs are modeled as operational and structural flexibility providers in ADN expansion planning problem. Then, the ADN expansion planning problem is modeled in an unbundled environment, where DNO and DER operator are two independent entities. It is worth mentioning that, expansion planning of network-driven flexibility providers is discussed in Chapter 12.

Note to mention, this chapter describes the basics of expansion planning of ADNs considering resource-driven flexibility providers. Additional information can be found in ref. [1–8]. Particularly, unbundled environment is examined in ref. [5].

## 6.2  CONVENTIONAL DISTRIBUTION NETWORK EXPANSION PLANNING VS. INTEGRATED EXPANSION PLANNING

In order to develop flexible ADN planning model, at first step it is necessary to understand the differences between conventional planning and flexible one. The differences can be highlighted by answering two main questions. The first one is that what are the main objectives that the decision-maker want to reach by optimal expansion planning? In fact, the answer to this question clarifies the borderlines of the planning problem. The second question is that how can the decision-making problem be solved? This question stems from the fact no matter how good a model is; it is useless if it cannot be solved. Moreover, the answer to this question clarifies the assumptions and approximations required for model development.

### 6.2.1  OBJECTIVES

In conventional planning models, network capacity is increased by its reinforcement/ expansion. In other words, reinforcement and expansion of the lines and the substation are the only expansion options, whereas in flexible one, in addition to these, expansion planning of DERs is considered as network capacity expansion options. Therefore, it is necessary to consider DER integrated planning of distribution network instead of separate planning of network and DERs in ADNs. Moreover, in addition to the main grid, DERs can supply the loads during scheduled or unscheduled outage events within islands in ADNs. As will be seen in sections 6.3 and 6.4, this important difference changes the problem modeling completely. Another important difference between conventional planning and flexible one stems from the uncertain nature of stochastic DERs. Generally, all decision-making problems are modeled in an environment having a level of uncertainty. However, it should be notified that DER inclusion in the network expansion problem increases the uncertainty level of the decision-making framework. As it will be explained later, this in turn, imposes the decision-maker to consider risk of the plans in addition to their expected cost. It will be shown that the main disadvantage of the ignoring risk in an uncertain environment is that the optimal values of the variables may lead to the minimum expected cost at the expense of experiencing very high costs in some unfavorable scenarios.

### 6.2.2 OPTIMIZATION TOOLS

Commonly, in order to solve conventional planning problem mathematical algorithms or heuristic-based ones are use. Such algorithm suitably can solve conventional planning problem in which only decision variables related to the network reinforcement/expansion exist. However, it should be noted in flexible planning environment, DER deployment is considered along with network reinforcement. This increases the number of decision variables which in turn increases dimension of decision space and complicates the mathematical model of the problem. Such high computational burden problems cannot easily be solved by heuristic or mathematical algorithms and hybrid algorithms should be developed. Moreover, as it will be discussed later in flexible environment, planning problem should be developed for both bundled and unbundled environments. It will be seen in the unbundled environment, planning problem is a multi-objective one which cannot be solved by preference-based multi-objective optimization algorithms in which the multi-objective problem is converted to a single-objective one using different methods. In such environment it is necessary that non-preference based algorithms are used to consider objectives of all players.

## 6.3   INVESTMENT ENVIRONMENT GOVERNING DER DEPLOYMENT

The first step to develop flexible planning problem, is to draw the investment environment governing DER deployment and to determine how DER investors interact with ADN decision-maker. In the next step, time frame of interactions should be clarified considering investment environment.

In a bundled environment, short-term and long-term behaviors of DERs are managed by a single entity and therefore, it is possible that these resources are used as operational and structural flexibility providers. It should be noted that different measures should be provided to reach this objective. Proper combination of DER units that complement each other is one of these measures.

In an unbundled environment, network investor and DER investors are independent entities which peruse their own objectives. Therefore, interactions practically are carried out in the long-term. In other words, in such environment, DERs generally contribute to structural flexibility enhancement of ADNs. In this environment, the expansion planning problem analysis considering different degrees of cooperation between players can provide appropriate perspective on creating incentives for the DER deployment. Hence, in this chapter, in addition to the bundled environment, planning problem is also discussed in an unbundled one.

## 6.4 FLEXIBILITY-BASED PLANNING IN A BUNDLED ENVIRONMENT

### 6.4.1 PROPOSED FRAMEWORK

In order to model the integrated expansion problem, it is necessary that first, utilized DER types are determined. In this section, renewable-based DERs and storage-based ones denoted by RDGs and ESSs, respectively, are considered.

From decision-maker's point of view, decisions can be distinguished between long-term and short-term. Network reinforcement/expansion and planning of DERs are long-term decision, whereas the optimal operational behaviors are decided in the short-term. Long-time decisions are made at the beginning of the planning horizon, while short-term decisions are made throughout. Therefore, as shown in Figure 6.1, the planning problem is modeled as a bi-level optimization problem. The first level (corresponding to the decision-making in the planning phase) comprises network reinforcement/expansion and planning of DERs. Decision variables include the optimal timing, location, and type/number of the system lines, substation and DER units to be upgraded or installed. All the candidate planning proposals are used as inputs in the next level. In its place, a two-stage optimization problem is defined for determination of the operational cost along with the energy not supplied (ENS) cost. In the first stage of the second level, DNO first combines different data which include

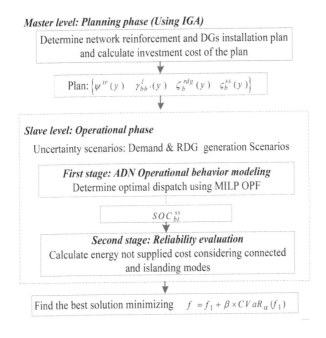

**FIGURE 6.1** Framework of the proposed bi-level ADN planning model.

network data received from the upper level, forecasted generation-demand data, as well as day-ahead market prices. Then, the optimal operational behavior is determined using combined data. The optimal operational behavior includes the optimal power generation of the DERs and the amount of the power purchased from the transmission network for each state of the network. These are achieved through the execution of the mixed integer linear programming (MILP)–based optimal power flow (OPF) which can consider environmental and economic objectives simultaneously. Meanwhile, the feasibility of the first-stage decisions is checked in the second stage, where the technically infeasible solutions are identified and discarded. In the second stage of the second level, the generated operational behavior is utilized to quantitatively analyze the reliability of the candidate planning proposal. Integration of RDGs as capacity expansion options in the ADN planning problem is a challenging issue due to the uncertain nature of the renewable generation. Therefore, the stochastic operational model is used to consider the effect of the uncertainty characterization of the RDGs on the optimal dispatch. Next to this step, the estimated operational cost, ENS cost along with the economic risk of the planning scheme will be fed back to revise previous planning design in the first level. With ongoing iterations, this procedure would finally reach the optimal plan.

## 6.4.2  Problem Formulation

The ADN planning problem is decomposed to an investment master problem and a two-stage operational slave problem is explained in the following section.

### 6.4.2.1  Investment Master Problem

The optimal design determination, which is a here-and-now decision, is made in the master problem by minimizing the net present value (NPV) of the total investment costs as presented in Equation (6.1). The first two lines in the objective function represent the substation/line expansion cost and the DER capital costs. Utilizing NPV as the objective function denotes that the planning horizon should be compatible with the lifetime of utilized expansion equipment. In order to ensure compliance with this requirement, the annualized costs are used in Equation (6.1). The last line of Equation (6.1) is related to the estimated operational cost along with the ENS cost that will be determined via solving the two-stage slave problem. In Equation (6.1), the investment costs are defined annually, while the operational costs are calculated hourly in the planning horizon, for each state of the system as discussed in the next section.

$$
\begin{aligned}
\min\{f_1\} = & \sum_y \frac{1}{(1+i)^y}\left(\sum_{y'\leq y}\sum_{bb'}IC^l(\gamma^l_{y',bb'})L_{bb'} + \sum_{y'\leq y}IC^{tr}(\psi^{tr}_{y'})\right) \\
& + \sum_y \frac{1}{(1+i)^y}\sum_{y'\leq y}\sum_b\sum_{der}IC^{der}\zeta^{der}_{y',b} \\
& + \sum_y\sum_\omega \frac{1}{(1+i)^y}\rho_\omega\Lambda_{y,\omega}
\end{aligned}
\tag{6.1}
$$

where $i$ is the discount rate, $\gamma^l_{y',bb'}$ is the investment decision variable in line $bb'$ and year y, $\psi^{tr}$ is the investment decision on the substation reinforcement in year y, $\zeta^{der}_{y',b}$ is the investment decisions for installation of DER unit in bus $b$, $IC^l$, $IC^{tr}$, $IC^{der}$ are the line reinforcement/installation cost, substation reinforcement cost, and DER investment cost, $\rho_\omega$ is the probability of scenario $\omega$, $\Lambda$ is the sum of the operational cost and ENS cost.

### 6.4.2.2 Operational Slave Problem

The main purpose of the slave problem is to project the optimal operational cost along with the ENS cost for each candidate planning proposal received from the master problem (6.2).

$$\Lambda_{y,\omega} = OC_{y,\omega} + ENC_{y,\omega} \tag{6.2}$$

where $OC$ and $ENC$ are the optimal operational cost and the ENS cost, respectively.

I. *First stage: Active network management (ANM) model*

The first stage of the slave problem corresponds to the determination of the optimal dispatch strategies, which are wait-and-see decisions. Consequently, these decisions depend on each realization vector of the uncertain parameters, which are made through applying the OPF for each uncertain scenario.

1. *Objective function*

The OPF objective function, which minimizes the expense for purchasing power from the grid, the energy loss cost, as well as the corresponding carbon tax created, is given in Equation (6.3).

It should be noted that if energy limited based DERs, such as ESSs is intended to be installed, these resources' depreciation cost arising from charging/discharging, should be considered in active network management (ANM) objective function. In ref. [9], this cost is considered.

$$\begin{aligned} \text{Min } OC_{y,\omega} = &\sum_t \sum_b \tau_{y,t} \pi^{grid}_{y,t} P^{grid}_{y,\omega,b,t} \\ &+ \sum_t \sum_{bb' \in \Omega^l} \tau_{y,t} \pi^{loss}_{y,t} R_{y,bb'} I^2_{y,\omega,bb',t} \\ &+ \sum_t \sum_b \tau_{y,t} \pi^{ct}_y P^{grid}_{y,\omega,b,t} E^{grid} \end{aligned} \tag{6.3}$$

where $P^{grid}$ is the active power injected from the substation in year y, scenario $\omega$, and time period $t$ (MW), $E^{grid}$ is the main grid emission factor (Ton $CO_2$/MWh), $\tau_{y,t}$ is duration of time period $t$ in year y (hour), $R_{bb'}$ is the resistance of line $bb'$, $I_{bb't}$ is the current of line $bb'$ in time period $t$ ($10^3$A), $\pi^{grid}$, $\pi^{loss}$, $\pi^{ct}$ are electricity market price ($/MWh), loss cost ($/MWh), and emission tax rate ($/Ton $CO_2$), respectively.

## 2. *Constraints*

OPF model should satisfy the following constraints:

a. *Power flow:* Power flow equations are given by Equations (6.4)–(6.7). In reactive power balance constraint, the reactive power consumption by DERs such as ESS should be considered.

$$P_{bt}^{grid} = P_{bt}^{d} - \sum_{der \in \Omega^{der}} P_{bt}^{der} + \sum_{bb' \in \Omega^{bb'}} P_{bb't}^{l} + \sum_{bb' \in \Phi^{bb'}} R_{bb'} \times I_{bb't}^{2}, \quad \forall\ t \in T,\ b \in \Omega^{b} \quad (6.4)$$

$$Q_{bt}^{grid} = Q_{bt}^{d} + \sum_{der} Q_{bt}^{der} + \sum_{bb' \in \Omega^{bb'}} Q_{bb't}^{l} + \sum_{bb' \in \Phi^{bb'}} X_{bb'} \times I_{bb't}^{2}, \quad \forall\ t \in T,\ b \in \Omega^{b} \quad (6.5)$$

$$V_{bt}^{2} - 2\left(R_{bb'} \times P_{bb't}^{l} + X_{bb'} \times Q_{bb't}^{l}\right) - Z_{bb'}^{2} I_{bb't}^{2} - V_{b't}^{2} = 0, \quad \forall\ t \in T,\ bb' \in \Omega^{l} \quad (6.6)$$

$$V_{b't}^{2} \times I_{bb't}^{2} = \left(P_{bb't}^{l}\right)^{2} + \left(Q_{bb't}^{l}\right)^{2}, \quad \forall\ t \in T,\ bb' \in \Omega^{l} \quad (6.7)$$

where $P_{bt}^{der}$ is the active power generated by the DER in bus $b$ and time period $t$, respectively, $P_{bt}^{d}$, $Q_{bt}^{d}$ are the active and reactive demands in bus $b$ and time period $t$, $P_{bb't}^{l}$, $Q_{bb't}^{l}$ are the active and reactive power flows through line $bb'$ in time period $t$, $T$ is the set of time periods, $Q_{bt}^{grid}$, $Q_{bt}^{der}$ are the reactive powers injected from the substation, and consumed by the DER, respectively, $V_{bt}$ is the voltage of bus $b$ in time period $t$, $X_{bb'}$, $Z_{bb'}$ are the reactance and impedance of line $bb'$, $\Omega^{bb'}$, $\Phi^{bb'}$ are the set of lines connected to bus $b$ and set of lines which their power losses are concentrated on bus $b$, respectively.

b. *Nodal voltage limit:* The voltage magnitude of each bus should be kept between the permissible operating limits as represented by (6.8).

$$\underline{V}^{2} \leq V_{bt}^{2} \leq \overline{V}^{2}, \quad \forall\ t \in T,\ b \in \Omega^{b} \quad (6.8)$$

where, $\underline{V}$ and $\overline{V}$ are the minimum and maximum of voltage levels.

c. *Thermal capacity limits of the lines and substation:* To sustain the security of the lines and the substation, the flow of current/power passing through them should not exceed the lines/substation thermal capacity limit as shown in (6.9) and (6.10).

$$-\overline{I}_{bb'} \leq I_{bb't} \leq \overline{I}_{bb'}, \quad \forall\ t \in T,\ bb' \in \Omega^{l} \quad (6.9)$$

$$\left(P_{t}^{grid}\right)^{2} + \left(Q_{t}^{grid}\right)^{2} \leq \left(\overline{S}^{tr}\right)^{2}, \quad \forall\ t \in T \quad (6.10)$$

where, $\overline{I}_{bb't}$, $\overline{S}^{tr}$ are the thermal capacity limits of line $bb'$ and substation, respectively.

d. *Power flow of the substation:* The reverse power flow through the substation is not allowed as denoted by (6.11). Moreover, the substation power factor should be kept in the specified range as presented in (6.12).

$$P_t^{\text{grid}} \geq 0 \tag{6.11}$$

$$-\tan(\varphi) \times P_t^{\text{grid}} \leq Q_t^{\text{grid}} \leq \tan(\varphi) \times P_t^{\text{grid}}, \quad \forall \ t \in T \tag{6.12}$$

where, $\varphi$ is the permissible power angle of the substation.

e.  *DER operational limits:* These types of constraints mainly depend on the DER types used as expansion options. Generally these limits include the following constraint set:

   i.  **DER power limit:** These constraints originate from the DER's primary energy resource limit or from its thermal capacity limit. Regardless of origination nature, (6.13) can be used to model active power limits. DER capacity limit, i.e. $\bar{P}_{bt}^{\text{der}}$, can be time dependent due to the primary resources limits of some DERs such as renewable-based ones. Therefore, for DERs don't have such limitation, e.g. ESSs, $t$ subscript should be eliminated from $\bar{P}_{bt}^{\text{der}}$. Note to mention, for DERs which have not power consumption capability, another constraint shown in (6.14) should be considered.

$$-\bar{P}_{bt}^{\text{der}} \leq P_{bt}^{\text{der}} \leq \bar{P}_{bt}^{\text{der}}, \quad \forall \ t \in T, \ b \in \Omega^b, \ \text{der} \in \Omega^{\text{der}} \tag{6.13}$$

$$0 \leq P_{bt}^{\text{der}} \leq \bar{P}_{bt}^{\text{der}}, \quad \forall \ t \in T, \ b \in \Omega^b, \ \text{der} \in \Omega^{\text{der}} \tag{6.14}$$

   ii.  **Reactive power model of DERs**: Depending on the DER type, there are different reactive power model such as dependent-Q (constant power factor), and circular thermal capacity curve which are given in (6.15) and (6.16), respectively. Therefore, reactive power model of DER should be determined before model development. More information on reactive power model of DERs can be found in refs. [10, 11].

$$Q_{bt}^{\text{der}} = \tan(\phi) \times P_{bt}^{\text{der}}, \quad \forall \ t \in T, \ b \in \Omega^b, \ \text{der} \in \Omega^{\text{der}} \tag{6.15}$$

$$\left(P_{bt}^{\text{der}}\right)^2 + \left(Q_{bt}^{\text{der}}\right)^2 \leq \left(\bar{P}_b^{\text{der}}\right)^2 \tag{6.16}$$

   iii.  **Energy limited-based DERs:** In addition to the above mentioned limits, another set of constraint exit for energy limited DERs. This set includes the following items:

- Energy stored in an ESS called as state of charge (SoC), should be kept in the acceptable range as presented in (6.17).
- The linking constraint of an ESS SoC along the time steps of a day should be satisfied as shown in (6.18).

- Constraint (6.19) defines the losses related to ESS active and reactive power output.
- For a sustainable operation of an ESS, its SoC at the end of a day should be equal to its initial SoC as given in (6.20).

$$0 \leq \text{SOC}_{bt}^{ss} \leq \bar{E}_b^{ss}, \quad \forall \quad t \in T, \ b \in \Omega^b \tag{6.17}$$

$$\text{SOC}_{bt}^{ss} = \begin{cases} \text{SOC}^{in} - P_{bt}^{der} - \text{loss}_{bt}^{ss} & \text{if } t = 1 \\ \text{SOC}_{bt-1}^{ss} - P_{bt}^{der} - \text{loss}_{bt}^{ss} & o.w \end{cases}, \quad \forall \quad t \in T, \text{der} \in ss, \ b \in \Omega^b \tag{6.18}$$

$$\text{loss}_{bt}^{ss} = r^{ss} \times \frac{\left(P_{bt}^{der}\right)^2 + \left(Q_{bt}^{der}\right)^2}{\left(V^{base}\right)^2}, \quad \forall \quad t \in T, \ b \in \Omega^b, \text{der} \in ss \tag{6.19}$$

$$\text{SOC}_{bt}^{ss} = \text{SOC}^{in} \quad \forall \quad t = \bar{T}, \ b \in \Omega^b \tag{6.20}$$

where $\bar{E}_b^{SS}$ is the energy capacity limit of a given ESS in bus $b$, $\text{SOC}_{bt}^{SS}$ is the SOC for a given ESS in bus $b$ and time period $t$, $\text{SOC}^{in}$ is the initial SOC of the ESS, $\text{loss}_{bt}^{SS}$ is the resistive losses of the ESS in bus $b$ and time period $t$, $r^{SS}$ is the resistance of an ESS, $V^{base}$ is the base voltage.

II. *Second stage: Reliability evaluation model*

As discussed earlier, DER capability in power restoration to the interrupted loads in the form of islands make a dramatic difference between conventional planning model and flexible ones. This obligates the decision-maker to consider reliability indices in decision-making process. Hence, at this stage, the reliability index which is ENS cost, is assessed considering both the connected and island mode operations. Successful islanding has special conditions. In this chapter, it is assumed that if the following two conditions are satisfied the islanding will be successful:

1. During island mode operation, relying on renewable-based DERs only may cause stability problems regarding voltage and frequency variations. Hence, the presence a dispatchable DER such an ESS, is assumed to be the primary condition for successful islanding.
2. The DER sources should have enough power injection capability to support the load points within the zone. In some cases, in order to satisfy this constraint, some of the loads would need to be curtailed. Furthermore, during this constraint checking, power generation limit of energy limited nature DERs should be considered.

Different DER integrated reliability evaluation models such as zoning-based ones [12–14] and energy injection-based model [15] are used in the planning. In this chapter, comprehensive zoning-based model in which the ENS cost assessment procedure includes seven steps is used, as shown in Table 6.1.

**TABLE 6.1**
**Procedure to Assess ENS Cost**

| Step | Brief description |
|------|-------------------|
| 1 | Zoning the network |
| 2 | Iterating steps 3–6 |
| 3 | Clustering fault-prone elements |
| 4 | Checking the primary condition for successful island formation |
| 5 | Calculating ENS cost of a zone in disconnected mode |
| 6 | Calculating ENS cost of a zone in island mode |
| 7 | Calculating the ENS cost of the network |

*Step 1: Zoning the network*

In this step, the system is divided into several zones considering the locations of protective devices. A zone is a group of components whose entry element is a protective device, as shown in Figure 6.2.

*Step 2: Iterating steps 3–6*

The ENS cost of the system should be computed considering all operational behaviors and zones. Consequently, steps 3–6 described in the following should be repeated a number of times proportional to the numbers of generated operational behaviors and zones.

*Step 3: Clustering fault-prone elements*

In this step, two sets are determined for each zone: set (a) that comprises all elements within the zone whose failures cause the outage of all loads within the zone and set (b) that includes all elements outside the zone whose failures cause the outage of the part of the loads. In other words, if a fault occurs in each element of set (b), the zone can be operated in the island mode.

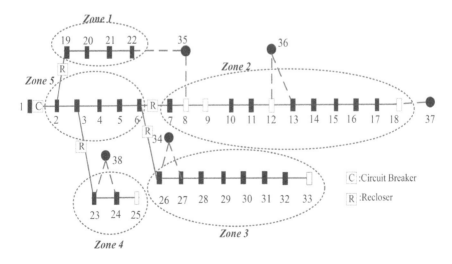

**FIGURE 6.2**   The 33-bus test distribution network.

*Step 4: Checking the primary condition for successful island formation*

In this step, the existence or non-existence of a dispatchable DER such as an ESS in the zone is checked. If there is dispatchable unit in the zone, the zone cannot be operated in the island mode even if a fault occurs in elements of set (b). Therefore, all elements of set (b) should be considered as elements of set (a). Hence, the ENS cost of zone z related to the outages in set (b) will be zero.

*Step 5: Calculating energy not supplied cost for a zone in disconnected mode*

The ENS cost of the zone for outage events in the elements of set (a) is calculated by Equation (6.21). The equation states that for each contingency in the elements of set (a), all loads within the zone are shed. Therefore, the total ENS cost of a zone in the disconnected mode due to a contingency in an element of set (a) is proportional to the outage time of that element and the total curtailed load cost in that zone. It should be notified that the failure rate of an element multiplied by the repair time of that element shows the annual outage time of the element.

$$\text{ENC}^{\text{dsc}}_{y,\omega,z,t} = \sum_{k \in \text{dsc}_z} \sum_{b \in \Omega_z} \frac{\lambda^f_{y,k}}{8760} t^{\text{rep}}_{y,k} P^d_{y,\omega,b,t} CD_b \tau_{y,t} \tag{6.21}$$

where $\text{ENC}^{\text{dsc}}$ is the zth zone ENS cost related to the outages in elements of set (a) in time period t, scenario ω, and year y, $\lambda^f_{y,k}$ is the failure rate of element k in year y, $t^{\text{rep}}_{y,k}$ is the repair time of element k in year y, $CD_b$ is the cost damage function of the curtailed load in bus b, $\Omega^b_z$ is the set of network buses in zone z.

*Step 6: Calculating energy not supplied cost for a zone in island mode*

For each outage event in the elements of set (b), it is checked whether the DER units have enough capacity to supply all load points within the zone. For this purpose, the maximum power injection capability of DERs is evaluated using (6.22). It should be noted that power injection capability of energy limited DERs can be calculated using Equation (6.23). If constraint (66.24) is satisfied, no load is cut. If not, some load must be curtailed. The load shedding policy utilized here is to curtail the loads with low ENS cost. The load shedding procedure continues until constraint (6.24) is satisfied. In other words, the loads in the zone are cut until the power generated by DER units will be enough to supply the loads that are not shed. Finally, ENS cost within the island is calculated utilizing Equation (6.25).

$$P^{\text{isl}}_{y,\omega,m,t} = \sum_{b \in \text{BMG}_m} \sum_{der \in \Omega^{\text{der}}} P^{\text{der}}_{y,\omega,b,t} \tag{6.22}$$

$$P^{\text{der}}_{y,\omega,b,t} = \frac{\eta^{\text{dis}} \text{SOC}_{y,\omega,b,t}}{t^{\text{rep}}_{y,k}} \tag{6.23}$$

$$P^{\text{isl}}_{y,\omega,m,t} \geq (1 + \xi^{\text{res}}) \sum_{b \in \text{BMG}_m} P^d_{y,\omega,b,t} - LS^m_{y,\omega,b,t} \tag{6.24}$$

$$\text{ENC}^{\text{isl}}_{y,\omega,z,t} = \sum_m \sum_{k \in \text{isl}_{m,z}} \sum_{b \in \Omega_z} \frac{\lambda^f_{y,k}}{8760} t^{\text{rep}}_{y,k} LS^m_{y,\omega,b,t} CD_b \tau_{y,t} \tag{6.25}$$

$P^{isl}$ is the active power generated by DER units in an FMG, $LS^m$ is the amount of the load shedding in bus $b$ and time period $t$ within islanded system, BMG is the set of an FMG buses, $\Omega_z$ is the set of network zones, $\xi^{res}$ is the required reserve margin for island operation, and $\text{ENC}^{isl}$ is $z$th zone ENS cost related to the outages in elements of set (b).

*Step 7: Calculating the energy not supplied cost of the network*

As seen, the ENS cost of a zone is separated into two distinct parts, i.e. the ENS cost in the island mode and the one in the disconnected mode. Therefore, the cost of the interruption in the ADN can be evaluated by summing these parts as given in Equation (6.26).

$$\text{ENC}_{y,\omega} = \sum_t \sum_z \text{ENC}^{isl}_{y,\omega,z,t} + \text{ENC}^{dsc}_{y,\omega,z,t} \qquad (6.26)$$

### 6.4.3 RISK CONTROL IN THE PROPOSED STOCHASTIC ADN PLANNING

As discussed earlier, in integrated planning, the probabilistic parameters including generation and demand put the system at risk that should be evaluated. In order to perform a correct comparison of the solutions, a term modeling the risk of the variability corresponding to the total cost is included in the formulation of the problem. This chapter takes into account the conditional value at risk (CVaR), a quite popular risk index used in financial research, to measure the related risk. The CVaR, regarding a predefined probability level $\alpha$ percent, is defined as the expected value which should be paid, in addition to the expected cost, during the $1-\alpha\%$ worst scenarios [16]. In other words, first, the expected cost in the $1-\alpha\%$ worst scenarios of the cost distribution is computed. Then, $\text{CVaR}_\alpha$ is calculated by subtracting this cost from the expected cost.

The CVaR can be incorporated into the main objective function as follows:

$$\text{Min } f = f_1 + \beta \times CVaR_\alpha(f_1) \qquad (6.27)$$

where $\beta \in [0,\infty)$ is a weighting coefficient utilized to make a tradeoff between the expected cost and the risk aversion strategy. If $\beta = 0$, the risk term in the objective function is disregarded. Thus, the resulting problem becomes the risk-neutral one. As $\beta$ increases, the expected cost term becomes less significant than the risk term.

### 6.4.4 OPTIMIZATION TOOL

Mathematically, the proposed planning model is a bi-level mixed-integer non-linear programming (MINLP) problem. Decision variable of the first stage are $\{\gamma^l, \psi^{tr}, \zeta^{der}\}$. For each candidate solution of these variables, the first stage objective function can be figured out only by calculating the second level main decision variables which are $\{P^{grid}, Q^{grid}, P^{der}, Q^{der}\}$. Therefore, as stated before, optimal expansion plan problem of ADN contains a huge number of decision variables which complicates the problem. In other words, proposal of a solution method calculating the optimal value of these variables simultaneously practically is impossible. There are different methods to solve such complex problems. One of which is to decompose the main problem into several subproblems. Bi-level hybrid heuristic-mathematical algorithm optimization methods are widely used to solve such problems. In ref. [9], immune-genetic algorithm (IGA) combined with linear programming-based algorithm has been proposed to solve such planning model.

## 6.4.5 Numerical Results

As an illustrative example, the planning methodology is applied to the IEEE 33-bus distribution test network depicted in Figure 6.2. It is assumed that circle-shape nodes belong to the new load points predicted to be expanded after 3 years. Moreover, new candidate corridors are shown with dashed-lines. Candidate buses for installation of DERs depend on technical, environmental, and economic factors, which are shown with the white rectangles in Figure 6.2. The capacity of the substation transformer that can be utilized to reinforce the substation is 4 MVA. Wind-based DERs and ESS-based ones are considered. The wind turbines have a capacity of 0.5 MW. The energy reservoir capacity and rated power of ESSs are assumed to be 0.6 MWh and 0.075 MW, respectively. The planning horizon includes 4 years. The other simulation parameters are presented in Table 6.2.

### 6.4.5.1 Considered Cases

Based on the considered objectives and expansion options, four different cases are studied.

Case A: In this case, DERs is not considered. No OPF is performed and only a load flow problem is solved. In other, words, this case examines simulation results in a passive network and a conventional environment. Moreover, in

## TABLE 6.2
## Data Used in the Case Studies

| Parameter | Unit | Value |
|---|---|---|
| Load growth rate | % | 3.5 |
| $E^{grid}$ | Ton $CO_2$/MWh | 0.91 |
| $\lambda^{ct}$ | $/ Ton $CO_2$ | 30 |
| $r^{cal}$ | Per day | 1 |
| $S, R'$ | | 4, 20 |
| $\alpha, \beta$ | | 0.8, 1 |
| $N_p$ | | 20 |
| $\kappa$ | % | 5 |
| $\rho_{m/c}^{max}$ | | 0.2, 0.8 |
| $UC^{tr}$ | M$ | 0.2 |
| $UC_{bb'}$ | M$/km | Type 1: 0.1, Type 2: 0.15 |
| $IC^{rdg}$ | M$ | 0.6135 |
| $IC^{ss}$ | M$ | 0.15 |
| $Cos(\phi)$ | | 0.99 |

this case neither carbon tax nor risk index is considered in the expansion planning procedure of passive distribution networks.

*Case B:* This case is similar to the previous case. The only difference is that the carbon tax and risk index are considered as the components of the objective function.

*Case C:* In this case, the ENS cost and carbon tax are considered as the components of the objective function in ADN in which DERs are installed.

*Case D:* This case embeds the ENS cost, carbon tax, and risk index into the objective function in active environment. It should be notified that all of the above-mentioned cases consider the capital investment cost of the corresponding expansion options and the cost of the purchased power from the main grid as the components of the objective function.

### 6.4.5.2 Results and Comparison

The NPVs of the optimal plan costs are presented in Table 6.3 for each case.

*Case A:* As shown in Table 6.4, the NPV of the total cost is 5.817 M$ in case A, which is the carbon-intensive conventional distribution network planning case. It should be noted that this cost ignores the created carbon tax, which is 1.363 M$.

*Case B:* For case B, the outcomes of the planning problem are found to converge to case A. However, the value of risk in case B is different from that one in case A. The reason is that the carbon tax minimization is not considered as a planning objective in case A.

*Case C:* In this case, the carbon tax is reduced up to 54.95% compared to the previous case. Nevertheless, the outcome of the simulation shows the highest value for $CVaR_{0.8}$ in case C in which the planning problem is solved from a risk-neutral DNO's viewpoint, which minimizes only the expected cost. Although the expected cost decreases of 1.44 M$ in C, $CVaR_{0.8}$ increases of

**TABLE 6.3**
**NPVs of Different Cost Terms in Different Cases (M$)**

| Cost terms | | | A | B | C | D |
|---|---|---|---|---|---|---|
| Investment costs considering salvage value | | Substation | 0.032 | 0.032 | 0.032 | 0.123 |
| | | Line | 2.189 | 2.189 | 1.378 | 1.323 |
| | | Wind | 0 | 0 | 1.848 | 0.546 |
| | | ESS | 0 | 0 | 0.084 | 0.094 |
| Total investment cost | | | 2.221 | 2.221 | 3.343 | 2.092 |
| Purchased power cost | | | 2.508 | 2.508 | 1.054 | 1.844 |
| ENS cost | | | 1.035 | 1.035 | 0.692 | 0.692 |
| Loss cost | | | 0.052 | 0.0516 | 0.034 | 0.051 |
| Carbon tax[*] | | | 1.363 | 1.363 | 0.614 | 1.059 |
| $CVaR_{0.8}$ | | | 0.415 | 0.568 | 2.158 | 1.086 |
| Total planning cost | | | 5.817 | 7.178 | 5.738 | 5.739 |

**TABLE 6.4**

**NPV of Planning Cost**

| Cost Terms ($) | Cost (Solution 25) |
|---|---|
| Network upgrade cost($10^6$) | 1.015 |
| Loss cost ($10^4$) | 4.652 |
| ENS cost ($10^6$) | 0.996 |
| $CVaR_{0.8}$ ($10^4$) | 7.805 |
| Total planning cost ($10^6$) | 2.058 |
| Total planning cost + CVaR ($10^6$) | 2.136 |
| GPLO's profit ($10^5$) | 0.942 |

1.59 M$ in comparison to B. Indeed, the average ADN planning cost in the worst scenarios is 7.896 M$.

*Case D:* The outcomes of the simulation show the expected total cost in case D is slightly higher than the one in case B. This increase in the expected total cost is due to the considered objective function that is minimizing the total cost along with the plan risk. Therefore, although the expected cost increases in D, the results show 49.68% CVaR decrease in comparison to C. In other words, the plan cost in the 20% of the worst scenarios in case C is considerably higher than the one in case D. This result can be verified by observing the cost PDFs of the optimal plans in cases C and D presented in Figures 6.3 and 6.4, respectively. As can be seen, compared to case C, the cost PDF of the optimal plan is compressed in case D.

Compared to B, the ENS cost savings in case D is 33.14%, which means that the flexibility of the network for supplying loads under outage events increases of 33.14%

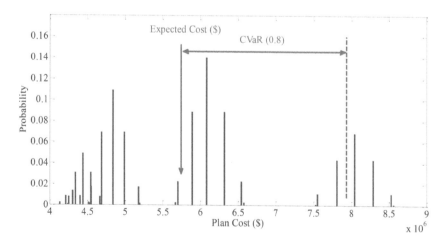

**FIGURE 6.3**   Probability distribution of the plan cost in case C.

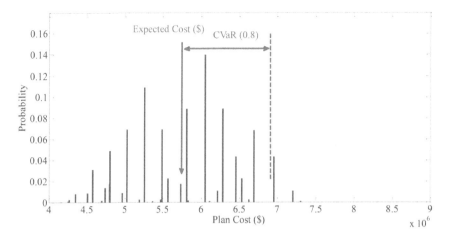

**FIGURE 6.4**   Probability distribution of the plan cost in case D.

by DERs. The network structure in the planning horizon for case D is depicted in Figure 6.5. As can be observed, ESSs with the reactive power capability are installed in zones 2–4, which enables the island mode operation for these zones.

## 6.5   FLEXIBILITY-BASED PLANNING IN AN UNBUNDLED ENVIRONMENT

In this section, flexibility-based ADN expansion planning problem is investigated in an unbundled environment. In this section, among different type of DERs, gridable parking lots (GPLs) are considered in the expansion problem. The first reason

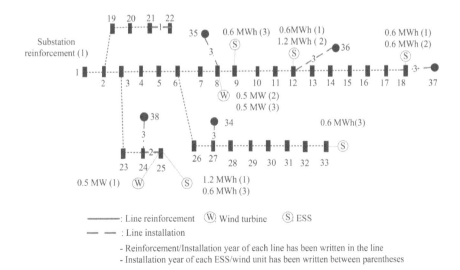

**FIGURE 6.5**   Optimal ADN expansion plan in D.

for this is that in most countries, GPLs are installed by private entities, i.e. entities other than DNOs. The second reason is related to the dual producer-consumer (prosumer) behavior of these resources which complicates the expansion model. This prosumer behavior creates both challenges and opportunities in the ADN expansion planning. On one hand, the stored electrical energy in vehicle batteries can be considered as an alternative supplying source and increase the ADN flexibility to restore the interrupted loads. On the other hand, the electrical energy required for vehicle battery charge may change the direction and magnitude of current flow through network substation and feeders and consequently, may change the characteristics of equipment needed for network capacity expansion. Regarding that the GPLs installation plan are determined by their owners and network expansion decision are made by DNO, two main question arise: (1) how the ADN expansion plan should be modeled in the presence of these prosumers?, and (2) considering that GPLO seeks their own objectives, can DNOs design incentive mechanism to benefit from these resources' flexibility resulting from their prosumer behavior? In this section, an expansion planning model is proposed regarding these two main questions. It should be mentioned that the model is a general one and other DER type can be incorporated in the model.

### 6.5.1  PROPOSED FRAMEWORK

In this section, DNO and DER owners' (called here in after GPL owner (GPLO)) behaviors are modeled in an interactive environment. The main objective of these parties is to maximize their profit depending on the investments made for distribution system expansion and DER deployment in the planning horizon. Profit maximization by these parties creates conflict in decision-making environment which leads each party tracks his own planning strategy. This conflict can be modeled by different mathematical approaches such as bi-level modeling. In the upper-level of the model, DNO seeks to make an optimal decision for system expansion, whereas in the lower-level GPLO determines the GPLs optimal allocation plan. The optimal decisions of these two levels are interdependent through decision variables related to allocation of GPLs.

The most important point is that the flexibility enhancement of ADN by utilizing electrical energy stored in GPLs for interrupted load restoration is considered in the bi-level model. In other words, the power injection capabilities of GPLs are considered as an alternative power supplying source when reliability of the network is jeopardized. More details are given in the next sections.

### 6.5.2  THE LOWER LEVEL: DER EXPANSION PLANNING MODEL

As mentioned earlier, the lower level is considered to model the GPL planning problem, aiming at maximizing the GPLO's benefit resulting from the financial interactions with the wholesale electricity market and plug-in electric vehicle (PEV) owners. At this level, GPLO decides about the allocation of GPLs and power trading with the wholesale electricity market and each GPL schedules the amount of charging/discharging power of PEV batteries.

### 6.5.2.1  Objective Function

As represented in Equation (6.28), the objective function contains two main terms: (1) short-term financial interactions and (2) long-term ones.

$$\text{Max}\left\{OF^{\text{GPL}}\right\} = OF^{\text{GPL},op} - OF^{\text{GPL},\text{inv}} \tag{6.28}$$

In this chapter, GPLs are modeled as aggregators which interact with wholesale market on one hand and with vehicles one the other hand. Therefore, the first term in Equation (6.28), which models the NPV of revenue resulting from the interactions with energy market as well as PEV owners, is presented in Equation (6.29). More detailed expression of the interactions is given in Equations (6.30) and (6.31). Equation (6.30) indicates that the income from selling energy back to the main grid minus the cost of purchasing energy from the grid is the net revenue obtained through the energy market interaction. In other words, according to the scenarios of availability of PEVs in parking lots and variations of electricity price, the sold power to the GPL and the sold power back to the main grid are computed in the energy market.

The revenues and the cost resulted from the interaction of GPLs with PEVs are given in Equation (6.31). The revenues are due to the charging of PEVs for driving and the parking fee. However, the cost comes from the depreciation of the vehicle batteries because of the operation in the lot-to-grid (L2G) mode.

$$OF^{\text{GPL},op} = \sum_y \sum_b \sum_t \sum_\omega \frac{1}{(1+i)^y} \rho_\omega \left( R^{\text{L2G}}_{y,\omega,b,t} + R^{\text{LEV}}_{y,\omega,b,t} \right) \tag{6.29}$$

$$R^{\text{L2G}}_{y,\omega,b,t} = P^{\text{dis}}_{y,\omega,b,t}\pi^{\text{grid}}_{y,t}\tau_{y,t} - P^{\text{ch}}_{y,\omega,b,t}\pi^{\text{grid}}_{y,t}\tau_{y,t} \tag{6.30}$$

$$R^{\text{LEV}}_{y,\omega,b,t} = \left(\text{SOC}^{fi}_{y,\omega,b,t}\pi^{\text{L2V}}_{y,t} + N^{\text{avil}}_{y,\omega,b,t}\pi^{\text{park}} - P^{\text{dis}}_{y,\omega,b,t}\pi^{\text{deg}}\right)\tau_{y,t} \tag{6.31}$$

The long term interactions, i.e. the second term of Equation (6.28), depends on parking capacity and includes the cost of purchasing charging facilities, the cost of land, and the cost of wiring which is formulated by Equation (6.32).

$$OF^{\text{GPL},\text{inv}} = \sum_y \sum_b \frac{1}{(1+i)^y}\left(IC^P_b\, CS_{y,b}\right) \tag{6.32}$$

### 6.5.2.2  Constraints

Physically, GPLs can be considered as a combination of an electric storage system with variable capacity and a semi-flexible load. Hence, similar to the conventional ESSs, charging strategy determination is the most important factor to model GPLs. Because of the comprehensiveness of smart charging strategy, this strategy in which the charging scheduling is determined based on electricity price, is used to model GPLs. Other PEV charging strategy can be considered in the model with slight modifications. Details on this can be found in ref. [12].

Charging mode of a GPL is the summation of available PEV charging modes as given in Equations (6.33) and (6.34). The maximum power of a PEV traded with network depends on its availability in the GPL and the rated power of a charging station as formulated in Equations (6.35) and (6.36). According to Equations (6.37) and (6.38), the number of PEVs that can inject/absorb power to/from the grid is limited to the number of installed charging stations. $\chi^{ch}$ and $\chi^{dis}$ are binary variables to ensure that even if a PEV is charged with a fraction of the rated power, a station will be occupied as formulated in Equations (6.39) and (6.40). In order to maintain the PEV battery life, upper and lower bounds of each battery SOC should be respected by GPLO as expressed in Equation (6.41). The PEV SOC at the arrival time of the lot equals to the initial SOC of the vehicle. At the other time steps, if the PEV is charged/discharged, its SOC will be increased/decreased considering charging/discharging efficiency. Therefore, the SOC of a PEV along the time steps of a generic day depends on the stored energy in the previous time step and the power traded with the main grid (6.42). Although the available PEVs in the GPLs can be operated as energy storage systems, the parking lot is required to charge them up to the requested SOC. Therefore, each PEV SOC at the departure time from the lot should be equal to the vehicle requested SOC (6.43). The final SOC of a GPL can be calculated using Equation (6.44). The aggregated number of available PEVs in the GPL at time step $t$ can be calculated by Equation (6.45). The total number of PEVs that uses the GPL installed in bus $b$ is less than the total number of available vehicles in the same bus, as given in Equation (6.46). Equation (6.47) ensures that the total number of PEVs that uses the GPLs in the network does not exceed the total number of available PEVs in the network.

$$P^{ch}_{y,\omega,b,t} = \sum_{n \in PC_{y,b}} p^{ch}_{n,\omega,t} \tag{6.33}$$

$$P^{dis}_{y,\omega,b,t} = \sum_{n \in PC_{y,b}} p^{dis}_{n,\omega,t} \tag{6.34}$$

$$p^{ch}_{n,\omega,t} \le P^{rated} n^{avil}_{n,\omega,t} \tag{6.35}$$

$$p^{dis}_{n,\omega,t} \le P^{rated} n^{avil}_{n,\omega,t} \tag{6.36}$$

$$\sum_{n \in PC_{y,b}} \chi^{ch}_{n,\omega,t} \le CS_{y,b} \tag{6.37}$$

$$\sum_{n \in PC_{y,b}} \chi^{dis}_{n,\omega,t} \le CS_{y,b} \tag{6.38}$$

$$\chi^{ch}_{n,\omega,t} \ge \frac{p^{ch}_{n,\omega,t}}{P^{rated}} \tag{6.39}$$

$$\chi^{dis}_{n,\omega,t} \ge \frac{p^{dis}_{n,\omega,t}}{P^{rated}} \tag{6.40}$$

$$soc^{min}_n \le soc_{n,\omega,t} \le soc^{max}_n \tag{6.41}$$

$$soc_{n,\omega,t} = \begin{cases} soc_{n,\omega}^{in} + p_{n,\omega,t}^{ch}\eta^{ch} - \dfrac{p_{n,\omega,t}^{dis}}{\eta^{dis}} & t = t_{n,\omega}^{ar} \\[2ex] soc_{n,\omega,t-1} + p_{n,\omega t}^{ch}\eta^{ch} - \dfrac{p_{n,\omega,t}^{dis}}{\eta^{dis}} & t_{n,\omega}^{ar} < t \le t_{n,\omega}^{dep} \\[2ex] 0 & \text{Otherwise} \end{cases} \tag{6.42}$$

$$soc_{n,\omega,t} = soc_{n}^{fi} \quad t = t_{n,\omega}^{dep} \tag{6.43}$$

$$SOC_{y,\omega,b,t}^{fi} = \sum_{n} soc_{n}^{fi} \,, \left\{ \forall n \in PC_{y,b} \,\middle|\, t_{n,\omega}^{dep} = t \right\} \tag{6.44}$$

$$N_{y,\omega,b,t}^{avil} = \sum_{n \in PC_{y,b}} n_{n,\omega,t}^{avil} \tag{6.45}$$

$$PC_{y,b} \le PC_{y,b}^{max} \tag{6.46}$$

$$\sum_{b} PC_{y,b} = PC_{y}^{N,max} \tag{6.47}$$

## 6.5.3 THE UPPER LEVEL: ADN EXPANSION PLANNING MODEL

An optimal distribution network plan is determined at the upper level by minimizing the NPV of the total planning cost as represented in Equation (6.48).

$$f_1 = UC + \sum_{y} \sum_{\omega} \frac{1}{(1+i)^y} \rho_\omega \left( LC_{y,\omega} + ENC_{y,\omega} \right) \tag{6.48}$$

### 6.5.3.1 Network Upgrading Cost

Similar to the bundled environment, the upgrading cost of the distribution network is the sum of the all costs paid for reinforcement of lines and substation as represented in Equation (6.49). As discussed before, this equation takes into account annualized costs.

$$UC = \sum_{y} \sum_{y' \le y} \frac{1}{(1+i)^y} \left( \sum_{bb'} IC^l (\gamma_{y',bb'}^l) L_{bb'} + IC^{tr} (\psi_{y'}^{tr}) \right) \tag{6.49}$$

### 6.5.3.2 Loss Cost

The equation representing the loss cost is given by Equation (6.50). In order to compute this cost term, a power flow should be solved. The power flow equations are same as constraints (6.4)–(6.10) in bundled environment. It should be noted that since in this section, GPLs are considered as DERs, the active power generated by the DER is calculated by (6.51).

$$LC_{y,\omega} = \sum_{t} \sum_{bb'} \pi_{y,t}^{loss} \tau_{y,t} R_{y,bb'} I_{y,\omega,bb',t}^2 \tag{6.50}$$

$$P_{b,t}^{der} = P_{b,t}^{dis} - P_{b,t}^{ch} \tag{6.51}$$

### 6.5.3.3   ENS Cost

Similar to the bundled environment, in the unbundled framework, electrical energy stored in GPLs are used as alternative power supply source. FMG boundaries depend on the locations of protective devices of ADNs. Hence, similar to the bundled environment, in order to assess reliability, network is divided into several zones. In order to restore interrupted loads the following conditions should be considered:

- In each FMG, at least one GPL should exist.
- Power supplying capability of GPLs depends on the energy stored in these resources, hence, load shedding in GPLs should be considered.

Reliability evaluation process in this section is similar to the process described in unbundled framework. The only difference is related to the amount of restored load, which depends on the electrical energy stored in GPLs. The electrical energy stored in GPLs is calculated using Equation (6.52). Other equations required to calculate ENS cost, are similar to the equations given in unbundled framework.

$$SOC_{y,\omega,b,t} = \sum_{n \in PC_{yb}} soc_{n,\omega,t} \tag{6.52}$$

According to the developed model, it can be stated that another part of the conflict in decision-making process of DNO and GPLO is due to the variable denoting stored energy in GPLs. This variable exists in both levels of the model.

### 6.5.4   Risk Control in the Proposed Planning Problem

The probabilistic parameters including the GPL behavior and demand put the network at risk. In order to perform a correct comparison of the solutions, similar to previous section, CVaR index is included in the formulation of the problem as follows:

$$OF^{DNO} = f_1 + \beta \times CVaR_\alpha(f_1) \tag{6.53}$$

### 6.5.5   Impact of Cooperative Behavior of GPL on Planning Problem

As seen in previous sections, GPLO and DNO set different goals in the decision-making process. However, the mentioned decision-makers are coupled by the location and capacity of GPLs with each other.

In order to model the impact of cooperative behavior of GPLs with DNO on the planning problem, the total saving accrued to DNO arising from the cooperation with GPLO is maximized. In other words, the value of the total planning cost is calculated two times: when GPLO follows only his/her goals in the decision-making process, which is $OF_a^{DNO}$ and when GPLO participates in the planning problem $OF_b^{DNO}$. The difference between these values represents the saving of DNO as given in (6.54).

$$Max\left\{OF_c^{DNO}\right\} = OF_a^{DNO} - OF_b^{DNO} \tag{6.54}$$

### 6.5.6 Optimization Tool

Mathematically, the proposed planning model is a bi-level MINLP problem in which two objectives, $\{OF_c^{DNO}, OF^{GPL}\}$, should be maximized, simultaneously. In other words, the lower level seeks to maximize the GPLO's profit, whereas the upper level tries to maximize the DNO's profit. The main decision variables in the upper level are $\left\{\gamma_{ybb'}^l, \psi_y^{tr}\right\}$. For any given set of these variables, secondary variables of the upper level, i.e. {network power flow variables, ENS cost, CVaR}, can only be figured out by solving the GPLO's profit maximization problem. In order to solve such bi-level problems various methods can be used. The Pareto optimal-based two-stage MILP-embedded heuristic method is one of the powerful methods to solve such problems. Reference [12] applies such a method to solve the planning problem. At the first stage of this method, the optimal Pareto front is determined using IGA. Then, at the second stage, the final solution is selected from this front by maximizing the minimum satisfaction among all objectives.

### 6.5.7 Numerical Results

The proposed planning methodology is applied to the IEEE 33-bus distribution test network. The charging rate of each charging station is assumed to be 11 kW per hour. It is assumed that in the first year of the planning horizon which includes 4 years, the maximum number of PEVs is 100, which increases 3.5% each year. The NPV of different cost terms is presented in Table 6.4. The resultant Pareto optimal front is shown in Figure 6.6, which contains 29 non-inferior solutions. Profits of GPLO and DNO are non-negative in all of the solutions. The values of objective functions and the ability of each solution in satisfying each decision-maker are given in Table 6.5. As mentioned earlier, the final solution is a solution with maximum of minimum

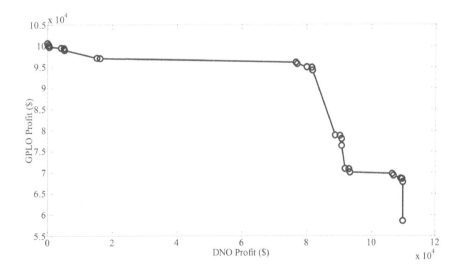

**FIGURE 6.6**  Pareto-optimal front.

**TABLE 6.5**

**Details of Pareto-Optimal Front**

| Solution Number | of$_1$ ($) | of$_2$ ($) | $\mu_{of_1}$ | $\mu_{of_2}$ |
|---|---|---|---|---|
| 1 | 5346 | 98945 | 0.049 | 0.961 |
| 2 | 5240 | 98975 | 0.048 | 0.961 |
| 3 | 5149 | 99419 | 0.047 | 0.972 |
| 4 | 5201 | 99103 | 0.047 | 0.964 |
| 5 | 0 | 100603 | 0.000 | 1.000 |
| 6 | 4250 | 99518 | 0.039 | 0.974 |
| 7 | 592 | 99737 | 0.005 | 0.979 |
| 8 | 486 | 99766 | 0.004 | 0.980 |
| 9 | 394 | 100210 | 0.004 | 0.991 |
| 10 | 16296 | 97015 | 0.148 | 0.915 |
| 11 | 15284 | 97114 | 0.139 | 0.917 |
| 12 | 109989 | 58533 | 1.000 | 0.000 |
| 13 | 109988 | 67777 | 1.000 | 0.220 |
| 14 | 109728 | 68510 | 0.998 | 0.237 |
| 15 | 106595 | 69694 | 0.969 | 0.265 |
| 16 | 109254 | 68609 | 0.993 | 0.240 |
| 17 | 106993 | 69301 | 0.973 | 0.256 |
| 18 | 90867 | 78018 | 0.826 | 0.463 |
| 19 | 90385 | 78752 | 0.822 | 0.481 |
| 20 | 88836 | 78850 | 0.808 | 0.483 |
| 21 | 90868 | 76413 | 0.826 | 0.425 |
| 22 | 93526 | 70066 | 0.850 | 0.274 |
| 23 | 93150 | 70799 | 0.847 | 0.292 |
| 24 | 91939 | 70898 | 0.836 | 0.294 |
| 25 | 81867 | 94180 | 0.744 | 0.847 |
| 26 | 81634 | 94913 | 0.742 | 0.865 |
| 27 | 76674 | 96097 | 0.697 | 0.893 |
| 28 | 80123 | 95012 | 0.728 | 0.867 |
| 29 | 77123 | 95705 | 0.701 | 0.884 |

satisfaction for both decision-makers. Simulation results in Table 6.5 reveal that solution 25 complies with this requirement. The planning scheme for solution 25 is presented in Figure 6.7.

## 6.6 SUMMARY AND CONCLUSION

In this chapter, flexible ADN planning frameworks are proposed, which consider different type of DERs as capacity expansion options and take into consideration island mode operation of the DERs to enhance the flexibility of the ADN. The models are developed in both bundled and unbundled environments. The model in unbundled environment can accommodate different cooperative behaviors between the

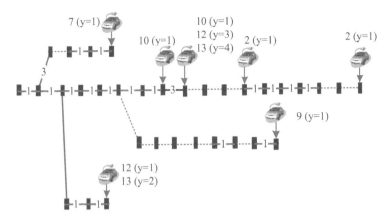

**FIGURE 6.7** Optimal distribution network scheme in the second case.

decision-makers in expansion planning problem. In order to deal with the economic risk corresponding to the uncertain parameters, a risk index is embedded in the objective function of the proposed ADN planning model. This framework can accommodate different risk aversion strategies of the decision-makers in ADN expansion planning problem. Simulation results demonstrate that the proposed method makes a number of contributions to the ADN expansion planning problem as follows:

1. Not only simultaneous DER and ADN expansion planning can defer investments in distribution network expansion, but also it can enhance the flexibility of the network for supplying loads in islands when the reliability of the network is jeopardized.
2. Considering that the proposed ADN expansion planning frameworks can present the risk level of the plans, they can provide helpful technical and economic signals for DNOs and be used as a guide to make a tradeoff between expected cost of the plans and payment in extreme scenarios.
3. Considering that the proposed multi-objective expansion framework in unbundled environment can present DNO's profit variations versus DER's ones (Pareto optimal front), it can provide helpful technical and economic signals for regulators and be used as a guide in designing incentive mechanisms so as to encourage investors for investing in DERs in a way that the location and capacity of DERs become more beneficial for the whole system.

## REFERENCES

[1] Naderi E, Seifi H, and Sepasian MS. A dynamic approach for distribution system planning. Considering distributed generation. IEEE Trans Power Deliv 2012; 27: 1313–1322.
[2] Ziari I, Ledwich G, Ghosh A, and Platt G. Integrated distribution systems planning to improve reliability under load growth. IEEE Trans Power Deliv 2012; 27: 757–765.
[3] Bagheri A, Monsef H, and Lesani H. Integrated distribution network expansion planning incorporating distributed generation considering uncertainties, reliability, and operational conditions. Electr Power Energy Syst 2015; 73: 56–70.

[4] Samper ME and Vargas A. Investment decisions in distribution networks under uncertainty with distributed generation—Part I: Model formulation. IEEE Trans Power Syst 2013; 28: 2331–2340.

[5] Soroudi A, Ehsan M, Caireand R, and Hadjsaid N. Hybrid immune-genetic algorithm method for benefit maximization of DNOs and DG owners in a deregulated environment. IET Gener Transm Distrib 2011; 5: 961–972.

[6] Ameli A, Farrokhifard M, Ahmadifar A, and Haghifam MR. Distributed generation planning based on the distribution company's and the DG owner's profit maximization. Int Trans Electr Energy Syst 2015; 25: 216–232.

[7] Soroudi A, and Ehsan M. A distribution network expansion planning model considering distributed generation options and techo-economic issues. Energy 2010; 35(8): 3364–3374.

[8] Soroudi A, Caire R, Hadjsaid N, and Ehsan M. Probabilistic dynamic multi-objective model for renewable and non-renewable distributed generation planning. IET Gener Transm Dis 2011; 5(11): 1173–1182.

[9] Moradijoz M, Moghaddam MP, and Haghifam MR. A flexible active distribution system expansion planning model: A risk-based approach. Energy 2018; 145; 442–457.

[10] Nick M, Cherkaoui R, and Paolone M. Optimal allocation of dispersed energy storage systems in active distribution networks for energy balance and grid support. IEEE Trans Power Syst 2014; 29(5): 2300–2310.

[11] Zhao M, Chen Z, and Blaabjerg F. Probabilistic capacity of a grid connected wind farm based on optimization method. Renew Energy 2006; 31(13): 2171–2187.

[12] Moradijoz M, Moghaddam MP, and Haghifam MR. A flexible distribution system expansion planning model: A dynamic bi-level approach. IEEE Trans Smart Grid 2017; 9(6): 5867-5877.

[13] Bagheri A, Monsef H, and Lesani H. Evaluating the effects of renewable and non-renewable DGs on DNEP from the reliability, uncertainty, and operational points of view by employing hybrid GA and OPF. Electr Power Energy Syst 2015; 25: 3304–3328.

[14] Al Kaabi SS, Zeineldin HH, and Khadkikar V. Planning active distribution networks considering multi-DG configurations. IEEE Trans Power Syst 2014; 29: 785–793.

[15] Neyestani N, Damavandi MY, Shafie-Khah M, Contreras J, and Catalão JP. Allocation of plug-in vehicles' parking lots in distribution systems considering network-constrained objectives. IEEE Trans Power Syst 2015; 30(5): 2643–2656.

[16] Esmaeeli M, Kazemi A, Shayanfar H, Haghifam MR, and Siano P. Risk-based planning of distribution substation considering technical and economic uncertainties. Electr Power Syst Res 2016; 135: 18–26.

# 7 Distribution Network Emergency Operation in the Light of Flexibility

*Seyed Ehsan Ahmadi[1] and Navid Rezaei[1]*
[1]University of Kurdistan, Sanandaj, Iran

## CONTENTS

## 7.1  INTRODUCTION

The concept of flexibility has been described by researchers. At present, studies on power system flexibility can be categorized into short-term operational flexibility and long-term planning flexibility. However, despite above categorization, there is no comprehensive definition of power system flexibility. Each research group has presented its definition based on main field of the study approach [1]. Accordingly, in this chapter, flexibility has been defined as a network's ability to regulate generation and demand concerning intentional or unintentional deviations. From the operational viewpoint, flexibility can be addressed as the capability of a power system to be applied to a specific time under net demand alterations. In past few years, flexibility enhancement of distribution network for an intelligent emergency operation has become an essential concern for power and energy researchers. About more than 70% of power outages are due to contingencies in distribution systems (DSs) [2]. DSs are imperative for innovative emergency operation strategies which employ available resources in a more reliable and corporative way to improve flexibility of the system. The prevalent solutions such as integration of sustainable energy resources and storages, microgrid (MG) development, and power line strengthening can be employed in the current DSs. Out of above-noted prevalent solutions, applying the MGs for flexibility enhancement is more desirable. This is because of the facility of MGs to corroborate the integration of the renewable energy sources (RESs) and potential of islanding scheduling (i.e., potentiality to survive their critical loads in the contingencies). During emergency operation, the on-fault area can be isolated from the entire network and subdivided into some independent MGs through the tie-line switches [3].

Furthermore, MGs can be applied as a black-start resource to start the main disrupted local resources. Accordingly, the self-healing method can be simultaneously investigated at the distribution level to enhance the flexibility of the system. Self-healing is one of the progressive design tools in the emergency operation of active DSs that can isolate the network in the contingencies and restore it to the normal mode. A self-heal index reduces interruption of service delivery to consumers, thus minimizing the undesirable results of disturbances on normal-operated areas. Accordingly, MGs have a great perspective to enhance vital parameters of DSs such as flexibility, reliability, environmental benefits, and techno-economic performance. The distributed energy resources (DERs) and local loads within the MGs are managed by local controllers (LCs), in which the scheduling of the RESs is addressed to guarantee the power balance. Reconfigurable MGs, through switching among the available MGs, can be reconnected to or disconnected from each other to support the critical loads. During emergency operation, the on-fault area is isolated from the entire network and subdivided into some independent MGs through the tie-line switches. Also, the load priority is determined and chances of survival of critical loads are increased through the local power generations [4].

Connecting multiple islanded MGs is another plan to enhance the flexibility of DSs. Networked MGs (NMGs) that connect to each other and to the main grid is the modern configuration of MGs resulting in great flexibility of the system. The correlation of the MGs as a networked framework improves system's flexibility by taking advantage of notable aspects of NMGs. In fact, NMGs have a preferable economic dispatch and islanding control than the single MG framework. Besides, the optimal implementation of the DERs in each MG and the power support from the main grid and other MGs are the significant features of the NMG framework. This means that the normal-operated MGs can support the critical loads of the on-fault MGs of the network having insufficient supply. Here, MGs can either be used as a local resource or as a group resource. Therefore, the flexibility of the DSs can be improved by sharing of power support among MGs in case of occurrence of a probable fault [5]. On the other hand, assigning a proper energy management system (EMS) during emergency operation is a key factor to guarantee reliable cooperation in the DSs to enhance the flexibility, resilience, and sustainability of the whole network. Power system resilience is defined as the flexibility of the network that can be recovered immediately after an unpredicted fault.

Demand response programs (DRPs) can modify the profile of energy consumption to achieve the highest performance during emergency operations. Therefore, a set of strategies that can be implemented using DRPs to increase flexibility can be considered as a predictive measure of achieving a self-healing and resilient system. Furthermore, applying locally distributed sources for service restoration to critical load in DSs is a sufficient method to improve the flexibility of the power grid. Mobile power sources (MPSs), including electric vehicles (EVs), mobile energy storage systems (MESSs), and mobile emergency generators (MEGs), can provide structural flexibilities to improve the resilience of the DSs. They can perform as backup sources to ensure power supply for critical loads. Moreover, when an extreme event occurs, the MPSs provide a novel aspect of flexibilities that can be investigated via the active network [6]. On the other hand, compared to conventional explanations, a distributed battery energy storage system (BESS) presents increased flexibility serving grid support concerning such as congestion management, Volt/Var support, and backup capacity. BESS can also provide key self-healing functionalities, which enable the islanding operation of small sections of distribution network. Enabling islanded operation mainly relies on the BESS capacity to supply the critical load and regulate the network frequency and voltage during the contingencies. The dispatch strategy of mobile emergency resources in the DSs is important and should be considered when deciding the restoration methods [7]. Furthermore, the DRPs can modify the profile of the energy consumption to achieve the highest performance during emergency operations. Therefore, a set of strategies that can be implemented using DRPs to increase flexibility can be considered as a predictive measure of achieving a self-healing and resilient system [8].

In this chapter, the concept of flexibility in DSs, the occurrence of emergency, and general strategies for enhancing the flexibility of the DSs are reviewed at first. Then, the configuration of the MGs is investigated. The general roles and approaches applied by MGs for enhancing the flexibility of the DSs in single, networked, and multi-energy system frameworks, the corresponding communication flexibilities, and the decision-making framework in DS with MGs are presented separately in four sections. The islanding strategy and DRPs are also investigated for an emergency operation. Applying various mobile and stationary power generation sources, their impacts on the DS flexibility, and the corresponding constraints during emergency operation are evaluated in a series of sections. The emergency power dispatch and balancing, and restoration method are also presented in these sections. Then, the IEC-61850-based EMS during emergencies is analyzed as a communication standard. Finally, the chapter is concluded with some points.

## 7.2 NETWORK FLEXIBILITY

The existing DSs are considered to be reliable since they can supply reliable power during normal and abnormal but predictable contingencies. The advanced DSs still require the flexibility and resiliency characteristics, since they are inefficient to preserve significant extreme events. Accordingly, the flexibility and resiliency enhancement investigations are very valuable. In this section, emergency modeling and flexibility enhancement strategies are presented.

### 7.2.1 EMERGENCY MODELING

Emergency modeling is the basis of the flexibility investigations, and DS studies should model the appropriate implementation of the network during the emergency operation. To illustrate the performance of the network during the extreme event, a flexibility trapezoid can be suggested in the same way as introduced by ref. [9], as demonstrated in Figure 7.1. Diverse objectives for different phases of the event can be illustrated from the suggested trapezoid. It is clear from Figure 7.1 that a flexible network desires to reduce the resisting, adjustment, and recovery intervals compared to a conventional network without flexibility characteristics [10].

### 7.2.2 FLEXIBILITY-ENHANCEMENT STRATEGIES

As illustrated in Figure 7.2, flexibility-enhancement strategies of DSs are divided into three main stages according to the emergency and clearance point. In the

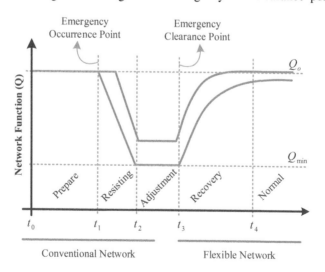

**FIGURE 7.1** Conventional versus flexible network performance during emergencies.

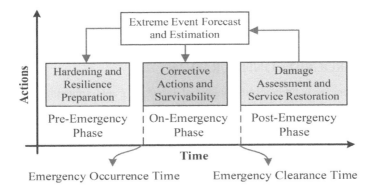

**FIGURE 7.2** Flexibility enhancement in diverse phases of emergency.

pre-emergency stage, the contingency situation in the network is evaluated and the network is provided for any further circumstances applying diverse strategies. In the on-emergency stage, the network is reconfigured, emergency generators and storages are utilized, the existing resources are rescheduled, and emergency load shedding may be conducted to prevent the network from effects of the contingency situation. This stage is comprehensively described in the further sections. Eventually, in the post-emergency stage, the network is recovered to the normal (or almost normal) situation through reconfiguring the network and regenerating the network (e.g., black-start). The network flexibility-enhancement schedules can be extensively allocated to long-term and short-term schedules as demonstrated in Figure 7.3. The long-term schedules enclose both transmission and distribution level improvement, sensor placement, system redundancy, and redeployment of resources. The short-term

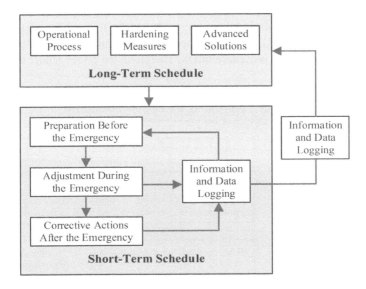

**FIGURE 7.3** Long- and short-term flexibility enhancement schedules.

schedules enclose MG operating, DRPs, distributed generations (DGs), developed monitoring, forecasting methods, and distributed control approaches [10].

## 7.3  MICROGRID CONFIGURATION

MGs can ensure economic advantages of integrating and adjusting the localized RES. MGs can be classified into simple self-sufficient, intelligent self-sufficient, and networked MGs according to their ability to meet local demands completely during a 24-hour scheduling time horizon of a day. The implementation of energy storages at distinct MGs or entire DSs could provide multiple advantages. Among several resources in an emergency operation, MGs have arisen as a beneficial resource for preserving or reducing the effects of extreme events that result in an emergency operation. MGs can strengthen the flexibility of the systems, especially at the distribution level, due to their capability of intentional islanding and integration of DERs. The islanding facility can separate normal-operated areas of the network from on-fault areas. DERs can also supply the critical loads locally during the emergency operation in the lack of an upstream network. The MGs scheduling for enhancing the flexibility of the DSs can be categorized into three main groups (i.e., local support, network support, and black-start support), as shown in Figure 7.4. In case of local support, DGs are employed as a backup power supply for only one formation or a group of a few formations. However, the DS can be reconfigured to support one MG for supplying the demand of the critical loads in another MG (i.e., MGs as network support). Eventually, MGs may support the setting up of the bulk generators that have deactivated because of an extreme event (i.e., MGs as black-start support). Accordingly, MGs not only enhance the flexibility of the network but also enhance the flexibility of the whole power system/regions by preserving the critical facilities.

During contingencies, the purpose is to instantaneously return a reliable power supply to loads. The prevalent restoration procedure is normally conducted in a centralized approach (i.e., black-start support adopted by transmission network, and then distribution network). In this manner, the loads can only receive power support after the restoration procedure in DS. Therefore, the power failure duration of

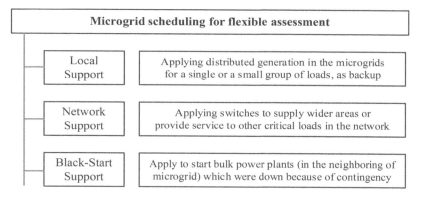

**FIGURE 7.4**  Performance of microgrid as a flexible resource.

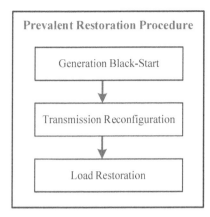

**FIGURE 7.5** Prevalent and microgrid restoration procedures.

critical loads is lengthened. However, if an MG is impaired during an extreme event, the MG can realize the self-healing directly at the local level to sustain maximum feasible critical loads. At the same time, the restoration procedure will be developed from the upstream grid toward the distribution level. Consequently, the full restoration procedure could be conducted simultaneously (i.e., a downward procedure from the upstream grid and an upward procedure from the downstream grid). The comparison of prevalent and MG restoration procedures is summarized in Figure 7.5.

In the long-term schedules of flexibility enhancement, complementary tie-switches and DERs are mostly utilized to enhance network flexibility. Accordingly, to modify a typical DS into MGs during the emergency operation, the optimal allocation of tie-switches is desired. Furthermore, the capacity of DERs should be optimally determined and allocated at the optimal location to decrease the power losses in the network while increasing the flexibility of the network through sited DERs. It should be noted that during an emergency operation, it may not be feasible to completely supply loads of the network. Hence, the priority of loads can be proposed to provide a reliable power supply to the maximum possible critical loads.

Dynamic MG configuration in response to the extreme event can feasibly enhance the flexibility of the on-fault area by providing the independence of the local power supply. Accordingly, reconfiguration of the present MGs during emergency operation can also enhance the flexibility of the MGs. This purpose can be assorted into the short-term schedules of flexibility enhancement. In this term, software methods are applied to enhance the network flexibility through controlling the tie-switches (allocated schedules of flexibility enhancement) and modifying network configuration.

On the other hand, neighboring local MGs can be networked to provide power support for each other during the emergency operation, as demonstrated in Figure 7.6 (a). However, because of the fault extension issue, the generation resources of one or some MGs can also be jeopardized. For this matter, other normal-operated MGs can support the priority loads of the on-fault MGs, as demonstrated in Figure 7.6 (b). Accordingly, NMGs can enhance the power supply reliability as a broadened structure of the single MG framework.

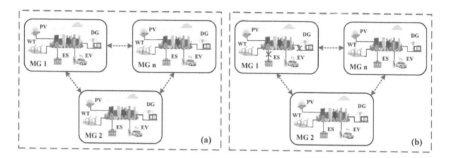

**FIGURE 7.6** Networked microgrids (MGs) in islanded mode: (a) normal-operated MGs; (b) MG1 in emergency operation.

### 7.3.1 MICROGRIDS AS ENERGY HUBS

During extreme events, not only power electrical facilities are impaired but other facilities may also be affected. The facility impairment may either be because of the event or maybe because of the power outage. However, the outage of other facilities may critically affect the power grid. Likewise, other energy grids may also be disabled because of the power outage. Accordingly, an energy hub can be proposed to provide the reliable islanded operation of the MG under diverse uncertainties of the generation and consumption, as shown in Figure 7.7, including different energy

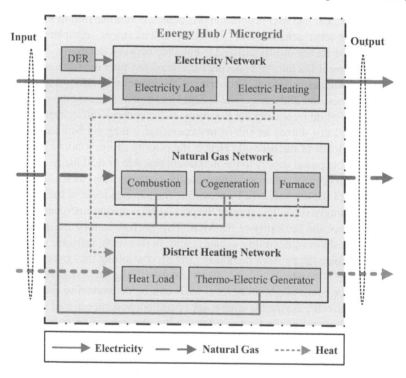

**FIGURE 7.7** Microgrid as an energy hub.

carriers, i.e., electricity, heat, natural gas, which can be optimized simultaneously for assuring particular intentions [11].

Inside an energy hub, all the conversion, transmission, distribution, and consumption of various energy carriers are strongly conjugated. It should be noted that the enhancement of MGs toward realizing energy hubs complies with the sight of smart cities, where the management of various associated energy carriers is complemented for realizing greater reliability in urban areas. Accordingly, the power system flexibility will not depend only on the electric power infrastructure but is also a function of the flexibility in other infrastructures where the energy can be converted to electric power and applied in an appropriate manner.

### 7.3.2 Approaches Applied by Microgrids for Flexibility Enhancement

MGs can reduce the outage duration in the network by supplying local power support during network contingencies. Similar to the network flexibility approaches, the flexibility enhancement approaches for MGs can also be classified into three main groups, as demonstrated in Figure 7.8. The classification is according to the contingency situation and fault elimination durations.

MGs require to develop for extreme events not later than the occurrence of the events by operating their facilities in a reasonable approach through a proactive operation approach. Previous operating information can be applied to forecast the occurrence of a specific event and the normal operation of MGs can be adjusted through the supporting block. Supporting block is mainly suggested for ESSs and reserve facilities to sustain a prescribed amount of power support to enhance the flexibility of the network during emergency operations. The supporting block does not authorize the EMS to discharge ESSs lower than a prescribed amount or restrict the current power consumption higher than a prescribed amount. Figure 7.9 illustrates the paradigm for the flexibility enhancement of MGs in the grid-connected mode, not later than the occurrence of the events. These operation approaches are recognized as proactive operation approaches based on their application.

During the emergency operation, MGs provide a reliable power supply of priority loads in the process of changing from grid-connected to islanded mode through

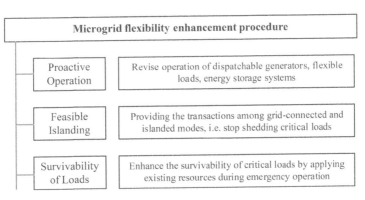

**FIGURE 7.8** Overview of microgrid flexibility enhancement procedure.

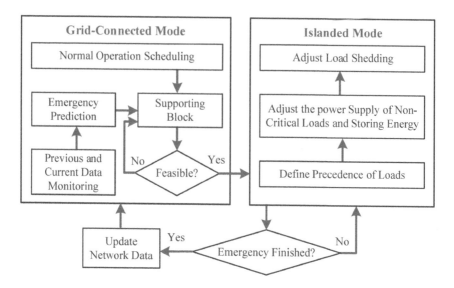

**FIGURE 7.9**   Flexibility enhancement in normal and emergency modes.

feasible islanding. This feasible islanding can be realized by employing the supporting block as demonstrated in Figure 7.9. Therefore, priority of loads is determined considering that whole loads may not be fed during emergency operations. According to the decision-makings, the load shedding value at diverse demand amounts is determined. After the contingency elimination, MG can switch back to the normal-operation mode. In addition to the abovementioned approaches, ESSs, DRPs, multi-agent frameworks, and artificial intelligence (AI) methods can also be applied as flexibility-enhancement approaches.

Besides, energy management approaches can affect flexibility enhancement due to the proposed central and distributed controllers. Figure 7.10 illustrates various

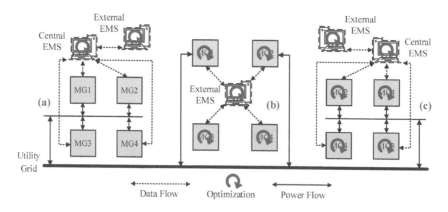

**FIGURE 7.10**   Overview of energy management frameworks: (a) centralized; (b) decentralized; (c) hierarchical.

energy management approaches in MG operation during normal and emergency operations. In a centralized EMS, the whole data is collected by a central controller. In the decentralized approach, the central controller is not obligatory and MGs can communicate straightforward with the external controller. Also, in the hierarchical approach, on a par with decentralized EMS, each MG is locally optimized. Then, the data of power surplus/deficiency is revised to the allocated central controller. The central controller, on a par with centralized EMS, can communicate with the external EMS. It should be noted that although the decentralized approach does not require to go through the whole decision-making procedure of the entire network, centralized EMS are proposed by various researchers to enhance the flexibility of the MGs due to their excellent capability to operate the network facilities [12].

### 7.3.3 COMMUNICATION FLEXIBILITY OF MGS

To provide network flexibility, a stable and flexible communication framework is majorly desired. Accordingly, software-defined networks can be suggested to enhance the flexibility of the communication network in MGs [10]. A summary of the software-defined network-based multi-layer framework for flexibility enhancement of MGs is illustrated in Figure 7.11.

### 7.3.4 DECISION-MAKING FRAMEWORK IN DS WITH DERS AND MGS

In this section, the decision-making framework in DS with MGs is presented to address the responsibility of different entities in the DS emergency operation model. In this framework, a wholesale energy market (WEM) organized by the market

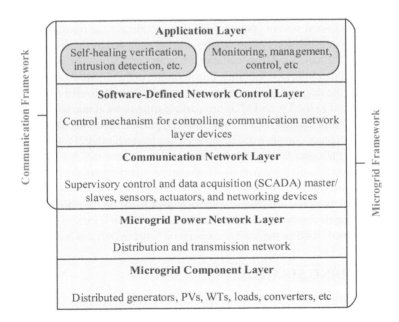

**FIGURE 7.11**   Software-defined microgrid framework.

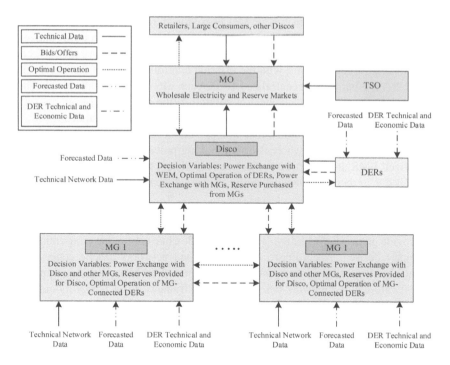

**FIGURE 7.12**   Decision-making framework in distribution system with distributed energy resources and microgrids.

operator (MO) at the transmission system level is also proposed. The MO collaborates with the transmission system operator (TSO) that evaluates the transmission system flexibility. The transmission system supplies the DS managed by the distribution company (Disco). Besides, the Disco collaborates with the distribution system operator (DSO) that evaluates the DS flexibility. Furthermore, in this framework, Disco can be considered as the decision-maker interacting with MGs, and diverse players such as DERs connected to the DS and MGs are participating in local markets to create the prices. The DER interacted with a single MG is regulated inside the MG.

The DS emergency operation can be modeled as a hierarchical decision-making problem, in which the upper level problem represents the decision-making problem of the Disco and the lower level problem represents the decision-making problem of each MG, as demonstrated in Figure 7.12. In this framework, Disco, MG operators, and DER operators exchange required energy and reserve with each other based on the price signals. These price signals can be assumed as fixed or variable prices. When price signals are variable, they are estimated based on the feasible cooperation between decision-makers [13, 14].

## 7.4   ISLANDING STRATEGY

Islanding or loss of main (LOM) is determined when an area of the DS is energized by the DG after being isolated from the rest of the DS. Taking into account the diverse impacts islanding may lead, IEEE 929 [15] and IEEE 1547 [16] approved that islanding needs be impeded. However, the advantages of the DG will not be greatly investigated if the DG regularly requires to inactivate in case the power supply in the DS is deficient. Accordingly, applying an intentional islanding scheduling of DG will define the strength of the power supply in such a way that the DG should be feasible to deal with the local power generating in the originated island. Therefore, the intentional islanding improves the flexibility and reliability of the power supply during the emergency operation in the DS. Noticing the advantages that intentional islanding could provide, IEEE 1547 society has advanced an outline. This outline will assist DSs to practice an intentional islanding scheduling. Accordingly, the DG's controllers should be configured particularly to schedule the distribution level in two operation modes including grid-connected and islanded mode. The network can be subjected to high risks, especially when it is not being developed to support it [17]. Some corresponding problems regarding unintentional islanding can be described as follows:

1. Power quality is the most significant desire for DSs in providing their customers with an appropriate power supply. However, the voltage and frequency can be considered alternative and the probabilities to be out of the allowed limits are very significant during unintentional islanding. This issue can deliver high risks to the local loads and other DGs.
2. Being out of synchronism is probably one of the most complicated problems to be mentioned during emergency operation due to the islanding formation. For this reason, most of the substations in DSs substation employ auto-reclosers in their protection system to enhance flexibility and reliability of power supply. In the islanding occurrence, the auto-recloser will try to reconnect the islanded area to the network a few times. It should be noted that the value of the voltage and frequency have to be within the desirable limit before the reconnection can happen.
3. Applying the various types of DGs in the DS leads to critical effects on the protection system coordination. Therefore, protection systems on the islanded area are improbable to be coordinated as well. The DG should prevent the available protection coordination with the alteration in fault current during the emergency operation.

There are also essential obligations in designing appropriate and flexible islanding detection techniques. The techniques must be reliable, safe, and quick response in a way that can detect all types of islanding occurrence. Furthermore, developing the islanding detection techniques greatly relies on the DG unit features.

## 7.5 DEMAND RESPONSE PROGRAM

DRP is one of the concepts which has been improved with the approaches in MG operation. Based on the United States Department of Energy (DOE), DRP is the energy adjustment of consumer's demand using diverse strategies such as financial incentives and performance modification via education. The applying of DRPs in the MGs gives rise to increasing the MG flexibility as well as controlling the alternative effects of RES. The mentioned requirements can be realized via the load reduction during emergency operation (i.e., particular hours during an energy scheduling in which the energy price has its highest rate or the reserve limit is bounded because of the failure of a central power generator, outage of transmission lines, and the disconnection from the main grid). In this regard, DRPs have been applied to sustain the system reliability and flexibility during peak load hours for the sake of great demand increment problems in California and the western United States area. Moreover, during emergency operations (e.g., outage of power generators or transmission lines), DSs have many issues in supplying power support through the rest of the power generators. Accordingly, the DS reliability is compromised during contingencies. An impressive technique is to apply demand-side management (DSM) programs to deal with all referred challenges in DSs. The DSM assigns strategies that can be implemented on the demand side of a power system to enhance its features. The definition of DSM demonstrates that the objective of these programs is to support DSs peak load hours or contingencies [18].

### 7.5.1 CLASSIFICATION OF DRPs

DRPs can be subdivided into two categories: (I) incentive-based DRPs and (II) price-based DRPs. In the first category, the consumers are granted incentives for modifying their load profile according to the intention of the supply side. In the second category, the consumers are charged with diverse prices at distinct demand times. There are various classifications of incentive-based DRPs such as direct load control (DLC), load curtailment, demand bidding, and emergency load reduction programs. In accordance with the emergency load reduction programs, the consumers are given a significant incentive for decreasing their consumption. These programs may support the DS to strengthen its flexibility. Furthermore, there are also different classifications of price-based DRPs including time-of-use pricing, real-time pricing, inclining block rate, and critical peak pricing. The critical peak pricing is similar to the time-of-use pricing besides for the time when the reliability of systems is compromised and the regular peak price is altered by an extremely upper price. It should be noted that this program is merely applied for a few hours per year and improves the flexibility of the system during emergency operations.

### 7.5.2 PERFORMANCE OF DRPs

To apply an appropriate DRP, advanced metering infrastructures (AMI) are set up on the demand side. These infrastructures can measure and store energy consumption in diverse terms and communicate with the utility to control the present consumption data. Applying advanced technologies such as smart meters, communication systems, and energy controllers are essential for the beneficial performance of DRPs.

## 7.6 ELECTRIC VEHICLES AS AN EMERGENCY RESOURCE

EVs are promising factors to reduce greenhouse gas emissions and support e-mobility via the high penetration of RES. Technology-based improvements in EVs have caused the novel concept of the EV aggregator in the network, especially at the distribution level, providing advantageous controlled charging-discharging methods. An EV can operate as an ESS to shift load peak to off-peak hours to support the acceptable emergency operation. Vehicle-to-grid (V2G) facility acts as a storage device that supplies the power support during peak demand in the emergency operation. The V2G strategy is convenient for large-capacity provisions in the DS, supports a smart grid approach to deal with the uncertain parameters. The main objective of the V2G facility is to comprehend charging/discharging regulation and sustain a proper charging scheme to overcome the operation conditions, stability, power balance, emergency charging, and unexpected power deficiency in V2G applications. To this end, collective decision-making has made suitable EV regulations for charging/discharging. Therefore, central regulation of V2G facility is taken into account practically due to its impressive performance.

V2G regulation is more beneficial than grid-to-vehicle (G2V) at the present moment; however, considerable challenges still exist. For instance, in wide DS, the current V2G framework cannot immediately and frequently supply the high-power support without taking into account the essential aspect such as the power market, network operation control, and demand management of EVs. Furthermore, the EV holder is not flexible to supply power into the network during emergencies. This may cause voltage deviations and reductions in voltage stability and network flexibility, resulting in problems with placed battery efficiency. Therefore, a multi-disciplinary data-processing model is required to guarantee EV aggregator regulation for realizing appropriate energy management. Consequently, the DSO can apply the stationary ESSs to deal with the alternative nature of RESs. Applying EVs is an advantageous solution for the fluctuation issue of the renewable energies in MGs. Unlike EVs, mobile ESS is a utility-scale storage bank contained and strongly controlled by the utility corporation. Accordingly, the storage is mobilized by a big truck and connected to the network at distinct stations. The suitability for transport is the capability to provide the local reactive power support, voltage regulation, and distributed RES integration [19].

On the other hand, the requirement of fast charging of EVs has led to the extensive structure of the EV fast DC charger substructures. Accordingly, feasible benefits of utilizing V2G facility have been greatly investigated, where bidirectional power exchange is the key factor. However, the EV batteries are mostly high-priced and the V2G facility is still not cost-effective to vindicate the bidirectional EV charging feasibility as it shortens the battery cycle life. Consigning some EVs to plug into the bidirectional charger substructures in a localized area during emergency operation could supply enough power support to compensate for the power deficiency, especially for the islanded networks. This would conveniently eliminate the requirements of backup diesel generators and the battery-based emergency power generation (EPG). Compared to the V2G facility, the EPG assignment is more feasible for bidirectional charging. During the emergency operation, if the state-of-charge of the battery and the power ramp of the EV connected to the charger complies with the EPG essentials, it operates in voltage source mode to support the critical loads [20].

The EPG design with the capability of bidirectional charging is illustrated in Figure 7.13. When a fault occurs in the network, it applies the EV batteries to supply power support to the on-fault area with emergency operation. Many grid-connected EPG merely employ a single-stage AC/DC connected with the integrated batteries.

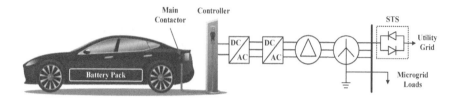

**FIGURE 7.13**   Proposed bidirectional electric vehicle charger based EPS.

However, to follow the wide range of the EV battery packs as a charger, an additional DC/DC conversion is necessary. The high voltage side of the bidirectional DC/DC conversion is the regular DC bus and the low voltage side couples with the EV battery through the charger station. An important point is to set a charger controller to interact with the battery management system (BMS) of the battery pack onboard an EV and manage the bidirectional power exchange, and the operation frameworks of the AC/DC and DC/DC.

## 7.7   MOBILE EMERGENCY POWER SOURCES

EPG system is a self-sufficient power supply facility to generate power support for the critical load during power deficiency and enhance the network flexibility for controlling the load disturbances. Mobile power sources (MPSs), including MEGs, MESSs, and plug-in EVs, can ensure great flexibilities to enhance the DS flexibility. The MPSs are an essential flexibility facility for prompt power restoration in the DSs. The MPSs are truck-mounted generators with the significant advantages of flexibility and great potential, up to a few MVA. They are one of the major impressive response facilities during emergency operations. The MPSs can act as an initial power source for reliable EPG. Moreover, the road network is significantly flexible based on the perfectly meshed configuration and overlapping paths. Consequently, the road network maintains the priority loads by allocating paths to transport flexibilities during the emergency operation.

In the past few years, technological improvements of the RES and ESS, diverse energy resources have the major capability for applying in MPSs during the emergency operation to enhance the flexibility of the power network. Accordingly, diverse hybrid energy generation and storage methodologies, such as photovoltaic (PV) and wind turbine (WT), and battery, etc., have provided the basis for mobile hybrid energy generating (HEG) facilities as backup EPG design. The power generation outline and optimal sizing of sources in the HEG facilities are crucial for the cost-effective sustainability of backup EPG. The available optimal sizing strategy of energy sources mainly emphasizes the backup of islanded MG to reduce implementation costs while enhancing the reliability of the power supply. However, the increasing implementation of the mentioned sizing strategy to the mobile HEG facilities is a complex issue until now. Besides, the alternative nature of the RES including PV and WT should be also taken into account. To optimally implement the RES, the applicable scheme of the mobile multi-energy framework becomes a significant issue according to the uncertainties of the RES and emergency power supply [21].

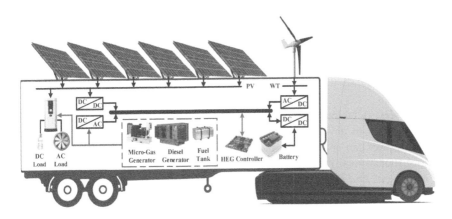

**FIGURE 7.14** The framework of a mobile HEG facility.

### 7.7.1 HEG Framework

The various multi-energy frameworks of HEG facilities can realize the fulfillment of the green and non-green energies to specify the flexibility, reliability, and techno-economical constraint of the particular energy resource. The main segments in the HEG facilities can be modeled using PV, WT, battery pack, micro-gas, and diesel generators, fuel tanks, central controller, and AC/DC, DC/DC, and DC/AC converters [22]. Figure 7.14 illustrates a representative scheme of the mobile HEG facility. In the mobile HEG framework, the PV and WT are coupled to a common DC bus through the DC/DC and AC/DC converters, and the battery pack is also coupled to the common DC bus via a DC/DC converter for saving the surplus power from PV and WT. The common DC bus can be converted to feed AC loads through the DC/AC inverter, and micro-gas and diesel generators are employed as the primary backup EPG to supply the emergency power support straightforwardly to the AC loads. The control mechanism should be applied through the HEG central controller to adjust the power generations and provide the optimal state-of-charge value of the batteries. In this scheme, the green and non-green energy resources operate jointly to supply emergency power support. In the case of surplus power generation of PV and WT, the power can be stored in the battery for further utilization. Besides, in case of deficient renewable generation, battery pack with micro-gas and diesel generators will be discharged to feed existing power demand.

#### 7.7.1.1 Multi-Energy MPS for Emergency Operation

*Photovoltaics*: Solar energy is one of the greatest renewable energies with the capability for mobile power supply. Mobile PV arrays provide adjustable and flexible employment of solar energy with diverse formations that range from 10 W to 40 kW. In unpleasant weather, the implementation of PV arrays coupled with battery energy storage can supply electric power to local priority loads for enhancing the network flexibility during the emergency operation. Accordingly, with the strong points of adjustable scale, portable, no fuel request, and techno-economic, the PV arrays can be integrated into the HEG facility [23].

*Wind Turbine*: Wind power is another infinite renewable energy applied in the HEG facility. The correspondent alternation of PV-WT energy structure enables to enhance the network efficiency and flexibility with more cost-effective and eco-friendly profits since higher wind power can be derived commonly in the overcast weather and nightly without solar irradiance to deal with the undersupply of the PV arrays [24].

*Batteries*: A battery pack is required for the HEG facility to reduce the uncertainty of RES to develop the power supply service. The required battery of the PV-WT power can enhance the network flexibility to further discharge once these are insufficient solar irradiation and wind speed [25].

*Micro-gas and Diesel Generators*: The mobile HEG facility should contain the micro-gas and diesel generators according to their economical foundations compared to the renewable resources. The micro-gas and diesel generators have swift startup qualification and strong disposability of fuel tank and are mostly applied as the crucial reserve power. Also, the backup fuel tanks are obligatory for refueling the micro-gas and diesel generators. Besides, the power density of these sources is considerable. [26]

### 7.7.1.2 Implementation Requirements of HEG Facility

It is required to investigate the implementation of the HEG facility for an appropriate assessment of energy investments. The following two requirements can be defined to evaluate the implementation of the HEG facility for mobile EPG application.

*Power Density Measurement*: Power density measurements of HEG facility consist of volumetric power density, $\rho_{vol}$, and gravimetric power density, $\rho_{gra}$, point out the measure of the energy generated through mobile EPG in per unit [27]. The EPG with high power density implies the loading potential to provide great load consumption. Accordingly, the power density measurement of the HEG facility can be indicated numerically in kW / m$^3$ or kW / kg as presented in Eqs. (7.1–7.3).

$$\rho_{vol} = \frac{P_{HEG}^{cap}}{v_{PV} + v_{WT} + v_{BT} + v_{DG} + v_{MG} + v_{FU}} \tag{7.1}$$

$$\rho_{gra} = \frac{P_{HEG}^{cap}}{m_{PV} + m_{WT} + m_{BT} + m_{DG} + m_{MG} + m_{FU}} \tag{7.2}$$

$$P_{HEG}^{cap} = P_{PV}^{cap} + P_{WT}^{cap} + P_{BT}^{cap} + P_{DG}^{cap} + P_{MG}^{cap} \tag{7.3}$$

where, $P_{HEG}^{cap}$ is the highest power output of the HEG facility. $P_{PV}^{cap}$, $P_{WT}^{cap}$, $P_{BT}^{cap}$, $P_{DG}^{cap}$, and $P_{MG}^{cap}$ are the power capacities of PVs, WTs, batteries, micro-gas, and diesel generators, respectively. $v_{PV}$, $v_{WT}$, $v_{BT}$, $v_{DG}$, $v_{MG}$, and $v_{FU}$ demonstrate the volume of the PVs, WTs, batteries, micro-gas and diesel generators, and fuel tanks, respectively. $m_{PV}$, $m_{WT}$, $m_{BT}$, $m_{DG}$, $m_{MG}$, and $m_{FU}$ demonstrate the mass of the PVs, WTs, batteries, micro-gas and diesel generators, and fuel tanks, respectively.

*Energy Density Measurement*: Energy density is the value of existing energy saved in a network per unit mass or volume. In mobile EPS implementation, the energy density corresponding to mass and volume of HEG facility and the energy density of a hybrid energy framework demonstrate the value of the energy potential of the power generation and traffic related to its mass and volume. Therefore, the volumetric and gravimetric energy density measurement of the HEG facility, $\varepsilon_{vol}$ and $\varepsilon_{gra}$, are presented in Eqs. (7.4–7.6) to indicate the value of the embedded energy per unit volume or weight in the mobile HEG facility [27], normally demonstrated in kW / m$^3$ or kW / kg.

$$\varepsilon_{vol} = \frac{E_{HEG}^{cap}}{V_{PV} + V_{WT} + V_{BT} + V_{DG} + V_{MG} + V_{FU}} \tag{7.4}$$

$$\varepsilon_{gra} = \frac{E_{HEG}^{cap}}{m_{PV} + m_{WT} + m_{BT} + m_{DG} + m_{MG} + m_{FU}} \tag{7.5}$$

$$E_{HEG}^{cap} = T_E P_{PV}^{cap} + T_E P_{WT}^{cap} + E_{BT}^{cap} + \eta_{DG} E_{DG}^{cap} + \eta_{MG} E_{MG}^{cap} \tag{7.6}$$

where, $E_{HEG}^{cap}$ is the predicted energy output of the HEG facility. $E_{BT}^{cap}$, $E_{DG}^{cap}$, and $E_{MG}^{cap}$ are the mobile energy capacities of batteries, micro-gas, and diesel generators, respectively. $\eta_{DG}$ and $\eta_{MG}$, demonstrate average energy conversion efficiency of micro-gas and diesel generators, respectively. In this study, mobile energy capacity of micro-gas and diesel generators is calculated according to the derivable electrical energy from the generator and backup fuel tanks. The highest value of the energy that a battery can discharge to the network is defined as battery energy capacity.

### 7.7.2 RESTORATION FRAMEWORK FOR MOBILE EMERGENCY RESOURCES

To consider the dispatch aspects of mobile emergency resources in the transmission system, a bi-stage restoration framework can be suggested to make decisions on restoration methods at the distribution level. According to the diverse decision procedure of the transmission and DSs, the full decision-making method can be subdivided into two stages:

1. Assess moving time and pathway of mobile emergency resources by taking into account the active traffic allocation in the transmission system.
2. According to the results obtained from stage I, determine the emergency load restoration method in the DS.

The decision-making scheme of the bi-stage restoration framework is presented in Figure 7.15.

In the restoration framework, the moving time of the mobile power sources determines when they can be applied for the service restoration and how to assign the optimal pathway in the transmission system. Besides, the moving time of the repair crews and the time of the repairing process for restoring the faulted segment are important when the associated program is provided for the load restoration [28]. The

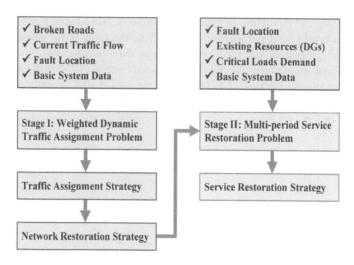

**FIGURE 7.15** The strategy of the suggested restoration framework.

moving time of the mobile power sources and the repair crews presented in stage I are the inputs of the DS restoration framework in stage II, as shown in Figure 7.15. In stage I, the active traffic allocation in the transmission system should be estimated to minimize the total mowing time of a diverse group of vehicles.

## 7.8 BATTERY ENERGY STORAGE SYSTEMS

BESS can enhance the power grid integration potential of RES at the distribution level. Besides, BESS can ensure major self-healing features which will provide an appropriate emergency operation of the DS during extreme events according to the principles of MG and NMGs. In case of unexpected network outages, an area of DS can be scheduled self-sufficiently, imparting to enhance the flexibility and survivability of the network. Providing an acceptable islanded scheduling during the emergency operation essentially depends on the BESS reserve potential to support the critical loads and adjust the voltage and frequency of the system. It is worth mentioning that primary frequency control will only provide the power balance if there is quite enough reserve potential and if the inverters' highest possible power margins are not overtaken [29].

### 7.8.1 Constraints of ESSs During Emergency Operation

As mentioned previously, the MG emergency operation has two major objectives according to the adjustment of the ESSs and feasible islanding:

1. Develop the part of the network which is scheduled to operate in islanding mode according to the forecasted generation and consumption for the following hours and the present state-of-charge of ESSs.

2. Evaluate the ESS emergency dispatch method which optimized the total accessible MG power and energy reserve potential. Accordingly, in case of insufficient ESS power or energy reserve potential, the EMS will apply an auxiliary controlling procedure to deploy the load flexibility.

The real-time EMS will determine the MG emergency schedule regularly for a specified duration, taking into account the decision-making time of the data predicting (per 15 mins). In the first stage, the EMS verifies the feasibility of developing islanding, taking into account the potential of the ESS with the islanding capacities. Conversely, in the case of insufficient power reserve, a network islanding procedure will be investigated. It should be mentioned that both mobile and stationary energy storages can be considered as ESSs. Accordingly, PEV aggregator and MESSs can be defined as ESS within the network.

According to the devised island, the EMS during emergency operation establishes the ESS emergency dispatch approach. As demonstrated in Figure 7.16, the emergency operation can be arranged into two major schedule stages, i.e., emergency power dispatch and emergency energy balancing [7].

The first stage schedules the first dispatch problem according to the ESS power reserve potential, and the second stage reschedules the dispatch method evaluating the unused energy potential of the whole ESSs after the islanding procedure.

### 7.8.1.1 Emergency Power Dispatch

As previously noted, the first stage schedules the first dispatch problem to provide a MG power balance and minimize the charging or discharging amount of the existing ESSs. The EMS will regulate a short-term schedule for diverse durations according to the ESS power and energy reserve. The unbalancing among generation and consumption will be dispatched among the existing ESS based on the proposed weights

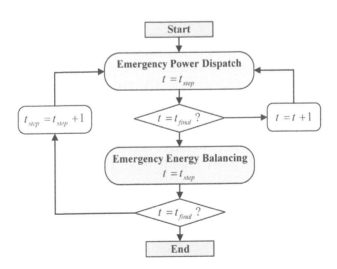

**FIGURE 7.16** Emergency operation schedule.

in Eq. (7.7), in which the power unbalance is dispatched correspondingly to the ESS taking into account their power constraints. Furthermore, the short-term performance of each ESS will be defined as in Eq. (7.8).

$$
\begin{cases}
\omega_{(s,t)} = \min\left( \dfrac{C_{(s)} - E_{(s,t)}}{\Delta t}; P_{(s)}^{ch} \right) & NL_{(t)} < 0 \\[3mm]
\omega_{(s,t)} = \min\left( \dfrac{E_{(s,t)} - E_{(s)}^{\min}}{\Delta t}; P_{(s)}^{dch} \right) & NL_{(t)} > 0
\end{cases}
\tag{7.7}
$$

$$
P_{(s,t)} = -\frac{\omega_{(s,t)}}{\sum \omega_{(s,t)}} \times NL_{(t)}
\tag{7.8}
$$

where, $\omega_{(s,t)}$ is the proposed weight, $E_{(s,t)}$ is the amount of energy in the proposed duration, $E_{(s,t)}^{\min}$ is the lowest level of allowed energy, $C_{(s)}$ is the defined energy potential, $P_{(s)}^{ch}$ and $P_{(s)}^{dch}$ are the maximum charging and discharging value, respectively, all for the $s$th ESS, $NL_{(t)}$ is the net predicted demand, and $\Delta t$ is the proposed time duration. $P_{(s,t)}$ shows the scheduled active power of the $s$th ESS. The net demand will be positive when corresponding amount of generation be lower than the present consumption. In this case, the proposed weight of a particular ESS is equal to the highest discharging amount of $P_{(s)}^{dch}$ which the ESS can feed in the proposed time duration. Besides, the highest power amount is bounded not only by maximum discharging value of the ESS but also by the existing energy potential for the proposed time duration $\left( E_{(s,t)} - E_{(s)}^{\min} / \Delta t \right)$. This term will refrain ESS from completely discharging in the proposed time duration. On the other hand, when net demand is negative, the proposed weight of the ESS is equal to the maximum charging value, $P_{(s)}^{ch}$ bounded by the defined ESS energy potential.

### 7.8.1.2 Emergency Energy Balancing

In this part, the emergency energy balancing will be conducted to regulate the shares of power dispatch of each ESS to balance the energy potential during the islanded scheduling caused by an emergency operation. According to the bounded ESS energy potential, the current demand of the network, and the state-of-charge of the ESS, in case of islanding scheduling, it is feasible that various available storages completely discharge their reserved energy. The EMS will initially determine the highest energy potential of the ESS after performing the islanded scheduling and then regulate the residual power and energy reserve for the proposed time duration according to the emergency power dispatch results. The residual reserve can be formulated as follows.

$$
R_{(HE,t)} = \min\left( \frac{E_{(HE,t)} - E_{(HE)}^{\min}}{\Delta t}; P_{(HE)}^{dch} \right) - P_{(HE,t)}
\tag{7.9}
$$

where, the $HE$ demonstrates the ESS with the highest energy and $R$ is its calculated residual power reserve.

The residual power reserve, $R_{(HE,t)}$, will be shared among other ESSs according to their energy potentials. To identify an appropriate decision of the EMS, an auxiliary constraint is also applied to avoid extravagant discharging of the defined *ME* ESS, as in Eq. (7.10).

$$E_{(HE,t)} + \left[ P_{(HE,t)} + R_{(HE,t)} \right] \Delta t \geq E_{(LE,t)} + \left[ P_{(LE,t)} + R_{(HE,t)} \omega_{(LE,t)} \right] \Delta t \qquad (7.10)$$

$$R_{(HE,t)} \geq \frac{E_{(HE,t)} - E_{(LE,t)} + \left[ P_{(LE,t)} - P_{(HE,t)} \right] \Delta t}{\left[ 1 + \omega_{(LE,t)} \right] \Delta t} \qquad (7.11)$$

where, the *LE* demonstrates the ESS with the lowest energy and $\omega_{(LE,t)}$ id the proposed weight for the residual power reserve shares. Therefore, the power regulation in each ESS can be determined by Eq. (7.12).

$$\Delta P_{(OT,t)} = \min\left( \left| P_{(OT,t)} - R_{(HE,t)} \cdot \omega_{(OT,t)} \right|; P_{(OT)}^{ch} \right) \qquad (7.12)$$

where, *OT* demonstrates the whole ESSs except the *HE* one. Using the power regulation for the whole ESSs, the strategy will be iterated with the difference that the *HE* ESS is omitted. This procedure will be terminated when only a single ESS lasts.

## 7.9    IEC-61850-BASED EMS DURING EMERGENCIES

During emergency operation, it is intended that the existing wireless networks and spectrum could be operated to set up communication. However, common wireless networks mostly operate in a specified spectrum range assigned by controllers. Therefore, applying diverse existing wireless networks is comparatively complicated. In this section, this complicated assignment is comprehended by implementing the cognitive radio (CR) design which employs existing spectrum and communication networks during emergency operations. Furthermore, the IEC 61850 standard does not assign any specific convention to be applied at information and power exchange levels. Accordingly, design of CR can be directly implemented on IEC 61850 communication without any problems of consistency and sustainability. Hence, the application of IEC 61850 communication based is investigated to figure out the management assignment during emergency operation. Besides, the implementation assessment of the suggested IEC 61850 communication based for energy management during emergency operation applying CR methodology is demonstrated [30].

In the emergency operation, EMS applies all stationary and mobile energy generators to supply a stable power output. To regulate the power balance between generation and consumption, it is required to implement intelligent EMS. Recent occurrences indicate that phone services and email are restricted while social media is accessible most of the time. Although there may be some problems, social media can be employed as an appropriate system to notify and manage the community during emergency operations. This is a communication framework that needs to

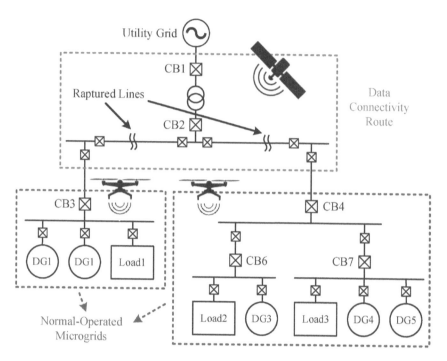

**FIGURE 7.17**  An energy management framework for achieving connectivity during emergency operation.

(i) ensure the internet connection, (ii) ensure the connection among fault-occurred areas which may be in islanded mode during the emergency operation. Also, with the assist of drones and satellites, the emergency data can be transfer from one network to another and to emergency center control at the end. The dispatchable DGs, RES-based units, and ESSs are connected via a communication network using this strategy. Figure 7.17 shows a full electrical network with some islanded areas during emergency operation.

For operating this network with EMS during emergency operation as described previously, the communication network is very critical and should be flexible. Consequently, the communication network of the EMS during emergency operations can be designed based on IEC 61850. Besides, it is essential that such a framework is attentive to its area and can apply diverse existing spectrums and link any available wireless networks (i.e., Wi-MAX, Wi-Fi, GSM, and radio networks). The mentioned assignments during emergency operations are comprehended by CR methodology. If a contingency situation happens, the EMS during emergency operations will be motivated in the prevalent DGs. The significant function of the EMS during emergency operation is to adjust the power flow in the islanded areas. According to the latest data, CR design a scheme that determines the connection network that will be employed. The EMS during emergency operation investigates for existing active power support and locality DGs be sending the present condition signals. Accordingly, circuit breakers or switches on the power transmission lines react to the signals from EMS, which renew the state of the power lines whether they are stable after the event. The emergency operation decisions made by EMS are dispatched to the mobile energy generators to send them to critical load areas to supply the power support [31].

It should be noted that EMS can be employed individually as a local distributed controller in the islanded area using existing facilities. When local DGs and ESs are not acceptable to supply the loads in the area with an emergency operation, it is required to apply extra MESSs detected in the vicinity. MESSs can communicate with the EMS and receive the required data to power support. On the other side, EMS during the emergency operation can collect the detailed location and the state-of-charge of each MESSs. Once the real-time data is collected from all resources, EMS adjusts the optimal power flow in the network. Accordingly, existing PEVs are requested to the emergency area to attach a particular EV group. The PEVs act as the generation unit under EV group assignment. After eliminating the emergency, the required data will be sent, and MESSs will be facilitated by their functions.

## 7.10 CONCLUSION

Power systems are always planned and operated in such a way that they provide a sufficient level of flexibility to confirm the power balance at any time. High penetration of RES and their fluctuation and uncertainty have enhanced the illustrious role of flexibility in active distribution networks. In this chapter, the flexibility enhancement methods for distribution networks were analyzed. The role of MGs in enhancing the flexibility of DSs by forming self-sufficient MGs, NMGs, multi-energy MGs, and energy hubs were presented. Furthermore, the flexibility of the communication networks in MG configuration is also investigated. In addition to MG configuration, strategies applied by MGs for enhancing their flexibility during the diverse emergency operation phases are also analyzed. The most important fact in the flexibility enhancement regarding the MG studies is the restoration procedure during or after the emergency operation, which can be realized by emergency operation strategies to enhance the flexibility of the load recovery congestion and decrease the restoration duration.

For a more secure and reliable DS emergency operation, more flexible services are required to deal with either forecasted or unexpected aberrations in net demand. Therefore, diverse resources for flexibility were suggested which will result in an advancement of DS flexibility. Each resource might enhance physical or structural flexibilities up to a particular level. Accordingly, the performance of the emergency resources including islanding strategies, DRPs, EVs, BESSs, MPSs, and HEG facility were investigated for the generation and demand reduction portfolios of the network. Besides, some generic assessment measures were formulated to designate the performance of the HEG facility. Furthermore, the emergency approaches in the light of the flexibility need proper communication and control among diverse resources. Accordingly, the IEC 61850 standard based scheduling was investigated to provide different communication facilities. The mentioned standardized EMS during emergency operations is majorly essential to minimize the impacts of the emergencies in the DSs.

## REFERENCES

[1] A. Akrami, M. Doostizadeh, and F. Aminifar, "Power system flexibility: an overview of emergence to evolution," *Journal of Modern Power Systems and Clean Energy*, vol. 7, no. 5, pp. 987–1007, 2019/09/01 2019, doi: 10.1007/s40565-019-0527-4.

[2] S. Y. Yun, J. C. Kim, J. C. Bae, Y. J. Jeon, S. M. Park, and C. H. Park, "Reliability Evaluation of Power Distribution Systems Considering the Momentary Interruptions," *IFAC Proceedings Volumes*, vol. 36, no. 20, pp. 761–766, 2003/09/01/ 2003, doi: https://doi.org/10.1016/S1474-6670(17)34563-9.

[3] S. Mishra, K. Anderson, B. Miller, K. Boyer, and A. Warren, "Microgrid resilience: A holistic approach for assessing threats, identifying vulnerabilities, and designing corresponding mitigation strategies," *Applied Energy*, vol. 264, p. 114726, 2020/04/15/ 2020, doi: https://doi.org/10.1016/j.apenergy.2020.114726.

[4] S. E. Ahmadi, N. Rezaei, and H. Khayyam, "Energy management system of networked microgrids through optimal reliability-oriented day-ahead self-healing scheduling," *Sustainable Energy, Grids and Networks*, vol. 23, p. 100387, 2020/09/01/ 2020, doi: https://doi.org/10.1016/j.segan.2020.100387.

[5] S. E. Ahmadi and N. Rezaei, "A new isolated renewable based multi microgrid optimal energy management system considering uncertainty and demand response," *International Journal of Electrical Power & Energy Systems*, vol. 118, p. 105760, 2020/06/01/ 2020, doi: https://doi.org/10.1016/j.ijepes.2019.105760.

[6] S. Lei, C. Chen, H. Zhou, and Y. Hou, "Routing and scheduling of mobile power sources for distribution system resilience enhancement," *IEEE Transactions on Smart Grid*, vol. 10, no. 5, pp. 5650–5662, 2019, doi: 10.1109/TSG.2018.2889347.

[7] J. Gouveia, C. Gouveia, J. Rodrigues, L. Carvalho, C. L. Moreira, and J. A. P. Lopes, "Planning of distribution networks islanded operation: from simulation to live demonstration," *Electric Power Systems Research*, vol. 189, p. 106561, 2020/12/01/ 2020, doi: https://doi.org/10.1016/j.epsr.2020.106561.

[8] N. Rezaei and M. Kalantar, "Smart microgrid hierarchical frequency control ancillary service provision based on virtual inertia concept: An integrated demand response and droop controlled distributed generation framework," *Energy Conversion and Management*, vol. 92, pp. 287–301, 2015/03/01/ 2015, doi: https://doi.org/10.1016/j.enconman.2014.12.049.

[9] M. Panteli, P. Mancarella, D. N. Trakas, E. Kyriakides, and N. D. Hatziargyriou, "Metrics and quantification of operational and infrastructure resilience in power systems," *IEEE Transactions on Power Systems*, vol. 32, no. 6, pp. 4732–4742, 2017, doi: 10.1109/TPWRS.2017.2664141.

[10] A. Hussain, V.H. Bui, and H.M. Kim, "Microgrids as a resilience resource and strategies used by microgrids for enhancing resilience," *Applied Energy*, vol. 240, pp. 56–72, 2019/04/15/2019, doi: https://doi.org/10.1016/j.apenergy.2019.02.055.

[11] Z. Li, M. Shahidehpour, F. Aminifar, A. Alabdulwahab, and Y. Al-Turki, "Networked microgrids for enhancing the power system resilience," *Proceedings of the IEEE*, vol. 105, no. 7, pp. 1289–1310, 2017, doi: 10.1109/JPROC.2017.2685558.

[12] Z. Cheng, J. Duan, and M. Chow, "To centralize or to distribute: that is the question: a comparison of advanced microgrid management systems," *IEEE Industrial Electronics Magazine*, vol. 12, no. 1, pp. 6–24, 2018, doi: 10.1109/MIE.2018.2789926.

[13] S. Bahramara, A. Mazza, G. Chicco, M. Shafie-khah, and J. P. S. Catalão, "Comprehensive review on the decision-making frameworks referring to the distribution network operation problem in the presence of distributed energy resources and microgrids," *International Journal of Electrical Power & Energy Systems*, vol. 115, p. 105466, 2020/02/01/ 2020, doi: https://doi.org/10.1016/j.ijepes.2019.105466.

[14] P. Siano, G. D. Marco, A. Rolán, and V. Loia, "A survey and evaluation of the potentials of distributed ledger technology for peer-to-peer transactive energy exchanges in local energy markets," *IEEE Systems Journal*, vol. 13, no. 3, pp. 3454–3466, 2019, doi: 10.1109/JSYST.2019.2903172.

[15] "IEEE Recommended Practice for Utility Interface of Photovoltaic (PV) Systems," *IEEE Std 929-2000*, p. i, 2000, doi: 10.1109/IEEESTD.2000.91304.

[16] "IEEE Application Guide for IEEE Std 1547(TM), IEEE Standard for Interconnecting Distributed Resources with Electric Power Systems," *IEEE Std 1547.2-2008*, pp. 1–217, 2009, doi: 10.1109/IEEESTD.2008.4816078.

[17] H. Mohamad, H. Mokhlis, A. H. A. Bakar, and H. W. Ping, "A review on islanding operation and control for distribution network connected with small hydro power plant," *Renewable and Sustainable Energy Reviews*, vol. 15, no. 8, pp. 3952–3962, 2011/10/01/ 2011, doi: https://doi.org/10.1016/j.rser.2011.06.010.

[18] A. R. Jordehi, "Optimisation of demand response in electric power systems, a review," *Renewable and Sustainable Energy Reviews*, vol. 103, pp. 308–319, 2019/04/01/ 2019, doi: https://doi.org/10.1016/j.rser.2018.12.054.

[19] H. S. Das, M. M. Rahman, S. Li, and C. W. Tan, "Electric vehicles standards, charging infrastructure, and impact on grid integration: a technological review," *Renewable and Sustainable Energy Reviews*, vol. 120, p. 109618, 2020/03/01/ 2020, doi: https://doi.org/10.1016/j.rser.2019.109618.

[20] H. Afrakhte and P. Bayat, "A contingency based energy management strategy for multi-microgrids considering battery energy storage systems and electric vehicles," *Journal of Energy Storage*, vol. 27, p. 101087, 2020/02/01/ 2020, doi: https://doi.org/10.1016/j.est.2019.101087.

[21] Y. Xu, Y. Wang, J. He, M. Su, and P. Ni, "Resilience-oriented distribution system restoration considering mobile emergency resource dispatch in transportation system," *IEEE Access*, vol. 7, pp. 73899-73912, 2019, doi: 10.1109/ACCESS.2019.2921017.

[22] B. Zhou *et al.*, "Multiobjective generation portfolio of hybrid energy generating station for mobile emergency power supplies," *IEEE Transactions on Smart Grid*, vol. 9, no. 6, pp. 5786-5797, 2018, doi: 10.1109/TSG.2017.2696982.

[23] D. Venkatramanan and V. John, "A reconfigurable solar photovoltaic grid-tied inverter architecture for enhanced energy access in backup power applications," *IEEE Transactions on Industrial Electronics*, vol. 67, no. 12, pp. 10531–10541, 2020, doi: 10.1109/TIE.2019.2960742.

[24] C. Nguyen, H. Lee, and T. Chun, "Cost-optimized battery capacity and short-term power dispatch control for wind farm," *IEEE Transactions on Industry Applications*, vol. 51, no. 1, pp. 595–606, 2015, doi: 10.1109/TIA.2014.2330073.

[25] J. Shi, W. Lee, and X. Liu, "Generation scheduling optimization of wind-energy storage system based on wind power output fluctuation features," *IEEE Transactions on Industry Applications*, vol. 54, no. 1, pp. 10–17, 2018, doi: 10.1109/TIA.2017.2754978.

[26] H. Rezk, M. Al-Dhaifallah, Y. B. Hassan, and H. A. Ziedan, "Optimization and energy management of hybrid photovoltaic-diesel-battery system to pump and desalinate water at isolated regions," *IEEE Access*, vol. 8, pp. 102512–102529, 2020, doi: 10.1109/ACCESS.2020.2998720.

[27] B. Whitaker *et al.*, "A high-density, high-efficiency, isolated on-board vehicle battery charger utilizing silicon carbide power devices," *IEEE Transactions on Power Electronics*, vol. 29, no. 5, pp. 2606–2617, 2014, doi: 10.1109/TPEL.2013.2279950.

[28] L. Che and M. Shahidehpour, "Adaptive formation of microgrids with mobile emergency resources for critical service restoration in extreme conditions," *IEEE Transactions on Power Systems*, vol. 34, no. 1, pp. 742–753, 2019, doi: 10.1109/TPWRS.2018.2866099.

[29] D. Zarrilli, A. Giannitrapani, S. Paoletti, and A. Vicino, "Energy storage operation for voltage control in distribution networks: a receding horizon approach," *IEEE Transactions on Control Systems Technology*, vol. 26, no. 2, pp. 599–609, 2018, doi: 10.1109/TCST.2017.2692719.

[30] S. M. S. Hussain, M. A. Aftab, I. Ali, and T. S. Ustun, "IEC 61850 based energy management system using plug-in electric vehicles and distributed generators during emergencies," *International Journal of Electrical Power & Energy Systems*, vol. 119, p. 105873, 2020/07/01/2020, doi: https://doi.org/10.1016/j.ijepes.2020.105873.

[31] M. H. Rehmani, A. C. Viana, H. Khalife, and S. Fdida, "A cognitive radio based Internet access framework for disaster response network deployment," in *2010 3rd International Symposium on Applied Sciences in Biomedical and Communication Technologies (ISABEL 2010)*, 7-10 Nov. 2010 2010, pp. 1–5, doi: 10.1109/ISABEL.2010.5702851.

# 8 Three-Layer Aggregator Solutions to Facilitate Distribution System Flexibility

*Xiaolong Jin[1], Saeed Teimourzadeh[2],*
*Osman Bulent Tor[2], and Qiuwei Wu[1]*
[1]Center for Electrical Power and Energy (CEE),
Department of Electrical Engineering, Technical
University of Denmark, Kongens Lyngby, Denmark
[2]EPRA Electric Energy Co. Ankara, Turkey

## CONTENTS

## NOMENCLATURE

### *Set:*

| | |
|---|---|
| $T, T^{\mathrm{c}}, T^{\mathrm{rb}}$ | Sets of the day-ahead scheduling hours, congestion hours, and allowable payback hours |
| $N^{\mathrm{nd}}, N^{l}$ | Sets of nodes and transmission lines |
| $N^{\mathrm{ev}}, N^{\mathrm{hp}}$ | Sets of EVs and HPs |
| $N^{\mathrm{ag}}$ | Set of aggregators |
| $N_{a,n}^{\mathrm{ev}}, N_{a,n}^{\mathrm{hp}}$ | Sets of EVs and HPs belonging to $a$th ($a \in N^{\mathrm{ag}}$) aggregator at node $n \in N^{\mathrm{nd}}$ |

### *Variables:*

| | |
|---|---|
| $f_{t,a,n}^{\mathrm{fle}}$ | The amount of available flexibility at hour $t$ stipulated in the bid provided by $a$th aggregator at node $n$ aggregator at node $n$ at hour $t$ |
| $r_{t,a,n}^{\mathrm{rb}}$ | The amount of payback power required by $a$th aggregator at node $n$ at hour $t$ |
| $f_{t,e}^{\mathrm{ev}}, f_{t,h}^{\mathrm{hp}}$ | The amount of flexibility provided by $e$th EV and $h$th HP at hour $t$, respectively |
| $r_{t,e}^{\mathrm{ev}}, r_{t,h}^{\mathrm{hp}}$ | The amount of payback power required by $e$th EV and $h$th HP at hour $t$, respectively |
| $p_{t,e}^{\mathrm{ev}}, p_{t,h}^{\mathrm{hp}}$ | Charging power of $e$th EV and power consumption of $h$th HP at hour $t$, respectively |
| $K_{t,h}^{\mathrm{h}}, K_{t,h}^{\mathrm{u}}$ | Household inside and outside temperature of the house with $h$th HP at hour $t$, respectively |
| $x_{t,a,n}^{\mathrm{rb}}$ | The binary variable representing the status of payback power, the $a$th aggregator at node $n$ can have payback power at hour $t$ if $x_{t,a,n}^{\mathrm{rb}} = 1$ |
| $x_{t,a,n}^{\mathrm{fle}}$ | The procurement percentage of the flexibility service bid provided by $a$th aggregator at node $n$ at hour $t$, $x_{t,a,n}^{\mathrm{fle}} \in [0,1]$ |
| $F_{t,l}$ | Active power flow of line $l \in N^{l}$ at hour $t$ |
| $p_{t,n}, q_{t,n}$ | Actual active and reactive power consumption at node $n$ at hour $t$, respectively |
| $f_{t,a,n}^{\mathrm{ag}}$ | The amount of flexibility purchased from $a$th aggregator and node $n$ at hour $t$ |
| $\lambda_{e}^{\mathrm{ev}}, \lambda_{h}^{\mathrm{hp}}$ | Increased day-ahead energy costs of $e$th EV and $h$th HP, respectively |

### *Parameters:*

| | |
|---|---|
| $e_{e}^{\mathrm{min/max}}$ | Lower and upper limits of the SOC level of $e$th EV |
| $e_{e,t0}$ | Initial SOC level of $e$th EV |
| $p_{e}^{\mathrm{ev,min/max}}$ | Lower and upper limits of $e$th EV charging power |
| $d_{t,e}$ | Driving distance of $e$th EV at hour $t$ |

| | |
|---|---|
| $s_{t,e}$ | The binary value representing the charging availability of $e$th EV at hour $t$ |
| $c_h^{\mathrm{cop}}$ | The performance coefficient of $h$th HP |
| $k_1, k_2$ | The heat transfer coefficients |
| $p_h^{\mathrm{hp,min/max}}$ | Lower and upper limits of $h$th HP power consumption |
| $K_h^{\mathrm{h,min/max}}$ | Lower and upper limits of the household inside temperature of the house with $h$th HP |
| $\lambda_{a,n}^{\mathrm{fle}}$ | The bidding price of the flexibility service bid provided by $a$th aggregator at node $n$ |
| $c_{t,e}^{\mathrm{ev}}, c_{t,h}^{\mathrm{hp}}$ | Cost coefficients of $e$th EV and $h$th HP at hour $t$, respectively |
| $\hat{p}_{t,e}^{\mathrm{ev}}, \hat{p}_{t,h}^{\mathrm{hp}}$ | Baseline charging power and power consumption of $e$th EV and $h$th HP at hour $t$, respectively |
| $\hat{p}_{t,n}$ | Baseline power consumption at node $n$ at hour $t$ |
| $\alpha$ | The upper limit of the payback percentage |
| $\lambda_t^{\mathrm{DA}}$ | The forecasted day-ahead energy prices at hour $t$ |
| $G_{n,l}$ | The element at $n$th row and $l$th column of the incidence matrix |
| $\cos(\phi_n)$ | The power factor at node $n$ |
| $F_l^{\mathrm{max}}$ | Capacity of line $l$ |
| $V^{\mathrm{s}}, V^{\mathrm{min}}$ | The voltage magnitude at the substation and the lower limit of the voltage magnitude |
| $z_{i,j}$ | The impedance of the transmission line between nodes $i$ and $j$ |
| $f_{t,a,n}^{\mathrm{pr}}$ | The amount of flexibility purchased from $a$th aggregator and node $n$ at hour $t$ after the LFM clearing |
| $\rho$ | Penalty parameter of the ADMM algorithm |
| $\tau, \zeta$ | Update parameters |

### Acronyms:

| | |
|---|---|
| DER | Distributed energy resources |
| EV | Electrical vehicle |
| HP | Heat pump |
| DSO | Distribution system operator |
| LFM | Local flexibility market |
| DT | Dynamic tariff |
| DS | Dynamic subsidy |
| RT | Real-time |
| **ESS** | Energy storage system |
| DG | Distributed generation |
| BRP | Balance responsible party |
| NLP | Nonlinear programming |
| MIQP | Mixed-integer quadratic programming |
| MILP | Mixed-integer linear programming |
| ADMM | Alternating direction method of multipliers |
| OPF | Optimal power flow |
| AC/DC | Alternating current/direct current |
| FRT | Flexibility requirement table |
| LP | Load point |

## 8.1    INTRODUCTION

Modern power systems continue to evolve under the influence of technological developments, regulatory policy modifications, and climate and environmental issues. The changes are more apparent at distribution systems level since distributed energy resources (DERs), such as electrical vehicles (EVs) and heat pumps (HPs), have been massively deployed in distribution networks. This leads to a significant paradigm shift of the network and poses more operational challenges to distribution system operators (DSOs) [1]. One major concern of DSO is the network congestion caused by the uncoordinated power consumption of DERs, e.g., simultaneous charging of EVs. In addition to network reinforcement, DSOs can use market-based demand response (DR) programs to utilize demand-side flexibility for resolving day-ahead congestion in an economically efficient way [2]. The DR program is established to change electricity consumption of end users in response to changes in electricity prices over time or given incentives [3]. DR benefits include, but are not limited to, improving the economic efficiency of electricity markets, enhancing the reliability of electric power systems, reducing peak demand, and alleviating price volatility. Depending on the design and structure of the DR program, DR could be captured from a large customer who meets the demand-side contribution requirements or an independent market player that aggregates small customers' contribution. Considering large number of small-scale costumers and owing to the proliferation of smart appliances at the household level, the DR aggregation could be acknowledged as an efficient solution to realize flexibility at distribution system level and increase the disclosure of consumers and prosumers to the wholesale electricity markets. A DR aggregator builds a bridge between the DSO and retail customers and takes the responsibility of aggregating and managing customer responses. The market-based DR programs can be roughly categorized into two types [4]: (1) price-based programs and (2) incentive-based programs.

### 8.1.1    LITERATURE REVIEW

#### 8.1.1.1    Price-Based Programs

In price-based programs, end users respond to changes in electricity prices by modifying power consumption patterns. In the dynamic tariff (DT)-based methods presented in refs. [5–9], the DSO publishes time-dependent DTs at different locations of the network, which leads to higher final electricity prices (base price plus DTs) at peak hours. As a result, the aggregators or end users shift power consumption from peak hours to off-peak hours to minimize energy costs, which consequently helps mitigate congestion. In ref. [5], a DT-based framework was developed to coordinate EV charging to alleviate congestion. However, due to the linear formulation of the DT optimization model, the model might have multiple optimal solutions in case of same energy prices at different hours, which will cause inconsistency between the solution at the DSO side and solutions at the aggregator side and finally lead to a failure of congestion management. In ref. [7], a quadratic programming (QP) formulation of the DT model was developed to resolve the multiple solution issue. The proposed DT frameworks in refs. [5, 6] rely on the DSO to accurately predict uncertain flexible demands. This is impractical due to inherent prediction errors. In ref. [7], an uncertainty management method based on the DT framework was developed to handle uncertainties of flexible demands. A sensitivity-based iterative procedure was used to quantify the probability distribution of forecast error in order to reserve a secure line capacity margin to deal with uncertainties. In ref. [8], a robust DT model was developed to account for uncertainties of flexible demands. To remove solution conservatism obtained with the robust DT model, a sensitivity-based real-time (RT) adjustment method was used in a receding horizon fashion to adjust the obtained DT in near RT. In order to relive the DSO from predicting information from the demand side, distributed optimizations for the DT framework were proposed in refs. [8] and [9], in which the DTs are obtained through an iterative procedure between DSO and aggregators such that DSO does not need to predict information at the demand side.

In ref. [10], a line shadow price method was proposed, which works similarly as the DT method with the shadow price obtained through an interactive process between the DSO and aggregators or end users. In the DT and shadow price methods, tariffs are published in peak hours to increase final prices. In contrast, a dynamic subsidy (DS) method was proposed in ref. [11], in which the subsidies are given to reduce final prices (base price minus subsidies) in the off-peak hours. The DS method shares a similar congestion management framework with the DT method but has a different model to calculate DSs.

### 8.1.1.2  Incentive-Based Programs

In the incentive-based programs, end users are rewarded by modifying power consumption profiles through local flexibility market (LFM) arrangements. A bilateral-contract-based LFM was proposed in ref. [12], in which trade processes are carried out between the DSO and aggregators and between the aggregator and end users. This market is based on a network of bilateral flexibility contracts with multiple contracts between each pair of agents that are able to trade with one another. In refs. [13–19], the pool-based LFM framework was studied, where flexibility sellers and buyers offer bids in the market and market operator clears the market for different operational purposes. The proposed LFM frameworks in refs. [13–19] share a similar operational mechanism where the aggregators or end users bid flexibility service in the LFM and then market operator clears the market to satisfy the DSO's flexibility requirements with different market-clearing models. An LFM framework for day-ahead and intraday congestion management was proposed in ref. [14], in which market participants consist of the DSO, market operator, balance responsible party (BRP), and aggregator. DSO is the only flexibility buyer in the day-ahead LFM for congestion management, whereas BRP participates in the intraday LFM and competes with the DSO for flexibility to minimize imbalance costs. Day-ahead and intraday market-clearing problems formulated as mixed-integer linear programming (MILP) models are solved by the market operator with the social welfare maximization. This LFM framework is complemented with an RT dispatching strategy in ref. [15]. If the market-based solution fails to resolve congestion, DSO takes over the control of flexible resources to adjust energy schedules. A similar LFM framework was proposed in ref. [16] to resolve day-ahead congestion with microgrids being flexibility service providers and DSO playing the role of the market operator.

The microgrid agent aggregates small size distributed generations (DGs), energy storage systems (ESSs), and customers to provide flexibility service. The market-clearing problem formulated as a nonlinear programming (NLP) model is solved by the DSO with minimum flexibility procurement cost.

Abovementioned LFM frameworks for day-ahead congestion management are executed before the operation of the day-ahead energy market. In such a case, LFM clearing solutions are used to adjust original energy schedules so that the resulting schedules after day-ahead energy market clearing lead to no congestion. In contrast, LFM frameworks proposed in refs. [17–19] operate after the day-ahead market clearing if the accepted energy schedules result in congestion. The market-clearing problems in refs. [17–19] formulated as NLP models are solved by the DSO to minimize flexibility procurement costs. Moreover, an RT LFM and the probability analysis-based method were developed in refs. [18] and [19] to deal with uncertainties of flexibility resources. However, in these LFM frameworks, the imbalance cost due to rescheduling energy schedules after the day-ahead energy market clearing is not considered.

### 8.1.2 MOTIVATIONS

There are issues with the price-based programs. In DT and shadow price methods [5–10], end users who help mitigate congestion need to pay congestion costs instead of getting rewards. In addition, in these price-based methods [5–11], the end users receive different tariffs or subsidies because of their locations in the network, which is against the nondiscrimination rule [11]. The LFM-based method can resolve the abovementioned issues. In the LFM framework, end users who help mitigate congestion receive rewards that are quantified by flexibility service provided. However, there are two major drawbacks when applying the current LFM frameworks to resolve day-ahead congestion.

The first one is privacy information protection. In existing LFM frameworks, it is assumed that the market operator has access to network parameters in order to ensure the market-clearing solution is technically feasible, i.e., the market-clearing solution satisfies network operation constraints, which may compromise the privacy protection [16–19]. The second issue is the optimal formulation of the flexibility service bid. One key element of the flexibility service bid is the energy payback condition, i.e., the power payback amount and hour [20]. However, these conditions are usually determined without modeling operation constraints of flexibility resources in the existing LFM frameworks [16–19]. Moreover, the flexibility cost, namely the payment to flexibility resources for providing flexibility, has not been considered in the bid formulation, which may make the end users or aggregators not willing to participate in the LFM because their revenues are uncertain.

This chapter aims to improve the current LFM framework. To improve privacy information protection, an alternating direction method of multipliers (ADMM)-based market-clearing strategy is proposed, in which the market operator communicates with DSO to clear the LFM so that the market-clearing solution respects network operational constraints without revealing network parameters to the market operator. The ADMM, as a distributed algorithm, has been widely used to solve convex problems in power system applications [21–26], such as load restoration [22], convexified optimal power flow (OPF) [23, 24], voltage control [25], and DR [26]. In ref. [22], a distributed load restoration strategy was proposed using ADMM algorithm. In the proposed strategy, the load restoration problem was decomposed into sub-problems for each node and solved by agents at each node through exchanging information between neighbor agents. In ref. [23], the convexified alternating current (AC)-OPF problem was decomposed and solved by ADMM algorithm in a distributed manner. In the algorithm, the closed-form solutions of sub-problems at each node are obtained, which significantly improve the computation efficiency. The direct current (DC)-OPF problem was solved by the consensus ADMM in ref. [24]. Based on the consensus ADMM framework, three distributed algorithms, i.e., distributed DC-OPF with a central controller, fully decentralized DC-OPF, and distributed DC-OPF with the accelerated ADMM, were developed and compared with respect to convergence performance and communication architecture. In ref. [25], a distributed reactive power control scheme based on the consensus ADMM was designed for voltage control of the wind farm cluster. Using the proposed distributed control scheme, the fair reactive power sharing among wind farms can be achieved to regulate voltages. Moreover, the computation burden of wind farm controllers and communication costs can be reduced. In ref. [26], an ADMM-based distributed residential load control strategy was proposed for peak load shaving considering operation constraints of unbalanced distribution networks and the privacy information protection. In addition, this chapter proposes an optimal bidding strategy for the aggregator under a given bidding price. The optimal bidding problem is formulated as an MILP that determines the energy payback condition considering operation constraints of flexibility resources and maximizes the aggregator's revenues with flexibility costs included. If the flexibility bid is accepted in the LFM, the aggregator and end users can receive revenues accordingly.

In summary, compared with the existing DT and shadow price methods [5–10], the end users who help mitigate congestion receive rewards instead of paying congestion costs in the proposed LFM framework. Compared with the DS method [11], the proposed LFM framework has no locational discrimination issue because the end users located in a network receive the same energy price. Compared with existing LFM frameworks [13–19], the proposed framework has two significant improvements: (1) the privacy information protection is improved by introducing an ADMM-based market-clearing strategy, in which the system parameters do not need to be revealed to the market operator; and (2) the energy payback conditions of flexibility resources and flexibility costs are considered and carefully modeled in the proposed optimal flexibility bidding strategy.

The contributions of this chapter are summarized as follows:

1. Propose a three-layer aggregator solution to provide flexibility for DSO by harvesting flexibility from DR aggregation.
2. Propose an optimal flexibility bidding strategy for aggregators in LFM, which carefully models energy payback conditions of flexibility resources and enables the aggregator to receive the maximum revenue with the flexibility cost included.

3. Develop an ADMM-based market-clearing strategy for the LFM, in which the market-clearing solution respects network operational constraints without revealing network parameters to the market operator.

The rest of this chapter is organized as follows: Section 8.2 presents the concept of the three-layer aggregator solutions; Section 8.3 presents the concept of the proposed LFM; Section 8.4 presents the proposed operational mechanism of day-ahead congestion management with the LFM; Section 8.5 presents three model formulations for the three-layer aggregator solution, including baseline energy schedule formulation, optimal flexibility bidding and rescheduling after flexibility activation for layer 2, and market-clearing problem for layer 3; Section 8.6 presents the proposed ADMM-based market-clearing strategy. Case study results are discussed in Section 8.7, followed by conclusions in Section 8.8.

## 8.2   INTRODUCTION OF THE THREE-LAYER AGGREGATOR SOLUTION

To address manifested merits of DR aggregation, this chapter studies harvesting flexibility from DR aggregation through a three-layer approach. Figure 8.1 depicts the outline of proposed three-layer aggregator solution to facilitate distribution system flexibility.

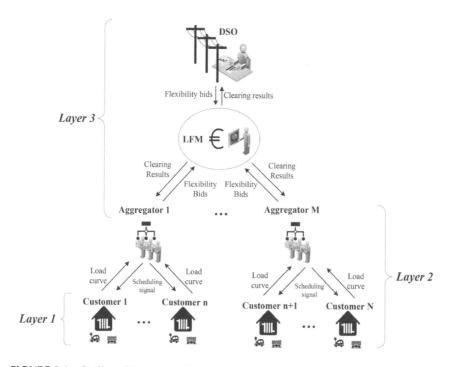

**FIGURE 8.1**   Outline of the proposed approach.

In Figure 8.1, the first layer deals with consumer/prosumer stage as the first step to attain benefits of DR programs is to increase awareness level of the customers. To do so, the advantages gained from altering utilization pattern of controllable loads are highlighted to the end users. The proposed approach acquires the end user's profiles pertaining to on-site generation, i.e., renewable-based generation, and consumption, i.e., fixed and flexible load; in turn, the proposed approach represents the cost difference between user-preferred and optimal scheduling of flexible appliance such as dishwasher, washer & dryer, pump, storage device, and EV. To investigate user-preferred operation condition, daily operation of a community aggregator is simulated where the controllable appliances are set by the end user considering comfort constraints. On the contrary, for offering the optimal operation, an optimization model is devised where the objective is to minimize operation cost of customer considering technical constraints pertaining to permissible operation range of equipment. Here, controllable loads are decision variables where their optimal setting is attained by solving the devised optimization model. An indicator is tailored to show the difference between usual and optimal operation and steer the user toward the optimal operation. More on consumer/prosumer stage DR and flexibility is presented in ref. [27].

As shown in Figure 8.1, the second layer deals with customers' demand control and flexibility management between aggregators and consumers. The aggregator acts as an intermediary between customers and the operator of LFM. Since the individual customer has limited negotiation power in LFM due to its small volume of flexibility, it is necessary to have an aggregator that can gather flexibility from customers, formulate flexibility service bids, and trade flexibility in LFM. The aggregator and customers reach an agreement that the aggregator can schedule flexibility sources of customers. An optimization model is developed in layer 2 for the aggregator to formulate flexibility service bids considering operational constraints of flexibility sources of customers, e.g., EVs and heat pumps, and submit the bids in LFM. The flexibility service bid stipulates the bidding price, amount and locations of flexibility provided at each congestion hour, energy payback hour and amount of payback power. After LFM clearing, the aggregators reschedule flexibility sources of customers to provide committed flexibility.

As shown in Figure 8.1, layer 3 provides a competitive trading platform, i.e., LFM. The LFM is utilized to trade flexibility as commodity between flexibility buyers (i.e., the DSO) and flexibility sellers (e.g., the aggregators representing customers). The DSO buys flexibility for voltage control, congestion management, and loss reduction. The aggregators sell flexibility for profits with customers' demand control and flexibility management between aggregators and consumers. A market-clearing model is developed in layer 3. The clearing model determines acceptable flexibility service bids of aggregators to satisfy DSO's flexibility requirements. To protect the privacy of network parameters, an alternating direction method of multipliers-based market clearing method is developed. With this market-clearing method, the operator of LFM communicates with DSO to clear the market such that the market-clearing solution respects network operation constraints without revealing network parameters to the market operator. After LFM clearing, DSO pays flexibility procurement costs and the aggregators receive payments according to bidding prices.

## 8.3 CONCEPT OF LOCAL FLEXIBILITY MARKET

LFM is a flexibility trading platform where different parties trade flexibility in a geographically limited area, such as a community or a city [28]. A schematic overview of an LFM is shown in Figure 8.2, where LFM participants are the aggregator, DSO, and market operator. The roles and responsibilities of the participants are described as follows:

1. *DSO*: The DSO is responsible for the secure operation of the distribution network and effective service delivery to customers. The DSO can procure flexibility in LFM for different operational purposes such as congestion management and line loss reduction.

2. *Aggregator*: The aggregator acts as an intermediary between customers and other LFM participants. Since the individual customer has limited negotiation power in LFM due to its small volume of flexibility, it is necessary to have an aggregator that can gather flexibility from customers, formulate flexibility service bids, and trade flexibility in LFM. The aggregator and customers reach an agreement that the aggregator can schedule flexibility sources of customers.

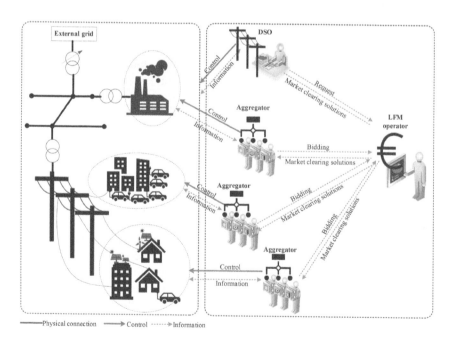

FIGURE 8.2    Schematic overview of a local flexibility market.

**FIGURE 8.3** Flexibility trading process in the local flexibility market.

3. *Market operator*: The LFM operator is an independent entity running LFM. The LFM operator is responsible for the market clearing process that determines trading results, i.e., the procurement cost and amount of flexibility traded.

The flexibility trading process is shown in Figure 8.3. First, after the DSO sends a flexibility request and publishes flexibility requirements, the aggregators formulate flexibility service bids and offer them in the LFM. Second, the market operator clears the LFM to determine accepted flexibility service bids to meet DSO's flexibility requirements. Third, according to the market-clearing solution, the aggregators schedule flexibility resources to provide committed flexibility. Finally, flexibility transactions are completed through financial settlement. DSO pays flexibility procurement costs while aggregators and customers receive revenues for providing flexibility service.

The capital flow of LFM is shown in Figure 8.4. As shown in the capital flow in Figure 8.4, customers receive payments from the aggregators because of providing flexibility and increase day-ahead energy costs. Therefore, customers do not need to pay extra money for increased energy costs and always have revenues when they provide flexibility. Aggregators receive flexibility selling revenues from the market operator or from the DSO directly and pay customers for providing flexibility and increased energy costs. Since the flexibility service bid is formulated with the maximization of the nonnegative revenue of the aggregator, it ensures that the aggregator always has revenues as long as its bid is accepted. Therefore, the proposed LFM is an attractive flexibility trading platform because it is profitable for the aggregators and customers.

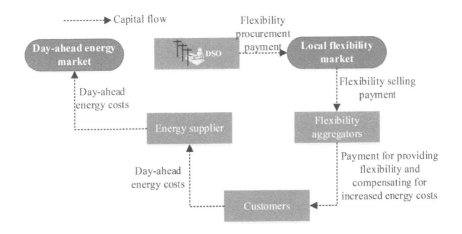

**FIGURE 8.4** Capital flow of financial settlement.

## 8.4 MECHANISM OF DAY-AHEAD CONGESTION MANAGEMENT WITH THE LFM

This study focuses on the application of the LFM to day-ahead congestion management in distribution networks. The proposed operational mechanism of day-ahead congestion management with the LFM is shown in Figure 8.5 and described as follows.

First, based on the forecasted day-ahead spot prices, energy suppliers make baseline energy schedules for customers with the minimum energy costs. The role of the energy supplier is to purchase energy in the day-ahead energy market on behalf of customers. It is assumed that the energy supplier and aggregator are two different entities; otherwise, the aggregator could intentionally make baseline energy schedules that result in congestion and then make profits by selling flexibility in FLM to deal with congestion that is originally caused by the aggregator itself.

Second, the baseline energy schedules are sent to DSO for technical validation. The DSO conducts the power flow analysis to examine if the schedules are technically feasible. Once DSO identifies there is congestion in the following day of operation, it sends a flexibility request to LFM and publishes a flexibility requirement table (FRT). The determination of the key parameters in the FRT is described below:

- Congestion hours and load points requested to provide flexibility
  After receiving the baseline energy schedules, the DSO can identify congestion hours and lines using the power flow analysis. According to the topology information, the DSO can determine those load points that are able to provide flexibility to resolve congestion.
- The minimum and maximum amount of flexibility required
  The minimum and maximum amount of flexibility required are approximate values. The two values provide a reference for aggregators to provide the appropriate amount of flexibility. The actual amount of flexibility required for resolving congestion is obtained after LFM clearing. For overloading

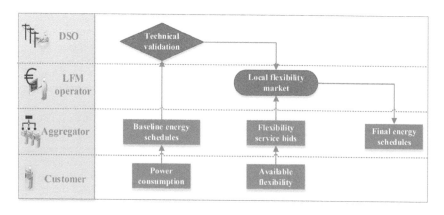

**FIGURE 8.5** The proposed operational mechanism of day-ahead congestion management with the local flexibility market.

management, the amount of flexibility required for resolving overloading issue is equal to the line overloading amount. Therefore, minimum and maximum amount of flexibility required for voltage management can be obtained according to the amount of flexibility required at each node.

- The upper limit of the energy payback percentage and maximum bidding price
  According to historical data of flexibility service bids, the DSO can stipulate the upper limits of the payback percentage and bidding price for the flexibility service bid.
- Allowable energy payback hours
  With the approximated maximum amount of flexibility required and the stipulated upper limit of the payback percentage, the approximated maximum amount of payback power can be calculated, based on which the DSO conducts the power flow analysis to select the allowable payback hours. In these payback hours, energy payback can occur without causing new congestion. To deal with approximation errors when selecting payback hours, a security margin can be reserved in the power flow analysis.

According to the FRT, the aggregators formulate flexibility service bids considering operation constraints of flexibility sources, e.g., EVs and HPs, and submit the bids in LFM. The flexibility service bid stipulates the bidding price, amount and locations of flexibility provided at each congestion hour, energy payback hour, and amount of payback power. For simplification of the bid formulation, the aggregator is assumed to have payback power at one hour only. After the flexibility bidding process, the market operator clears the LFM to determine accepted bids. With the trading result and baseline energy schedules, aggregators reschedule flexibility sources to provide committed flexibility. Then, the final energy schedules will be submitted to the day-ahead energy market. Finally, financial settlement regarding flexibility trading is carried out based on the pay-as-bid rule. The DSO pays flexibility procurement costs and the aggregators receive payments according to bidding prices. Moreover, the customers providing flexibility service receive payments from the aggregators accordingly.

As introduced in Section 8.2, layer 2 deals with customers' demand control and flexibility management between aggregators and consumers, while layer 3 provides a competitive trading platform, i.e., LFM. LFM is utilized to trade flexibility as commodity between flexibility buyers (i.e., the DSO) and flexibility sellers (e.g., the aggregators representing customers). Therefore, as shown in Figure 8.5, the modules of "Flexibility service bids" and "Final energy schedules" are conducted in layer 2 and the module of "Local flexibility market" is conducted in layer 3. The specific models of these modules for layer 2 and layer 3 will be presented in Section 8.5.

## 8.5 MODEL FORMULATIONS

This section presents four optimization models. First optimization model is to obtain the baseline day-ahead energy schedules for customers with minimal costs according to the forecasted day-head energy prices. Second optimization model is for the aggregator to formulate the optimal flexibility service bid considering operation constraints of flexible resources. The second model is used in layer 2. Third optimization model is the LFM-clearing model that determines acceptable flexibility service bids to satisfy the DSO's flexibility requirements. The third model is used for layer 3. Fourth optimization of for the aggregator to reschedule its contractual customers based on the trading results in the LFM. The fourth model is also used in layer 2.

### 8.5.1 Baseline Day-Ahead Energy Schedules

Due to the massive deployment of EVs and HPs at the residential buildings [29, 30] and the large potential to provide demand flexibility [31, 32], EVs and HPs are considered as flexibility providers in this study. Suppose that each customer owns an EV and a HP. The baseline energy schedules of EVs and HPs are obtained by minimizing day-ahead energy costs while satisfying operational constraints. The optimization model is formulated as below:

$$\text{Min} \sum_{t \in T} \lambda_t^{\text{DA}} \left( \sum_{e \in N^{\text{ev}}} p_{t,e}^{\text{ev}} + \sum_{h \in N^{\text{hp}}} p_{t,h}^{\text{hp}} \right) \tag{8.1}$$

$$e_e^{\text{min}} \leq \sum_{t_- \leq t} \left( p_{t_-,e}^{\text{ev}} - d_{t_-,e} \right) + e_{e,t_0} \leq e_e^{\text{max}}; \forall t \in T, e \in N^{\text{ev}} \tag{8.2}$$

$$s_{t,e} p_e^{\text{ev,min}} \leq p_{t,e}^{\text{ev}} \leq s_{t,e} p_e^{\text{ev,max}}; \forall t \in T, e \in N^{\text{ev}} \tag{8.3}$$

$$\sum_{t \in T} p_{t,e}^{\text{ev}} = \sum_{t \in T} d_{t,e}; \forall e \in N^{\text{ev}} \tag{8.4}$$

$$c_h^{\text{cop}} p_{t,h}^{\text{hp}} - k_1 \left( K_{t,h}^{\text{h}} - K_{t,h}^{\text{u}} \right) = k_2 \left( K_{t,h}^{\text{h}} - K_{t-1,h}^{\text{h}} \right); \forall t \in T, h \in N^{\text{hp}} \tag{8.5}$$

$$K_h^{\text{h,min}} \leq K_{t,h}^{\text{h}} \leq K_h^{\text{h,max}}; \forall t \in T, h \in N^{\text{hp}} \tag{8.6}$$

$$p_h^{\text{hp,min}} \leq p_{t,h}^{\text{hp}} \leq p_h^{\text{hp,max}}; \forall t \in T, h \in N^{\text{hp}} \tag{8.7}$$

The objective function (8.1) is to minimize day-ahead energy costs of EVs and HPs based on the forecasted day-ahead spot prices $\lambda_t^{\text{DA}}$, where $p_{t,e}^{\text{ev}}$ and $p_{t,h}^{\text{hp}}$ are $e$th EV and $h$th HP power consumption at hour $t$, respectively. Constraint (8.2) limits the state of charge (SOC) level of the EV, where $e_e^{\text{min}}$ and $e_e^{\text{max}}$ are the lower and upper bounds of $e$th EV's SOC level, $e_{e,t_0}$ is the $e$th EV's initial SOC level, and $d_{t,e}$ is the power consumption of $e$th EV at hour $t$. Constraint (8.3) limits the EV charging power considering the EV charging availability, where $p_e^{\text{ev,min}}$ and $p_e^{\text{ev,max}}$ are the lower and upper bounds of $e$th EV charging power, and $s_{t,e}$ is the charging availability of $e$th EV at hour $t$. Constraint (8.4) represents that the total charging power is equal to the total power consumption during the optimization horizon. Constraint (8.5) represents the household thermal balance equation, where $c_h^{\text{cop}}$ is the performance coefficient of $h$th HP, $k_1$ and $k_2$ are heat transfer coefficients, and $K_{t,h}^{\text{h}}$ and $K_{t,h}^{\text{u}}$ are household inside and outside temperature of the house with $h$th HP at hour $t$, respectively. Constraint (8.6) limits the household inside temperature within a thermal comfort range, where

$K_h^{\text{h,min}}$ and $K_h^{\text{h,max}}$ are lower and upper bounds of household inside temperature of the house with $h$th HP. Constraint (8.7) limits the HP power consumption, where $p_h^{\text{hp,min}}$ and $p_h^{\text{hp,max}}$ are lower and upper bounds of HP power consumption, respectively.

### 8.5.2 LAYER 2: OPTIMAL FLEXIBILITY BIDDING

Based on the baseline energy schedules and the FRT, the aggregators formulate flexibility service bids by rescheduling EVs and HPs, e.g., by shifting power consumption of EVs in peak periods to off-peak periods to provide flexibility. First, each aggregator sets a bidding price for its bid according to the maximum bidding price stipulated in the FRT and historical data. A large bidding price may make the bid unaccepted in the LFM, whereas a small price may denote a small revenue. Therefore, the bidding price should be determined properly to make the bid competitive in the LFM and to enable the aggregator to gain as many revenues as possible at the same time. This chapter focuses on the optimal flexibility service bid formulation under a given bidding price. The aggregator solves the following optimization model to optimally determine the amount of flexibility provided at each congestion hour, energy payback hour, and amount of payback power.

For aggregator $a$ at node $n$ with a bidding price $\lambda_{a,n}^{\text{fle}}$:

$$\text{Max} \sum_{t \in T^c} \left\{ \lambda_{a,n}^{\text{fle}} f_{t,a,n}^{\text{fle}} - \sum_{e \in N_{a,n}^{\text{ev}}} c_{t,e}^{\text{ev}} \left( f_{t,e}^{\text{ev}} \right)^2 - \sum_{h \in N_{a,n}^{\text{hp}}} c_{t,h}^{\text{hp}} \left( f_{t,h}^{\text{hp}} \right)^2 \right\} - \sum_{e \in N_{a,n}^{\text{ev}}} \lambda_e^{\text{ev}} - \sum_{h \in N_{a,n}^{\text{hp}}} \lambda_h^{\text{hp}} \quad (8.8)$$

$$\begin{cases} f_{t,e}^{\text{ev}} = \hat{p}_{t,e}^{\text{ev}} - p_{t,e}^{\text{ev}}; \forall t \in T^c, e \in N^{\text{ev}} \\ f_{t,h}^{\text{hp}} = \hat{p}_{t,h}^{\text{hp}} - p_{t,h}^{\text{hp}}; \forall t \in T^c, h \in N^{\text{hp}} \\ f_{t,e}^{\text{ev}} \geq 0, f_{t,e}^{\text{ev}} \geq 0; \forall t \in T^c, e \in N^{\text{ev}}, h \in N^{\text{hp}} \end{cases} \quad (8.9)$$

$$\begin{cases} r_{t,e}^{\text{ev}} = p_{t,e}^{\text{ev}} - \hat{p}_{t,e}^{\text{ev}}; \forall t \in T^{\text{rb}}, e \in N^{\text{ev}} \\ r_{t,h}^{\text{hp}} = p_{t,h}^{\text{hp}} - \hat{p}_{t,h}^{\text{hp}}; \forall t \in T^{\text{rb}}, h \in N^{\text{hp}} \\ r_{t,e}^{\text{ev}} \geq 0, r_{t,e}^{\text{ev}} \geq 0; \forall t \in T^{\text{rb}}, e \in N^{\text{ev}}, h \in N^{\text{hp}} \end{cases} \quad (8.10)$$

$$\begin{cases} f_{t,a,n}^{\text{fle}} = \sum_{e \in N_{a,n}^{\text{ev}}} f_{t,e}^{\text{ev}} + \sum_{h \in N_{a,n}^{\text{hp}}} f_{t,h}^{\text{hp}}; \forall t \in T^c, a \in N^{\text{ag}}, n \in N^{\text{nd}} \\ r_{t,a,n}^{\text{rb}} = \sum_{e \in N_{a,n}^{\text{ev}}} r_{t,e}^{\text{ev}} + \sum_{h \in N_{a,n}^{\text{hp}}} r_{t,h}^{\text{hp}}; \forall t \in T^{\text{rb}}, a \in N^{\text{ag}}, n \in N^{\text{nd}} \end{cases} \quad (8.11)$$

$$\begin{cases} \left| r_{t,a,n}^{\text{rb}} \right| \leq x_{t,a,n}^{\text{rb}} M; \forall t \in T^{\text{rb}}, a \in N^a, n \in N^{\text{nd}} \\ \sum_{t \in T^{\text{rb}}} x_{t,a,n}^{\text{rb}} = 1; \forall a \in N^a, n \in N^{\text{nd}} \\ x_{t,a,n}^{\text{rb}} \in \{0,1\} \end{cases} \quad (8.12)$$

$$r_{t,a,n}^{\mathrm{rb}} \leq \alpha \sum_{t^*} f_{t^*,a,n}^{\mathrm{fle}}; \forall t \in T^{\mathrm{rh}}, a \in N^a, n \in N^{\mathrm{nd}} \tag{8.13}$$

$$e_e^{\min} \leq \sum_{t_- \leq t} \left( p_{t_-,e}^{\mathrm{ev}} - d_{t_-,e} \right) + e_{e,t_0} \leq e_e^{\max}; \forall t \in T, e \in N^{\mathrm{ev}} \tag{8.14}$$

$$p_e^{\mathrm{ev,min}} s_{t,e} \leq p_{t,e}^{\mathrm{ev}} \leq s_{t,e} p_e^{\mathrm{ev,max}}; \forall t \in T, e \in N^{\mathrm{ev}} \tag{8.15}$$

$$c_h^{\mathrm{cop}} p_{t,h}^{\mathrm{hp}} - k_1 \left( K_{t,h}^{\mathrm{h}} - K_{t,h}^{\mathrm{u}} \right) = k_2 \left( K_{t,h}^{\mathrm{h}} - K_{t-1,h}^{\mathrm{h}} \right); \forall t \in T, h \in N^{\mathrm{hp}} \tag{8.16}$$

$$K_h^{\mathrm{h,min}} \leq K_{t,h}^{\mathrm{h}} \leq K_h^{\mathrm{h,max}}; \forall t \in T, h \in N^{\mathrm{hp}} \tag{8.17}$$

$$p_h^{\mathrm{hp,min}} \leq p_{t,h}^{\mathrm{hp}} \leq p_h^{\mathrm{hp,max}}; \forall t \in T, h \in N^{\mathrm{hp}} \tag{8.18}$$

$$\begin{cases} \lambda_e^{\mathrm{ev}} = \sum_{t \in T} \lambda_t^{\mathrm{DA}} \left( p_{t,e}^{\mathrm{ev}} - \hat{p}_{t,e}^{\mathrm{ev}} \right); \forall e \in N_{a,n}^{\mathrm{ev}} \\[2mm] \lambda_e^{\mathrm{hp}} = \sum_{t \in T} \lambda_t^{\mathrm{DA}} \left( p_{t,e}^{\mathrm{hp}} - \hat{p}_{t,e}^{\mathrm{hp}} \right); \forall h \in N_{a,n}^{\mathrm{hp}} \end{cases} \tag{8.19}$$

The objective function (8.8) is to maximize the aggregator's revenues if the bid is accepted in the DLM. The first term in (8.8) represents revenues of selling flexibility at the bidding price, where $f_{t,a,n}^{\mathrm{fle}}$ is the amount of available flexibility provided by $a$th aggregator at node $n$ at hour $t$. The second and third terms represent payments to customers for providing flexibility, where $f_{t,e}^{\mathrm{ev}}$ and $f_{t,h}^{\mathrm{hp}}$ are the amount of flexibility provided by $e$th EV and $h$th HP, respectively, and $c_{t,e}^{\mathrm{ev}}$ and $c_{t,h}^{\mathrm{hp}}$ are flexibility cost coefficients of $e$th EV and $h$th HP at hour $t$, respectively. It is assumed that the payment of one unit of flexibility has a linear relation to the total amount of flexibility [4]. Each customer can adjust coefficients $c_{t,e}^{\mathrm{ev}}$ and $c_{t,h}^{\mathrm{hp}}$ to change its willingness to provide flexibility. The last two terms are to compensate customers for their increased day-ahead energy costs due to energy rescheduling, where $\lambda_e^{\mathrm{ev}}$ and $\lambda_h^{\mathrm{hp}}$ are increased day-ahead energy costs of $e$th EV and $h$th HP, respectively. When a customer decreases its power consumption to provide flexibility at one hour, it requires payback power at another hour. Since the baseline energy schedule has the minimum energy cost, energy rescheduling will increase the energy cost of the customer. To guarantee that customers have positive net profits when they provide flexibility, the aggregators are required to compensate customers for the increased day-ahead energy costs.

The amount of flexibility provided by each EV and HP at each congestion hour is calculated in (8.9), where $\hat{p}_{t,e}^{\mathrm{ev}}$ and $\hat{p}_{t,h}^{\mathrm{hp}}$ are baseline energy schedules of $e$th EV and $h$th HP at hour $t$, respectively. The amount of payback power required by each EV ($r_{t,e}^{\mathrm{ev}}$) and HP ($r_{t,h}^{\mathrm{hp}}$) at each allowable payback hour is calculated in (8.10). The total amount of flexibility ($f_{t,a,n}^{\mathrm{fle}}$) and the total amount of payback power ($r_{t,a,n}^{\mathrm{rb}}$) are calculated in (8.11). Constraint (8.12) guarantees that the aggregator has payback power at one hour only, i.e., aggregator $a$ at node $n$ can have payback power at hour $t$ if $x_{a,n,t}^{\mathrm{rb}} = 1$, where $M$ is a very big number. Constraint (8.13) represents that the amount of payback power is constrained by the maximum payback percentage $\alpha$ stipulated in the FRT. Constraints (8.14–8.18) are the operation constraints of EVs and HPs, as described in (8.2–8.6). The increased day-ahead energy costs of EVs and HPs are calculated in (8.19).

## 8.5.3 Layer 3: LFM-Clearing Problem

After the flexibility service bidding process, LFM is cleared to determine the procurement percentages of flexibility service bids to resolve congestion with the minimum flexibility procurement cost. The market clearing problem is formulated as below:

$$\text{Min.} \sum_{t \in T, a \in N^{\text{ag}}, n \in N^{\text{nd}}} \lambda_{a,n}^{\text{fle}} f_{t,a,n}^{\text{ag}} \tag{8.20}$$

$$\sum_{l \in N^l} G_{n,l} F_{t,l} = p_{t,n}; \forall t \in T, n \in N^{\text{nd}} \tag{8.21}$$

$$p_{t,n} = q_{t,n} tg(\phi_n); \forall t \in T, n \in N^{\text{nd}} \tag{8.22}$$

$$\hat{p}_{t,n} = p_{t,n} + \sum_{a \in N^{\text{ag}}} f_{t,a,n}^{\text{ag}}; \forall t \in T, n \in N^{\text{nd}} \tag{8.23}$$

$$f_{t,a,n}^{\text{ag}} = x_{t,a,n}^{\text{fle}} f_{t,a,n}^{\text{fle}}; \forall t \in T, a \in N^{\text{ag}}, n \in N^{\text{nd}} \tag{8.24}$$

$$|F_{t,l}| \le F_l^{\max}; \forall t \in T, l \in N^l \tag{8.25}$$

$$\left(V^s\right)^2 - \text{Re}\left(\sum_{n^* \in N^{\text{nd}}} z_{n,n^*}(p_{t,n^*} - jq_{t,n^*})\right) \ge V^{\min}; \forall t \in T, n \in N^{\text{nd}} \tag{8.26}$$

The objective function (8.20) is to minimize the flexibility procurement cost, where $f_{t,a,n}^{\text{ag}}$ is the amount of flexibility purchased from $a$th aggregator at node $n$ at hour $t$. Constraint (8.21) represents the active power flow balance at each node at each hour, where $p_{t,n}$ is active power consumption at node $n$ at hour $t$, $F_{t,l}$ is the power flow on line $l$ at hour $t$, and $G_{n,l}$ is the mapping matrix. The reactive power is modeled in (8.22) to maintain a constant power factor at each node, where $q_{t,n}$ is the reactive power consumption at node $n$ at hour $t$, and $\cos(\phi_n)$ is the power factor at node $n$. Constraint (8.23) models the relation between baseline power consumption ($\hat{p}_{t,n}$) and final power consumption after providing flexibility. The amount of flexibility purchased from each flexibility service bid is calculated in (8.24), where $x_{t,a,n}^{\text{fle}}$ is the procurement percentage of the flexibility bid provided by $a$th aggregator at node $n$ at hour $t$. Constraint (8.25) represents line capacity limits, where $F_l^{\max}$ is the line capacity of $l$th line. Constraint (8.26) is the linearized voltage magnitude constraint [33], where $V^s$ is the voltage magnitude at the substation node, $V^{\min}$ is the lower bound of the voltage magnitude, Re() is the operation to extract the real part of a complex number, and $z_{n,n^*}$ is the element at the $n$th row and $n^*$th column of the inverse matrix of the partial nodal admittance matrix $Y_{ll}$, which is a submatrix of the full admittance matrix $Y$:

$$Y = \begin{bmatrix} Y_{00} & Y_{0l} \\ Y_{l0} & Y_{ll} \end{bmatrix}$$

As shown in (8.21)–(8.22) and (8.25)–(8.26), network operation constraints are incorporated into the market clearing problem in order to obtain a flexibility procurement solution that is technically feasible. Therefore, the market operator is assumed to know network parameters [16–19]. To relax this assumption, this chapter develops an ADMM-based market-clearing strategy, in which the market operator clears the market with communication with the DSO such that the market clearing solution respects network operation constraints without posing network parameters to the market operator.

### 8.5.4   Layer 2: Rescheduling after Flexibility Activation

After the LFM clearing, each aggregator reschedules its contractual EVs and HPs according to the amount of flexibility purchased in LFM. The rescheduling model is the optimal flexibility bidding model in (8.8–8.19) with the variable $f_{t,a,n}^{\text{fle}}$ fixed as the amount of flexibility purchased in LFM. The rescheduling model is formulated as below:

Min (8.8)

Subject to: (8.9–8.19); $f_{t,a,n}^{\text{ag}} = f_{t,a,n}^{\text{fle}}; \forall t \in T^c, a \in N^{\text{ag}}, n \in N^{\text{nd}}$

## 8.6   THE ADMM-BASED MARKET-CLEARING STRATEGY

In this section, an ADMM-based algorithm is developed first to solve the market-clearing problem with an iterative procedure, as shown in Table 8.1. Then, an ADMM-based market-clearing strategy is proposed based on the ADMM-based algorithm. Since the market-clearing problem in (8.20–8.26) is formulated as a convex model, the convergence of the ADMM algorithm is guaranteed [21].

---

**TABLE 8.1**

**Pseud Code of the ADMM-Based Algorithm**

|  |  |
|---|---|
| **ADMM-Base Algorithm for the Market-Clearing Model** | |
| **1: inputs** | Parameters of the system and flexibility bids |
| **2: outputs** | The amount of flexibility purchased from each flexibility service bid |
| **3: while** | $\sum\limits_{t \in T, a \in N^a, n \in N^{\text{nd}}} \left( \tilde{f}_{t,a,n}^{\text{ag},(k+1)} - f_{t,a,n}^{\text{ag},(k+1)} \right) > \sigma_1$; or $\sum\limits_{t \in T, a \in N^a, n \in N^{\text{nd}}} \left( \tilde{f}_{t,a,n}^{\text{ag},(k+1)} - \tilde{f}_{t,a,n}^{\text{ag},(k)} \right) > \sigma_2$ |
| **4: do** | |
| **5:** | Solve sub-problem I to update primal variabels **X** |
| **6:** | Solve sub-problem II to update primal variables **Y** |
| **7:** | Update dual variables using (8.29) |
| **8: end** | |

---

### 8.6.1   ADMM-BASED ALGORITHM

#### 8.6.1.1   Reformulation of Market-Clearing Model

Before solving the market-clearing model, a set of auxiliary variables $\tilde{f}_{t,a,n}^{ag}$ and a set of equality constraints are introduced as below:

$$f_{t,a,n}^{ag} = \tilde{f}_{t,a,n}^{ag}; \ \forall t \in T, a \in N^{ag}, n \in N^{nd} \tag{8.27}$$

By replacing the original variables $f_{t,a,n}^{ag}$ in (8.23) with auxiliary variables, the original market-clearing model is reformulated as below:

Min. (8.20)

Subject to: (8.21)-(8.22) and (8.24)-(8.27)

$$\hat{p}_{t,n} = p_{t,n} + \sum_{a \in N^{ag}} \tilde{f}_{t,a,n}^{ag}; \forall t \in T, n \in N^{nd} \tag{8.28}$$

#### 8.6.1.2   Augmented Lagrangian Function

The augmented Lagrangian is formulated by adding the equality constraint (8.27) into the objective function (8.20) through dual variables $\mu_{t,a,n}$, as below:

$$\text{Min.} \sum_{t \in T, a \in N^{ag}, n \in N^{nd}} \lambda_{a,n}^{fle} f_{a,n,t}^{ag} + \sum_{t \in T, a \in N^{ag}, n \in N^{nd}} \mu_{t,a,n} \left( \tilde{f}_{t,a,n}^{ag} - f_{t,a,n}^{ag} \right) + \frac{\rho}{2} \left\| \tilde{f}_{t,a,n}^{ag} - f_{t,a,n}^{ag} \right\|^2;$$

Subject to: (8.21–8.22) and (8.24–8.26), (8.28)

The augmented Lagrangian is optimized over two groups of primal variables $(\mathbf{X}, \mathbf{Y})$ and one group of dual variables $(\Lambda)$, as defined below:

$$\begin{cases} \mathbf{X} = \left[ \tilde{f}_{t,a,n}^{ag}, p_{t,n}, q_{t,n}, F_{t,l} \right]; \\ \mathbf{Y} = \left[ f_{t,a,n}^{ag}, x_{t,a,n}^{fle} \right]; \\ \Lambda = \left[ \mu_{t,a,n} \right]; \end{cases}$$

#### 8.6.1.3   Iterative Procedure

In the ADMM-based algorithm, primal variables and dual variables of the augmented Lagrangian are optimized with an iterative procedure. Two groups of primal variables are optimized in sub-problems I and II, respectively.

1. Sub-problem I

   In the sub-problem I, at the $(k+1)$th iteration, primal variables $\mathbf{X}^{k+1}$ are optimized with the values of $\mathbf{Y}^*$ and $\Lambda^*$ obtained at the $k$th iteration. The sub-problem I is formulated as below:

$$\text{Min.} \sum_{t \in T, a \in N^{ag}, n \in N^{nd}} \mu_{t,a,n}^{(k)} \tilde{f}_{t,a,n}^{ag,(k+1)} + \frac{\rho}{2} \left\| \tilde{f}_{t,a,n}^{ag,(k+1)} - f_{t,a,n}^{ag,(k)} \right\|^2;$$

Subject to: (8.21–8.22), (8.25–8.26), and (8.28)

2. Sub-problem II

In the sub-problem II, primal variables $\mathbf{Y}^{k+1}$ are optimized at $(k+1)$th iteration with the values of $\mathbf{X}^*$ and $\mathbf{\Lambda}^*$ obtained at the $k$th iteration. The sub-problem II is formulated as below:

$$\text{Min.} \quad \sum_{t\in T, a\in N^{\mathrm{ag}}, n\in N^{\mathrm{nd}}} \lambda_{a,n}^{\mathrm{fle}} f_{t,a,n}^{\mathrm{ag},(k+1)} - \sum_{t\in T, a\in N^{\mathrm{ag}}, n\in N^{\mathrm{nd}}} \mu_{t,a,n}^{(k)} f_{t,a,n}^{\mathrm{ag},(k+1)} + \frac{\rho}{2} \left\| \tilde{f}_{t,a,n}^{\mathrm{ag},(k)} - f_{t,a,n}^{\mathrm{ag},(k+1)} \right\|^2 ;$$

Subject to: (8.24)

3. Update of dual variables

After solving sub-problems I and II, dual variables are updated using the values of $\mathbf{X}^*$ and $\mathbf{Y}^*$ obtained at $(k+1)$-th iteration in (8.29).

$$\mu_{t,a,n}^{(k+1)} = \mu_{t,a,n}^{(k)} + \rho\left( \tilde{f}_{t,a,n}^{\mathrm{ag},(k+1)} - f_{t,a,n}^{\mathrm{ag},(k+1)} \right); \forall t \in T, a \in N^{\mathrm{ag}}, n \in N^{\mathrm{nd}} \qquad (8.29)$$

4. Convergence criteria

After optimizing primal variables and updating dual variables, the convergence conditions are checked. The optimizations of primal variables and update of dual variables are carried out iteratively until the aggregated primal and dual residuals (cr and cd) are lower than the specified thresholds, as shown in (8.30).

$$\begin{cases} \text{primal residual:cr} = \left| \sum_{t\in T, a\in N^{\mathrm{a}}, n\in N^{\mathrm{nd}}} \left( \tilde{f}_{t,a,n}^{\mathrm{ag},(k+1)} - f_{t,a,n}^{\mathrm{ag},(k+1)} \right) \right| \le \sigma_1; \\ \\ \text{dual residual:cd} = \left| \sum_{t\in T, a\in N^{\mathrm{a}}, n\in N^{\mathrm{nd}}} \left( \tilde{f}_{t,a,n}^{\mathrm{ag},(k+1)} - f_{t,a,n}^{\mathrm{ag},(k)} \right) \right| \le \sigma_2; \end{cases} \qquad (8.30)$$

5. Dynamic update of the penalty parameter

During the iteration procedure, the penalty parameter $\rho$ is updated at each iteration according to primal and dual residuals using (8.31) [21, 34], where $\zeta$ and $\tau$ are update parameters.

$$\rho^{k+1} = \begin{cases} \tau\rho^k & \text{if } \mathrm{cr} \ge \zeta\,\mathrm{cd} \\ \rho^k / \tau & \text{if } \mathrm{cd} \ge \zeta\,\mathrm{cr} \\ \rho^k & \text{otherwise} \end{cases} \qquad (8.31)$$

## 8.6.2 ADMM-BASED LFM-CLEARING STRATEGY

Based on the proposed ADMM-based algorithm, the proposed LFM clearing strategy is shown in Figure 8.6, where the market operator clears the market with

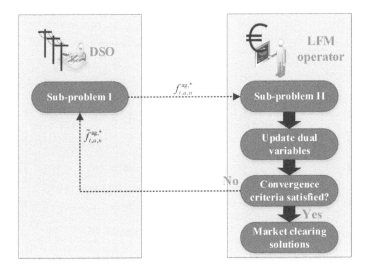

**FIGURE 8.6** The ADMM-based LFM clearing strategy.

communication with the DSO. In the interactive procedure, the parameters associated with the amount of purchased flexibility are exchanged between the market operator and the DSO. The market operator solves the sub-problem II to select flexibility service bids and calculate flexibility procurement costs, while DSO solves the sub-problem I to check if network operation constraints are violated with the selected flexibility service bids. In the proposed strategy, the market clearing solution respects network operation constraints without revealing network parameters to the market operator.

## 8.7 CASE STUDIES

### 8.7.1 CASE SETTING

Case studies were conducted on the Bus 4 distribution network of the Roy Billinton Test System (RBTS) [35] to demonstrate the effectiveness of day-ahead congestion management with LFM. The single-line diagram of the Bus 4 distribution network is shown in Figure 8.7. Line segments of feeder 1 are labeled as L1–L12 and load points are labeled as $LP_{1-7}$, $LP_{11-16}$, $LP_{18-25}$, and $LP_{32-38}$. The data of conventional load consumption profile and lines can be found in ref. [6]. The loading limits of L2 and L7, load point data, and key parameters of the EV and HP are listed in Table 8.2 and Table 8.3. Each residential load point has 200 customers, each of which owns an EV and HP to provide flexibility. It is assumed that there are four aggregators ($ag_1$, $ag_2$, $ag_3$, and $ag_4$) participating in LFM. At each residential load point, each of $ag_1$ and $ag_4$ has contracts with 40 customers, and each of $ag_2$ and $ag_3$ has contracts with 60 customers. The forecasted day-ahead spot price profile is

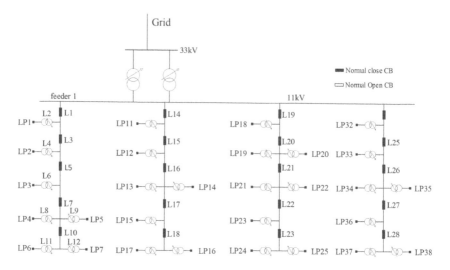

**FIGURE 8.7**   The single diagram of the Bus 4 distribution network.

shown in Figure 8.8. The lower limit of the voltage magnitude is set to be 0.95 p.u. is order to have a security margin of 0.01 p.u. compared to the assumed physical limit of 0.94 p.u. It is assumed that the bidding prices of aggregators and cost coefficients of EVs and HPs are produced randomly with the normal distribution.

## 8.7.2   SIMULATION RESULTS

### 8.7.2.1   Baseline Day-Ahead Energy Schedules
Based on the forecasted day-ahead spot prices, the energy suppliers obtain baseline day-ahead energy schedules and submit the schedules to the DSO for technical

## TABLE 8.2
## Key Parameters

| Parameters | Value |
|---|---|
| COP of HP | 3.0 |
| $k_1$ (5 types of houses) | 0.280/0.315/0.350/0.385/0.420 |
| $k_2$ (5 types of houses) | 3.638/4.093/4.548/5.002/5.457 |
| Lower and upper limits of SOC | 20–90% |
| Thermal comfort range | 20–24°C |
| Peak consumption power | 5 kW |
| EV battery size | 30 kWh |
| Energy consumption per km | 150 kWh/km |
| Peak charging power | 10 kW |
| Loading limit of L2/L7 | 1700/3700 kW |
| Resistance/reactance | 0.26/0.027 omh/km |

**TABLE 8.3**
**Load Point Data**

| Load Points (LP) | Types of Customers | Peak Conv. Load [kW] | Number of Customers Per LP |
|---|---|---|---|
| $LP_{1-4}$, $LP_{11-13}$, $LP_{18-21}$, $LP_{32-35}$ | Residential | 886.9 | 200 |
| $LP_5$, $LP_{14-15}$, $LP_{22}$, $LP_{36}$ | Residential | 813.7 | 200 |
| $LP_{23}$, $LP_{37}$ | Residential | 986.9 | 200 |
| $LP_{6-7}$, $LP_{16-17}$ | Commercial | 415.0 | 10 |
| $LP_{24-25}$, $LP_{38}$ | Commercial | 986.9 | 10 |

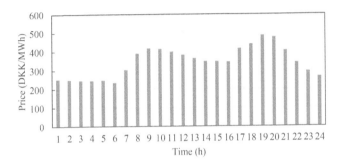

**FIGURE 8.8**   Forecasted day-ahead spot price profile.

validation. The baseline charging power profiles and SOC levels of an EV at $LP_4$ is shown in Figure 8.9. The baseline power consumption profiles of an HP at $LP_4$ and household inside temperature is shown in Figure 8.10. The DSO conducts the power flow analysis to identify if the line capacity and voltage magnitude constraints are violated. As shown in Figure 8.11, there is an overload of 132.8 kW at L2 at $t_6$, and there are overloads of 378.2 kW and 86.1 kW at L7 at $t_6$ and $t_{16}$, respectively.

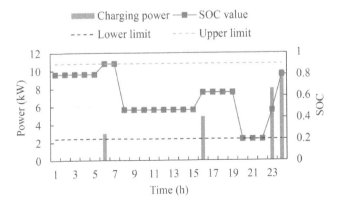

**FIGURE 8.9**   Baseline charging power profile and SOC values of an electric vehicle at $LP_4$.

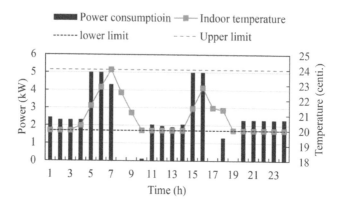

**FIGURE 8.10** Baseline power consumption profile of a heatpump at LP$_4$ and household inside temperature.

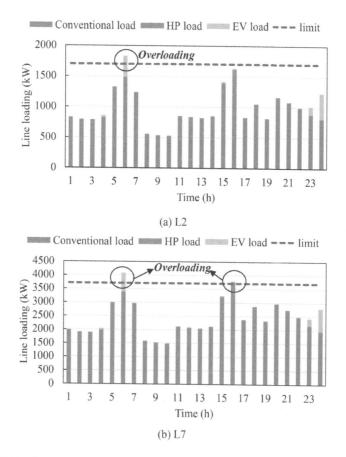

**FIGURE 8.11** Line loadings of L2 and L7.

**TABLE 8.4**
**Flexibility Requirement Table**

| Load Points | Minimum Amount of Flexibility Required [kW] | | Payback Hours | Upper Limit on Payback Percentage | Maximum Price [DKK/kW] |
|---|---|---|---|---|---|
| | $t_6$ | $t_{16}$ | | | |
| $LP_1$ | 132.8 | – | $t_{1-5}, t_{7-15}, t_{17-24}$ | 150% | 1.5 |
| $LP_{4-5}$ | 378.2 | 86.1 | $t_{1-5}, t_{7-15}, t_{17-24}$ | | |

To resolve these problems, DSO needs to procure flexibility in LFM to perform over-load and voltage management. After conducting the power flow analysis, the DSO submits a FRT shown in Table 8.4 to the LFM.

### 8.7.2.2 Results of Layer 2: Optimal Flexibility Bidding of Aggregators

Upon the flexibility request, each aggregator solves the optimal flexibility bid-ding model to formulate flexibility service bids according to the FRT. As listed in Table 8.5, four aggregators provide flexibility service bids at $LP_1$ and $LP_{4-5}$ that the DSO can use to resolve congestion on L2 and L7. Specifically, at $LP_1$, four aggrega-tors ($ag_1$–$ag_4$) provide flexibility at $t_6$ and choose to have payback power at $t_4$. At $LP_4$,

**TABLE 8.5**
**Flexibility Service Bids**

| Aggregator | Price (DKK/kW) | Load Point | Flexibility Amount (kW) | | Payback Hour and Amount (kW) | |
|---|---|---|---|---|---|---|
| | | | $t_6$ | $t_{16}$ | $t_4$ | $t_{10}$ |
| $ag_1$ | 0.75 | $LP_1$ | 60.896 | – | 66.478 | – |
| | 0.62 | $LP_4$ | 49.673 | 20.819 | – | 74.341 |
| | 0.70 | $LP_5$ | 52.857 | 22.692 | – | 79.820 |
| $ag_2$ | 0.58 | $LP_1$ | 87.408 | – | 94.763 | – |
| | 0.66 | $LP_4$ | 78.673 | 34.506 | – | 119.152 |
| | 0.63 | $LP_5$ | 89.463 | – | 97.047 | – |
| $ag_3$ | 0.53 | $LP_1$ | 83.661 | – | 90.853 | – |
| | 0.71 | $LP_4$ | 81.833 | 35.373 | – | 123.723 |
| | 0.59 | $LP_5$ | 84.901 | – | 92.302 | – |
| $ag_4$ | 0.84 | $LP_1$ | 67.318 | – | 73.219 | – |
| | 0.78 | $LP_4$ | 57.224 | 24.412 | – | 85.806 |
| | 0.73 | $LP_5$ | 54.810 | – | 59.565 | – |

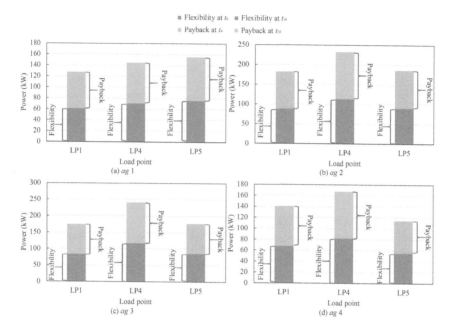

**FIGURE 8.12**   Flexibility service bids of aggregators.

$ag_1$–$ag_4$ provide flexibility at $t_6$ and $t_{16}$ simultaneously and require payback power at $t_{10}$, however, at $LP_5$, $ag_2$–$ag_4$ provide flexibility at $t_6$ only. The flexibility service bids of aggregators are also shown in Figure 8.12.

### 8.7.2.3   Results of Layer 3: LFM-Clearing Results

After the flexibility service bidding process, the market operator clears the LFM to obtain the procurement percentage of each bid, as listed in Table 8.6. With the market clearing solution, the aggregators reschedule EVs and HPs to provide committed flexibility and formulate the final day-ahead energy schedules. The resulting loadings of L2 and L7 are shown in Figure 8.13 after the LFM clearing. It can be seen that congestion on L2 and L7 is resolved and energy payback occurs at $t_4$ and $t_{10}$ without causing new congestion, which demonstrates the effectiveness of overload management with the LFM.

The market-clearing problem is solved by the proposed ADMM-based algorithm in an iterative manner. In the study, the primal and dual convergence thresholds are set as 0.00001 and the initial penalty parameter is set as 5. As shown in Figure 8.14, the aggregated primal and dual residuals can converge to the specified thresholds after 94 iterations. The iteration process of the objective value is shown in Figure 8.15, which demonstrates that the ADMM-based algorithm can reproduce the centralized solution. The centralized solution is obtained by directly solving the market-clearing model with the CPLEX solver.

**TABLE 8.6**
**Procurement Percentages of Flexibility Service**
**Bids and Total Amount of Purchased Flexibility**

| Aggregator | Load Point | Percentage | |
|---|---|---|---|
| | | $t_6$ | $t_{16}$ |
| $ag_1$ | $LP_1$ | 0 | 0 |
| | $LP_4$ | 1.000 | 1.000 |
| | $LP_5$ | 0.392 | 1.000 |
| $ag_2$ | $LP_1$ | 0.562 | 0 |
| | $LP_4$ | 1.000 | 1.000 |
| | $LP_5$ | 1.000 | 0 |
| $ag_3$ | $LP_1$ | 1.000 | 0 |
| | $LP_4$ | 0 | 0.230 |
| | $LP_5$ | 1.000 | 0 |
| $ag_4$ | $LP_1$ | 0 | 0 |
| | $LP_4$ | 0 | 0 |
| | $LP_5$ | 1.000 | 0 |
| total amount [kW] | | 511.040 | 86.135 |

### 8.7.2.4  Results of Layer 2: Rescheduling after Flexibility Activation

The baseline and final charging power profiles and SOC levels of an EV at $LP_4$ are shown in Figure 8.16. In order to provide flexibility at $t_6$ and $t_{16}$, the EV charges more power at $t_{10}$ so that it can have power consumption reductions at $t_6$ and $t_{16}$ while maintaining the sufficient SOC level for the daily driving consumption. The baseline and final power consumption profiles of a HP at $LP_4$ and its household inside temperature are shown in Figure 8.17. Similarly, the HP decreases power consumption at $t_6$ and $t_{16}$ to provide flexibility while having payback power at $t_{10}$ to maintain the household inside temperature within the thermal comfort range.

## 8.8  CONCLUSION

This chapter proposes a three-layer aggregator solution to provide flexibility for DSO by harvesting flexibility from DR aggregation. Layer 1 deals with consumer/prosumer stage, as the first step to attain the benefits of DR programs is to increase awareness level of the customers. Layer 2 deals with customers' demand control and flexibility management between aggregators and consumers. Layer 3 provides a competitive trading platform, i.e., LFM to facilitate flexibility trading between DSO and aggregators representing customers. In the existing LFM frameworks, there are three major drawbacks that hinder the application of the LFM to day-ahead congestion management: (1) the market operator is assumed to have access to network parameters such that the market clearing solution satisfies network operation

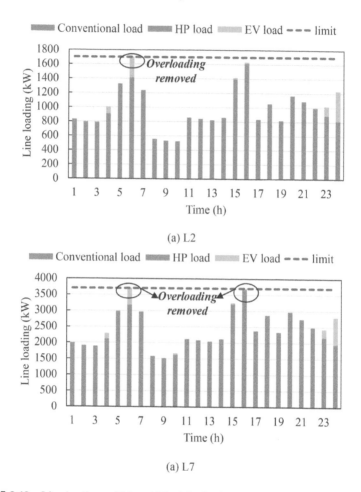

(a) L2

(a) L7

**FIGURE 8.13** Line loadings of L2 and L7 of the final day-ahead energy schedules.

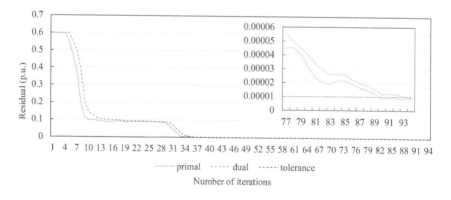

**FIGURE 8.14** Convergence process of the ADMM-based algorithm.

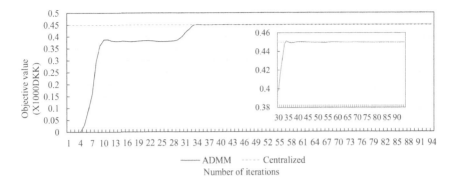

**FIGURE 8.15**  Convergence of the objective value.

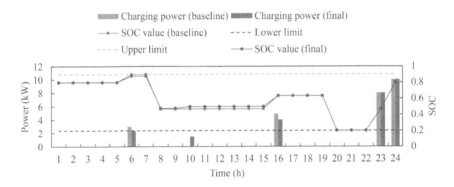

**FIGURE 8.16**  Baseline and final charging power profiles and SOC levels of an electric vehicle at LP4.

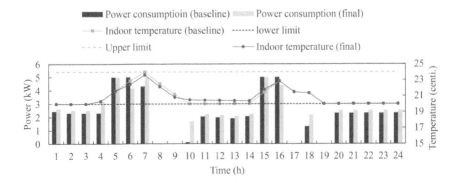

**FIGURE 8.17**  Baseline and final power consumption profiles of a heat pump at LP4 and household inside temperature.

constraints, which compromises the privacy information protection; (2) the energy payback conditions of flexibility resources are usually determined without considering operational constraints; and (3) the flexibility cost of flexibility resources has not been considered. To overcome first drawback, this chapter proposes an ADMM-based market clearing strategy, in which the market operator communicates with the DSO to clear the market such that the market clearing solution respects network operation constraints without revealing network parameters to the market operator. To overcome second and third drawbacks, this chapter proposes an optimal flexibility bidding strategy for aggregators. The optimal flexibility bidding problem is formulated as a mixed-integer quadratic programming model, which carefully models energy payback conditions and enables the aggregator to receive the maximum revenue with the flexibility cost considered. The case studies results demonstrate that the proposed LFM framework can perform effective day-ahead congestion management including overload and voltage management. The proposed ADMM-based market clearing strategy can efficiently solve the market clearing problem.

## ACKNOWLEDGMENT

This chapter presents the scientific results of the transnational project "Multi-layer aggregator solutions to facilitate optimum demand response and grid flexibility" (SMART-MLA Project No: 89029), co-financed by TÜBİTAK (Turkey), SWEA (Sweden), EUDP (Denmark), UEFISCDI (Romania), and RCN (Norway) under ERA-Net Smart Energy Systems, SG+ 2017 Program. The study is supported by "TÜBİTAK TEYDEB 1509 – Uluslararası Sanayi Ar-Ge Projeleri" program with the project number of 9180003.

## REFERENCES

[1] P. Hallberg, J. J. A. Rios, C. Bergerland, A. Blanquet, M. Cailliau, C. Clifford, P. D. Wit, E. Diskin, H. Feuk, B. Gouverneur, M. Kay, P. Lawson, M. Lombardi, P. Mandatova, R. Otter, P. S. Caballero, J. T. Guijarro, W. Tenschert, D. Trebolle, and S. Wanzek, "Active distribution system management—A key tool for the smooth integration of distributed generation", *Eurelectric*, Brussels, Belgium, Feb. 2013.

[2] K. Spiliotis, A. I. Ramos Gutierrez, and R. Belmans, "Demand flexibility versus physical network expansions in distribution grids", *Applied Energy*, vol. 182, pp. 613–624, Nov. 2016.

[3] U.S. Department of Energy, "Benefit of demand response in electricity market and recommendations for achieving them," Feb. 2006.

[4] F. Shen, Q. Wu, S. Huang, X. Chen, H. Liu, and Y. Xu, "Two-tier demand response with flexible demand swap and transactive control for real-time congestion management in distribution networks", *International Journal of Electric Power and Energy Systems*, in press.

[5] R. Li, Q. Wu, and S. S. Oren, "Distribution locational marginal pricing for optimal electric vehicle charging management," *IEEE Trans. Power Syst.* vol. 29, no. 1, pp. 203–211, Jan. 2014.

[6] S. Huang, Q. Wu, S. S. Oren, R. Li, and Z. Liu, "Distribution locational marginal pricing through quadratic programming for congestion management in distribution network," *IEEE Trans. Power Sys.*, vol. 30, no. 4, pp. 2170–2178, Jul. 2015.

[7] S. Huang, Q. Wu, L. Cheng, Z. Liu, and H. Zhao, "Uncertainty management of dynamic tariff method for congestion management in distribution networks," *IEEE Trans. Power Syst.*, vol. 31, no.6, pp. 4340–4347, Nov. 2016.

[8] S. Hanif, H. B. Gooi, T. Massier, T. Hamacher, and T. Reindl, "Distributed congestion management of distribution grids under robust flexible buildings operations," *IEEE Trans. Power Syst.*, vol. 32, no. 6, pp. 4600–4613, Nov. 2017.

[9] S. Huang, Q. Wu, H. Zhao, and C. Li, "Distributed optimization based dynamic tariff for congestion management in distribution networks," *IEEE Trans. Smart Grid*, vol. 10, no. 1, pp. 184–192, Jan. 2019.

[10] B. Biegel, P. Andersen, J. Stoustrup, and J. Bendtsen, "Congestion management in a smart grid via shadow prices," in *Proc. 2012 8th IFAC Symposium on Power Plant and Power System Control*, pp. 518–523.

[11] S. Huang and Q. Wu, "Dynamic subsidy method for congestion management in distribution networks," *IEEE Trans. Smart Grid*, vol. 9, no. 3, May 2018.

[12] T. Morstyn, A. Teyteloym, and M. D. McCulloch, "Designing decentralized markets for distribution system flexibility", *IEEE Trans. Power Syst.*, vol. 34, no. 3, May 2019.

[13] D. T. Nguyen, M. Negnevitsky, and M. de Groot, "Pool-based demand response exchange–concept and modeling," *IEEE Trans. Power Syst.*, vol. 26, no. 3, pp. 1677–1685, Aug. 2011.

[14] S. S. Torbaghan, N. Blaauwbroek, D. Kuiken, M. Gibescu, M. Hajighasemi, P. Nguyen, G. J. M. Smit, M. Roggenkamp, and J. Hurink, "A market-based framework for demand side flexibility scheduling and dispatching", *Sustain Energy, Grids Networks*, vol. 14, pp. 47–61, Jun. 2018.

[15] Torbaghan SS, Blaauwbroek N, Nguyen P, Gibescu M., "Local market framework for exploiting flexibility from the end users", *Int. Conf. Europe Energy Market*, Jul. 2016, pp. 1–6.

[16] E. Amicarelli, T. Q. Tran, and S. Bacha, "Flexibility service market for active congestion management of distribution networks using flexible energy resources of microgrids," *2017 IEEE PES Innovative Smart Grid Technologies Conference Europe*, Torino, 2017, pp. 1–6.

[17] A. Esmat, J. Usaola, and M. Á. Moreno, "Distribution-level flexibility market for congestion management", *Energies*, vol. 11, no. 5, Apr. 2018.

[18] A. Esmat, P. Pinson, and J. Usaola, "Decision support program for congestion management using demand side flexibility," in *Proceedings of IEEE Power Technical Conference Manchester*, Manchester, U.K., 2017, pp. 1–6.

[19] A. Esmat, J. Usaola, and M. Á. Moreno, "A decentralized local flexibility market considering the uncertainty of demand", *Energies*, vol. 11, no. 8, Aug. 2018.

[20] P. Olivella-Rosell, P. Lloret-Gallego, Í. Munné-Collado, R. Villafafila-Robles, A. Sumper, S. Ø. Ottessen, J. Rajasekharan, and B. A. Bremdal, "Local flexibility market design for aggregators providing multiple flexibility services at distribution network level", *Energies*, vol. 11, no. 44, Apr. 2018.

[21] S. Boyd, N. Parikh, E. Chu, B. Peleato, and J. Eckstein, "Distributed optimization and statistical learning via the alternating direction method of multipliers," *Foundations and Trends in Machine Learning*, vol. 3, no. 1, pp. 1–122, Jan. 2010.

[22] R. R. Nejad and W. Sun, "Distributed load restoration in unbalanced active distribution systems," *IEEE Trans. Smart Grid*, vol. 10, no. 5, pp. 5759–5769, Sep. 2019.

[23] Q. Peng and S. H. Low, "Distributed optimal power flow algorithm for radial network, I: Balanced single-phase case," *IEEE Trans. Smart Grid*, vol. 9, no. 1, pp. 111–121, Jan. 2018.

[24] Y. Wang, L. Wu, and S. Wang, "A fully-decentralized consensus-based ADMM approach for DC-OPF with demand response," *IEEE Trans. Smart Grid*, vol.8, no. 6, pp. 2637–2647, Nov. 2017.

[25] S. Huang, Q. Wu, Y. Gao, X. Chen, B. Zhou, and C. Li, "Distributed voltage control based on ADMM for large-scale wind farm cluster connected to VSC-HVDC," *IEEE Trans. Sustain. Energy,* vol. 11, no.2, pp. 584–594, Apr. 2020.

[26] W. Zheng, W. Wu, B. Zhang, and C. Lin, "Distributed optimal residential demand response considering operational constraints of unbalanced distribution networks", *IET Gener. Transm. Distrib.,* vol. 12, no. 9, pp. 1970–1979, May 2018.

[27] S. Teimourzadeh, O. B. Tor, M. E. Cebeci, A. Bara, S. V. Opera, S. M. Kisakurek, "Enlightening customers on merits of demand-side load control: a simple-but-efficient-platform", *IEEE Access,* (To be published).

[28] P. Olivella-Rosell, E. Bullich-Massagué, M. Aragüés-Peñalba, A. Sumper, S. Ø. Ottesen, J. A. Vidal-Clos, and R. Villafáfila-Robles, "Optimization problem for meeting distribution system operator requests in local flexibility markets with distributed energy resources", *Applied Energy,* vol. 210, pp. 881–895, Jan. 2018.

[29] U.S. Department of Energy, "Global plug-in light vehicle sales increased by about 80% in 2015," 2016. Available: https://www.energy.gov/eere/vehicles/fact-918-march-28-2016-global-plug-light-vehicle-sales-increased-about-80-2015

[30] European heat pump association, "The European heat pump market has achieved double-digit growth for the fourth year in a row," 2019. Available: https://www.ehpa.org/market-data/

[31] T. Wolf, *"Model-based assessment of heat pump flexibility,"* Master thesis, UPPSALA University, Apr. 2016.

[32] R. Deng, X. Yue, D. Huo, Y. Liu, Y. Huang, C. Huang, and J. Liu, "Exploring flexibility of electrical vehicle aggregators as energy reserve", *Electric Power Systems Research,* vol. 184, pp. 106305, Jul. 2020.

[33] S. Bolognani and S. Zampieri, "On the existence and linear approximation of the power flow solution in power distribution networks", *IEEE Transactions on Power Systems* vol. 31, pp. 163–172, Jan. 2016.

[34] M. Doostizadeh, F. Aminifar, H. Lesani, and H. Ghasemi, "Multi-area market clearing in wind-integrated interconnected power systems: a fast parallel decentralized method", *Energy Conversion and Management,* vol. 113, pp. 131–142, Apr. 2016.

[35] R. N. Allan, R. Billinton, I. Sjarief, L. Goel, and K. S. So, "A reliability test system for educational purposes-basic distribution system data and results," *IEEE Transactions on Power Systems,* vol. 6, no. 2, pp. 813–820, May 1991.

# 9 Decongestion of Active Distribution Grids via D-PMUs-based Reactive Power Control and Electric Vehicle Chargers

*Gabriel E. Mejia-Ruiz[1], Mario R. Arrieta Paternina[1], Juan M. Ramirez[2], Juan R. Rodríguez-Rodríguez[1], Romel Cárdenas-Javier[1], and Alejandro Zamora-Mendez[3]*
[1]National Autonomous University of Mexico, Mexico City, Mexico
[2]CINVESTAV, Jalisco, Mexico
[3]Michoacan University of Saint Nicholas of Hidalgo, Michoacan, Mexico

## CONTENTS

## 9.1   INTRODUCTION

The massive introduction of renewable energy sources (RESs) and electric vehicles (EVs) at the distribution level is playing an essential role in reducing greenhouse gas and polluting emissions [1–8]. Likewise, they impose techno-economic challenges to distribution system operators (DSOs), as the need for additional operational flexibility. In this context, it is necessary to analyze how they can contribute to enhancing the operational flexibility [9]. EV chargers working in the vehicle-to-grid (V2G) mode can extend the benefits of EVs in electrical power systems by providing ancillary services to the grid, such as voltage support through reactive power injection, peak demand reductions, power factor corrections, among others [9–11]. Distribution operators are required to maintain the voltage within ANSI C84.1 and IEC 60038 limits and interconnection reliability operating limits, as shown in Figure 9.1 [12, 13]. To quantify the impact of reactive power injection on the grid voltage regulation, a detailed analysis is needed that considers a proper model of the particular distribution grid. Moreover, a suitable model is necessary to represent the relevant dynamics and modal information of the network [14, 15].

Other technical challenges to DSOs are intermittent and stochastic nature of renewable energy systems, connection of nonlinear loads, and growing consumer demand [16–21]. They also increase the probability of conflict between control actions of the DSOs, the conventional control devices installed in distribution grids (DGs), and new renewable energy generator controllers [10, 22, 23]. EV chargers can increase the operational flexibility through their interactions with the network. Likewise, appropriate charging and discharging control strategies of EVs allow reducing the impact on the DG flexibility of renewable energy output fluctuations [24–29]. Furthermore, the reactive power injection from EV chargers can also be optimally controlled to improve the power quality, grid operation reliability, energy efficiency, and network voltage profile [30–32].

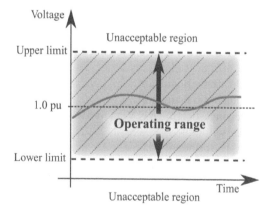

**FIGURE 9.1**   Regions of operation defined by ANSI C84.1 limits [13].

The abovementioned challenges are part of the additional operational flexibility that DGs need to increase so that their operational limits keep within bounds. Flexibility can be defined as the adaptability of generation and load units to rapidly adapt their behavior for supporting DGs based on external command/request signals that establish their behavior. The flexibility offered depends on the physical capabilities of each technology in terms of active and reactive powers, response time, rate of change, and available energy [33, 34].

Nowadays, active DGs are equipped with management systems to simultaneously control distributed energy resources (DERs), loads, and energy storage systems (ESSs) [35–40]. Such ESSs also have potential to provide flexibility supporting DGs [41]. In this context, the availability of flexible resources and dynamic system states must be precisely monitored and automatically controlled at the network level [33]. These tasks are tackled by intelligent and hierarchical control systems, working in coordination with the grid operator, which can optimally handle the energy resources available in the network to regulate the voltage variations and increase the flexibility of the system [10, 42].

In the past, electromechanical devices, such as shunt capacitor banks and under load tap-changers, have been widely used to perform voltage and reactive power control in DGs [43, 44]. However, a slow response is evident to rapid voltage fluctuations. Conversely, electronic power converters can act in short time frames due to their intrinsic nature, protecting sensitive loads and devices against those voltage variations.

As a solution to voltage fluctuations, parking lots composed of EV chargers that work within a coordinated hierarchical control scheme, as shown in Figure 9.2, with the ability to provide reactive power support, become a paramount solution to effectively mitigate changes in the grid voltage profile. Specially, the use of off-board chargers can supply reactive power ancillary services without degrading the batteries' life. The EV charger itself can provide voltage support, even when the EVs are disconnected, using the capability of its own DC-AC converter [45]. Prior to analyzing the impact of the reactive power injection on the grid voltage regulation, a detailed analysis considering the appropriate modeling of the particular DG is required, since this model has to represent the relevant dynamics and modal information [15, 46].

The development of new concepts and technologies that allow increasing the flexibility of modern electrical networks is a current topic. Different works that analyze the potential for flexibility of ESSs have been proposed [42] in topics as response to demand [47]; these are microgrids [48], multigeneration [43], and the interconnection of EVs to the grid [34]. Recently, the interest in providing EV-based voltage control has increased due to the growing availability of the EV chargers with the capacity to inject reactive power to the grid [48, 49].

Most of the previous works are limited to analyzing the grid operation programming in a 24-hour time horizon, obviating the short-time dynamic response. Thus, this work proposes a reactive power and voltage regulation framework for highly unbalanced DGs, using a hierarchical and coordinated closed-loop optimal control scheme by exploiting the eigensystem realization (ER)-based system identification technique and a linear quadratic Gaussian (LQG) control structure [50, 51]. The

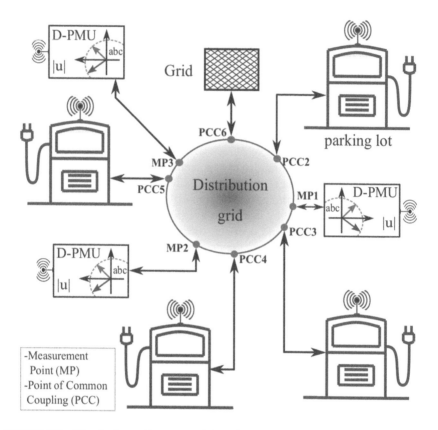

**FIGURE 9.2** Distribution grid concept with ancillary services supplied by electric vehicle battery chargers and remote measurements.

proposed framework incorporates the advance in time-synchronized measurement devices to compensate for the voltage in DSs remotely, injecting reactive power in a controlled manner. Accordingly, the capacity to transport active power is increased, making the network more flexible [52].

Thus, the significant contributions of the proposed approach are pointed out, as follows:

1. The work addresses a new reactive power flow management framework, including the individual injection capacity of every charger along with the DS. Consequently, the proposed system may intelligently handle voltage regulation and decongest the network increasing its flexibility; even when measurements are made at different nodes than the point of common coupling (PCC) between the chargers and the grid.
2. The effectiveness of the proposed control structure is evaluated using a study case with a modified 13-node IEEE test feeder, demonstrating that the proposal can compensate for voltage variations under highly unbalanced conditions in less than 205 ms.

3. The ER-based identification is introduced to identify power distribution networks interfaced with power electronic converters, by exploiting the intelligent phasor measurement units data (named as D-PMUs) and extending its applicability from low-frequency dynamics to higher frequencies [50, 53, 54].

4. The optimal control structure is in charge to face with the voltage compensation under unbalanced conditions, regulating the EV charger input given its multiple input multiple output (MIMO) characteristic, and overcoming the performance of single input single output (SISO) controllers.

The remaining sections of the chapter are structured as follows: The parking lot architecture and the EV battery charger topology together with its control system are described at the device-level in Section 9.3. Then, the ER approach via chirp modulations is presented in Section 9.4. The voltage controller design, based on Bellman's principle, is claimed in Section 9.5. In Section 9.6, the controller performance is tested under unbalanced conditions in distribution networks. Finally, concluding remarks are pointed out in Section 9.7.

## 9.2    HIERARCHICAL CONTROL ARCHITECTURE AND EV CHARGER TOPOLOGY

The high penetration of DERs and expected dynamic loads have imposed that the traditional slow-change DGs gradually turn to multi-source grid with faster dynamics. Consequently, new challenges arise and must be addressed advocating dynamic control structures. Thus, the proposed hierarchical control approach consists of multiple and remote measurement points. The phasor measurement units convey these at the distribution level (D-PMUs) [52, 55, 56], where numerous parking lots at different geographic locations on the grid are monitored (Figure 9.2) to promptly regulate the voltage through the reactive power injection on the network. Therefore, the network flexibility is feasible, thanks to the decongestion in the active power flow.

Thanks to the proposed architecture, a bidirectional information flow between the EV charger and an aggregator equipped with an optimal grid-side controller makes possible the smart management of the reactive power to be injected into the network. In this architecture, the key role of the grid-side controller is to calculate the magnitude of the reactive power required to improve the network voltage profile, based on the remote measurements made by D-PMUs. Subsequently, this controller requests the reactive power to the aggregator from each parking lot by sending operating commands. Then, the aggregator establishes the amount of reactive power requested from each charger based on their maximum operating limits. In this approach, the aggregator acts as an intermediary between the EV chargers and the grid-side controller, reducing the computing load for the control system and DSO. Figure 9.3 remarks the parking lot in the proposed architecture.

The hardware of each EV battery charger, summarized in Figure 9.4, is composed of two primary electronic converters: (i) a three-phase voltage source

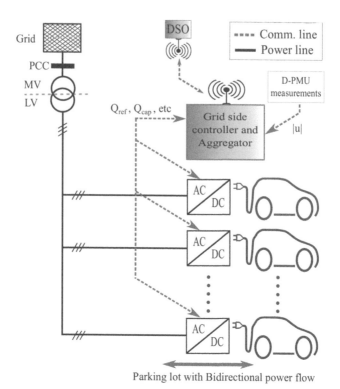

Parking lot with Bidirectional power flow

**FIGURE 9.3**   Parking lot architecture with electric vehicle battery chargers and D-PMUs.

converter (VSC) and (ii) an isolated dual active bridge (DAB) converter. These topologies have been extensively studied by the scientific community, allowing to operate in the four quadrants of $P-Q$ plane and enabling the bidirectional power flow [57–59]. In this topology, the VSC enables the active and reactive power exchange between the DC link capacitor $(C)$ and the utility grid. DAB regulates active power flow between DC link and battery pack inside the EVs. The ripple on the current injected into the network at the PCC is reduced by the inductance filter $(L)$ [59].

The VSC and DAB control system is achieved using PI controllers (Figure 9.4). This system considers the following charger operation modes: V2G and grid-to-vehicle (G2V). Battery charges employ both constant current and constant voltage (CC-CV) modes. VSC is controlled in the $d-q$ reference frame, regulating the direct current $(i_d)$ and the quadrature current $(i_q)$. This controller synchronizes the charger with the grid through the PLL. Meanwhile, the DAB power flow is regulated by two controllers: the first regulates the charge or discharge current $(i_{bat})$ and the second controls the battery voltage $(v_{bat})$. DAB and VSC average small-signal models are derived from refs. [57–59]. These models are defined in (9.1) and (9.2), respectively, enabling to tune the PI controllers' parameters inside the charger via the pole location method.

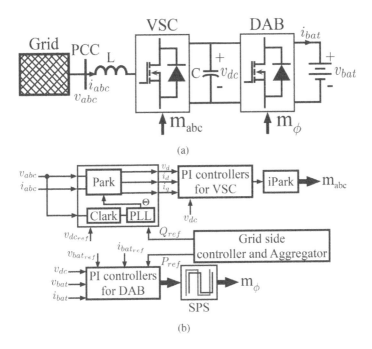

(a)

(b)

**FIGURE 9.4** Diagram of the electric vehicle (EV) battery chargers and its control architecture. (a) VSC-DAB-based EV Battery charger. (b) PI controllers for VSC and DAB converters.

$$
\frac{d}{dt}
\begin{bmatrix}
\Delta v_o^0 \\
\Delta i_t^{1R} \\
\Delta i_t^{1I}
\end{bmatrix}
=
\begin{bmatrix}
\dfrac{-1}{RC} & \dfrac{-4\sin(D\pi)}{\pi C} & \dfrac{-4\cos(D\pi)}{\pi C} \\
\dfrac{2\sin(D\pi)}{\pi L_t} & \dfrac{-R_t}{L_t} & w_s \\
\dfrac{2\cos(D\pi)}{\pi L_t} & -w_s & \dfrac{-R_t}{L_t}
\end{bmatrix}
\begin{bmatrix}
\Delta v_o^0 \\
\Delta i_t^{1R} \\
\Delta i_t^{1I}
\end{bmatrix}
+
$$

$$
\begin{bmatrix}
\dfrac{4}{C}\left(I_0^I \sin(\pi D) - I_0^R \cos(\pi D)\right) \\
\dfrac{2V_o^0}{L_t}\cos(\pi D) \\
\dfrac{-2V_o^0}{L_t}\sin(\pi D)
\end{bmatrix}
\Delta d
$$

(9.1)

$$
\frac{di_d}{dt} = Lw_0 i_q - R i_d - \hat{v}_{sd} + v_{td}
$$

$$
\frac{di_q}{dt} = -Lw_0 i_d - R i_q - \hat{v}_{sq} + v_{tq}
$$

(9.2)

## 9.3   EIGENSYSTEM REALIZATION MODEL VIA CHIRP MODULATIONS FOR DISTRIBUTION GRIDS

This work adopts the ER-based system identification technique to estimate the linear model representation for DGs. To this end, voltage magnitudes stemming from different D-PMUs in Figure 9.2 are established as the output sequence; meanwhile, the input sequence is established by the exponential chirp function that modulates the reference reactive power ($Q_{ref}$) at every parking lot. Therefore, its main foundations are described in the following.

To perform ER [60], a system identification problem is stated, taking into consideration input and output sequences. The input sequence is assumed as known $u(0), u(1), \cdots, u(N)$. The corresponding output sequence stems from time-simulations or even actual raw data from the distribution system response.

For multiple measurement channels [53], the output sequence is shaped in matrix form, such that each $m$ column arrays correspond to single channels, as follows:

$$\mathbf{Y}_m = [\mathbf{y}^{\{1\}} \, \mathbf{y}^{\{2\}} \cdots \mathbf{y}^{\{q\}} \cdots \mathbf{y}^{\{m\}}] \tag{9.3}$$

where, $\mathbf{Y}_m \in \mathfrak{R}^{N \times m}$ and the $q-th$ column is represented by $\mathbf{y}^{\{q\}} = [y(0) \, y(1) \cdots y(N-1)]^T$. Thus, the multivariate representation is defined by:

$$\begin{bmatrix} \mathbf{Y_0} \\ \mathbf{Y_1} \\ \vdots \\ \mathbf{Y_k} \end{bmatrix} = \begin{bmatrix} [y_0^{\{1\}} \ y_0^{\{2\}} \ \cdots \ y_0^{\{m\}}]^T \\ [y_1^{\{1\}} \ y_1^{\{2\}} \ \cdots \ y_1^{\{m\}}]^T \\ \vdots \quad \vdots \quad \vdots \quad \vdots \\ [y_{N-1}^{\{1\}} \ y_{N-1}^{\{2\}} \ \cdots \ y_{N-1}^{\{m\}}]^T \end{bmatrix} \tag{9.4}$$

Thus, the input/output measurement pairs allow expressing the output sequence for multiple channels as:

$$\begin{aligned} \mathbf{Y}_0 &= \tilde{\mathbf{D}} \\ \mathbf{Y}_1 &= \tilde{\mathbf{C}}\tilde{\mathbf{B}} \\ \mathbf{Y}_2 &= \tilde{\mathbf{C}}\tilde{\mathbf{A}}\tilde{\mathbf{B}} \\ &\vdots \\ \mathbf{Y}_{N-1} &= \tilde{\mathbf{C}}\tilde{\mathbf{A}}^{N-1}\tilde{\mathbf{B}} \end{aligned} \tag{9.5}$$

It is noteworthy that (9.5) represents the output sequences in terms of the Markov parameters for multiple channels which are termed as $\tilde{\mathbf{A}}$, $\tilde{\mathbf{B}}$, $\tilde{\mathbf{C}}$, and $\tilde{\mathbf{D}}$. Thereby, a Hankel matrix can be stated for multiple output channels as $\tilde{\mathbf{H}}(k) = \tilde{\xi}\tilde{\mathbf{A}}^{k-1}\tilde{\mathbf{B}}$. Afterwards, the block Hankel matrix becomes:

$$\tilde{\mathbf{H}}(k) = \begin{bmatrix} \mathbf{Y}_k & \mathbf{Y}_{k+1} & \cdots & \mathbf{Y}_{k+N} \\ \mathbf{Y}_{k+1} & \mathbf{Y}_{k+2} & \cdots & \mathbf{Y}_{k+N+1} \\ \vdots & \vdots & \ddots & \vdots \\ \mathbf{Y}_{k+N} & \mathbf{Y}_{k+N+1} & \cdots & \mathbf{Y}_{k+2N} \end{bmatrix} \tag{9.6}$$

By assuming $k = 1$ and $k = 2$ into (9.6), the Hankel matrices $\tilde{\mathbf{H}}(1)$ and $\tilde{\mathbf{H}}(2)$ are derived as:

$$\tilde{\mathbf{H}}(1) = \tilde{\xi}\tilde{\mathbf{B}}$$

$$\tilde{\mathbf{H}}(2) = \tilde{\xi}\tilde{\mathbf{A}}\tilde{\mathbf{B}}$$

(9.7)

Now, $\tilde{\mathbf{B}}$ can be obtained from $\tilde{\mathbf{H}}(1) \in \mathfrak{R}^{m(N/2-1)\times(N/2-1)}$, and $\tilde{\mathbf{A}}$ is derived from $\tilde{\mathbf{H}}(2) \in \mathfrak{R}^{m(N/2-1)\times(N/2-1)}$. Therefore, the Markov parameters for multiple output channels have the following form:

$$\tilde{\mathbf{A}} = \tilde{\mathbf{S}}^{-1/2}\tilde{\mathbf{P}}^T\tilde{\mathbf{H}}(2)\tilde{\mathbf{Q}}\tilde{\mathbf{S}}^{-1/2}$$

$$\tilde{\mathbf{B}} = \tilde{\mathbf{S}}^{1/2}\tilde{\mathbf{Q}}^T$$

$$\tilde{\mathbf{C}} = \tilde{\mathbf{P}}\tilde{\mathbf{S}}^{1/2}$$

$$\tilde{\mathbf{D}} = \mathbf{Y}_0$$

(9.8)

Until now, the distribution grid linear model represented for matrices $\tilde{\mathbf{A}}, \tilde{\mathbf{B}}, \tilde{\mathbf{C}}$, and $\tilde{\mathbf{D}}$ can be identified for multiple voltage channels coming from different locations, thanks to the availability of time-synchronize voltage measurements provided by D-PMUs, Figure 9.2.

### 9.3.1 Probing Signals for Input Modulations

In this section, the input sequence is rendered by assuming probing signals represented by an exponential chirp function that modulates the reactive reference power ($Q_{ref}$) at every parking lot. Thereby, the chirp function allows making a frequency sweep [61]. The probing signal stimulates the EV charger dynamics and is defined by:

$$u_i(t) = \alpha_i \sin\left(\frac{2\pi f_s(r_f^t - 1)}{\ln(r_f)}\right)$$

(9.9)

$$r_f = \left(\frac{f_e}{f_s}\right)^{1/T}$$

(9.10)

where, $\alpha_i$ is the amplitude, $T$ is the lasting signal time, $f_s$ and $f_e$ are respectively the starting and ending frequencies of the chirp signal.

### 9.3.2 Output Sequence Data Preparation

A time output sequence represented by the synchrophasor voltage magnitudes is required by the ER formulation, which is generated via time-domain simulations in response to the chirp modulations. That is, the system response is captured after the steady state is reached. Afterwards, Fourier spectra are applied to the output sequence signals, fft($y_m$), resulting in the frequency response $Y_m(s)$, for the $m$th output signal. Finally, a time output sequence per signal correlated with the $i$th input is obtained taking the inverse Fourier transform of the impulse response as [61]:

$$y_m(t) = F^{-1}\left(\frac{Y_m(s)}{U_i(s)}\right) \tag{9.11}$$

where, $U_i(s) = $ fft($u_i$); F = Fourier transform.

Once, the distribution system linear model is drawn, this is embedded into the optimal control strategy presented in Section 9.5.

## 9.4  VOLTAGE CONTROLLER DESIGN VIA EIGENSYSTEM REALIZATION AND OPTIMAL CONTROL

Once the system identification is carried out, the identified linear system is embedded into a linear control structure that is driven by an LQG controller aiming to regulate the reactive power supplied by the EV charger. This control has high flexibility for adapting to the state-space system in (9.8). It is composed of a linear quadratic regulator (LQR) that compensates the voltage droop/swell, taking into consideration the distribution system model drawn by the ER method through a frequency sweep in a selective frequency range. Thus, the Bellman's principle of optimality that makes up the LQR design is tackled in the following. Likewise, the LQG structure has an estimation stage known as the linear quadratic estimator (LQE). In the next section, both LQR and LQE are briefly discussed.

### 9.4.1  LINEAR QUADRATIC REGULATOR

The optimal formulation seeks that the LQR controller generates an optimal state-feedback control vector for regulating the deviation in the dynamic response of each state and output variables after disturbance in the infinite time-continuous system in (9.8). This is accomplished minimizing the performance function $J(u)$ whose optimal command that complies with the Bellman's principle is given by $u^*(x(t),t)$:

$$J(u) = \lim_{P \to +\infty} \frac{1}{2} \int_0^P \left(x_t^T Q_c x_t + u_t^T R_c u_t\right) dt \tag{9.12}$$

where, $Q_c$ is a positive semidefinite, i.e., $Q_c \geq 0$, and $R_c$ is a positive definite, i.e., $R_c > 0$. Both weighting matrices of LQR ensure a unique optimal solution. A proper selection of these matrices allows affecting the output and control signals giving them a degree of importance. Therefore, this criterion in (9.12) calculates the optimal command $u^*$ to minimize the states controlling the output $y$; it also decreases the effort to achieve this goal by reducing the amount of control.

Since a pair $(A, B)$ is controllable if and only if the pair $(A - BK)$ is controllable, the following optimal state-feedback control vector is defined:

$$u^* = -Kx \tag{9.13}$$

where, the optimal control gain $\mathbf{K}$ is defined as follows:

$$\mathbf{K} = \mathbf{R_c}^{-1}\mathbf{B}^T\mathbf{P_c} \tag{9.14}$$

and $\mathbf{P_c}$ is a unique symmetric and positive semidefinite solution which is obtained in the design of optimal control system by employing the algebraic Riccati equation (ARE) given by:

$$\mathbf{A}^T\mathbf{P_c} + \mathbf{P_c}\mathbf{A} - \mathbf{P_c}\mathbf{B}\mathbf{R_c}^{-1}\mathbf{B}^T\mathbf{P_c} + \mathbf{Q_c} = 0 \tag{9.15}$$

Then, the closed loop system $(\mathbf{A} - \mathbf{BK})$ is stable.

### 9.4.2 LINEAR QUADRATIC ESTIMATOR

It is not realistic to measure precisely all states $\mathbf{x}$. Sometimes they are not observable, or even it is only too expensive to measure them. Thus, one solution to cope with this problem and apply the optimal control $\mathbf{u} = -\mathbf{Kx}$ in (9.13) is to add an estimation of $\mathbf{x}$ in the control loop. Dealing with the proposed LQR presented in Section 9.5.1, the so-called LQE, based on the Kalman filter theory, is recommended.

Now let's consider the following state-space system hinged on the estimated states $\hat{\mathbf{x}}$:

$$\begin{cases} \dot{\hat{\mathbf{x}}} = \mathbf{A}\hat{\mathbf{x}} + \mathbf{Bu} + \mathbf{G}(\mathbf{y} - \hat{\mathbf{y}}) \\ \hat{\mathbf{y}} = \mathbf{C}\hat{\mathbf{x}} + \mathbf{Du} \end{cases} \tag{9.16}$$

which corresponds to a simulation $\dot{\hat{\mathbf{x}}} = \mathbf{A}\hat{\mathbf{x}} + \mathbf{Bu}$ corrected by the difference between estimated and measured output, weighted by the Kalman gain $\mathbf{G}$.

Then, the estimation error $\mathbf{e}(t) = \mathbf{x}(t) - \hat{\mathbf{x}}(t)$ can be introduced and the dynamic error $\dot{\mathbf{e}}(t)$ can be written as:

$$\dot{\mathbf{e}}(t) = \dot{\mathbf{x}}(t) - \dot{\hat{\mathbf{x}}}(t)(\mathbf{A} - \mathbf{GC})\mathbf{e}(t) \tag{9.17}$$

Thus, if the Kalman gain $\mathbf{G}$ is chosen such that the system $(\mathbf{A} - \mathbf{GC})$ is asymptotically stable, then $\mathbf{e}(t)$ will tend to 0.

Solving this problem means to find the optimal command to stabilize the system:

$$\dot{\mathbf{z}} = \mathbf{A}^T\mathbf{z} + \mathbf{C}^T\mathbf{h} \tag{9.18}$$

It has been proved in Section 9.5.1 that the optimal control $\mathbf{h}^*$ is calculated for $\mathbf{h}^* = -\mathbf{Gz}$ and the following analogies with Section 9.5.1 can be done:

$$\begin{aligned} \mathbf{A} &\rightarrow \mathbf{A}^T \\ \mathbf{B} &\rightarrow \mathbf{C}^T \\ \mathbf{K} &\rightarrow \mathbf{G}^T \end{aligned} \tag{9.19}$$

Thus, the optimal Kalman gain is obtained through:

$$\mathbf{G} = \mathbf{P}_o \mathbf{C}^T \mathbf{R}_o^{-1} \tag{9.20}$$

where, $\mathbf{P}_o$ is the semidefined solution of the ARE and $\mathbf{R}_0$ is one of the weight matrices appropriately chosen for the estimation.

## 9.5 ACTIVE POWER DECONGESTION IN DGs VIA D-PMUs AND EV CHARGERS

The straightforward implementation for the coordinated hierarchical voltage regulation control scheme that interfaces the DS in Figure 9.2 with the parking lot architecture in Figure 9.3 is established through the block diagram in Figure 9.5. The control inputs of EV chargers are feedback by the control actions that respond to the voltage magnitude variations measured by the D-PMUs, providing a remotely controlled system. The system identification provided by the ER method is attained running one offline simulation, selecting a frequency range from 1–100 Hz for the chirp function in (9.9)–(9.10) in turn, this is used as a probing signal to modulate the reference reactive power for every parking lot. Then, the LQG design is conducted by solving the discrete-time ARE to achieve the state feedback gain ($\mathbf{K}$) and Kalman gain ($\mathbf{G}$) matrices in (9.14) and (9.20), respectively. This is carried out using the Markov parameters provided by the ER-based identification via chirp modulations and the Matlab function *dlqr*. Dynamic time-domain simulations are performed in the Matlab & Simulink™ environment.

### 9.5.1 SYSTEM DESCRIPTION

The IEEE 13-node test feeder depicted in Figure 9.6 is used to demonstrate the feasibility and effectiveness of the proposed hierarchical control structure. The system is implemented in ref [62].

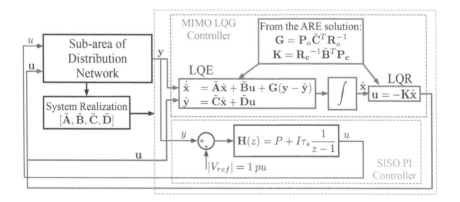

**FIGURE 9.5** Block diagram for the voltage controller implementation based on eigensystem realization and linear quadratic Gaussian.

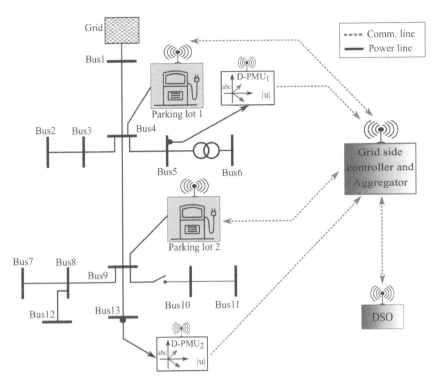

**FIGURE 9.6** IEEE 13-node test feeder in *Scenarios 1–3* [8].

As stated in Section 9.3, every charger is equipped with a VSC and a DAB as displayed in Figure 9.4, whose PI controllers are tuned by using the pole location method. Both VSC and DAB are implemented in accordance with the parameters in Table 9.1, yielding in Figure 9.7 the waveforms corresponding to the voltages, currents, and powers of the EV battery charger under changes in their references for

**TABLE 9.1**

**Simulation Parameters**

| Parameter | Value |
|---|---|
| Nominal voltage and rated battery capacity | 380 V, 5.4 × 70 Ah |
| DC link reference voltage ($v_{dcref}$) | 800 V |
| DC link capacitor bank ($C$) | 3600$\mu$F |
| DAB leakage inductance ($L_t$) | 1.5 $\mu$H |
| DAB transformer ratio | 400/800 V$_{rms}$ |
| VSC inductance filter ($L$) | 600 $\mu$H X3 |
| Grid side transformer ratio | 480/4160 V$_{rms}$ |
| Simulation sample time | 50$\mu$s |

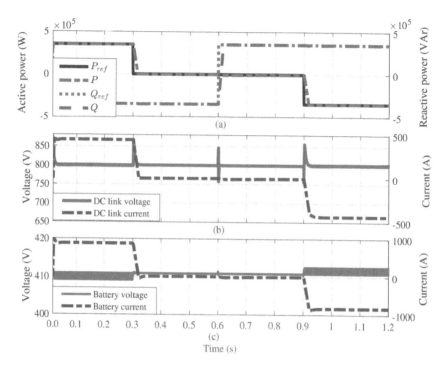

**FIGURE 9.7** Waveforms of voltages, currents, and powers for the electric vehicle battery charger.

reactive $Q_{ref}$ and active $P_{ref}$ powers. These changes are illustrated in Figure 9.7(a), where the PI controllers track the changes following the reference requested by the aggregator, injecting both powers into the PCC. The step-type changes in $P_{ref}$ exhibit the charge and discharge of the battery bank. Similarly, the step-type changes in $Q_{ref}$ confirm the charger's ability to inject inductive and capacitive reactive powers, working in the four quadrants of the $P-Q$ plane. Meanwhile, the currents at the DC bus ($i_{dc}$) and at the battery ($i_{bat}$) proportionally change with respect to the active power, as shown in Figure 9.7(b, c). Likewise, the VSC controller regulates the voltage on the DC link, keeping it around the point of operation (800 V) (Figure 9.7(b)).

For the simulation analysis, four scenarios are managed. They include two D-PMUs and two parking lots. In these cases, the switch installed between bus 9 and bus 10 closes at 0.5 s, adding the loads connected on buses 10 and 11 to the network. The parking lot 1 can inject up to 5.2 MVA (upon request) based on the connection of 15 EV chargers in shunt connection. Similarly, the parking lot 2 comprises 10 EV chargers with the capacity to provide up to 3.5 MVA (upon request). Each charger installed at the parking lots can supply 350 kVA at 420 $V_{LL}$. The parameters for each charger are presented in Table 9.1. At the PCC, a step-up power transformer adapts the voltage levels from 420 $V_{LL}$ up to 4160 $V_{LL}$, between the parking lots and the network feeder. As a reference, V3 supercharger from Tesla Motors can supply between

250 and 350kW, [63]. Similarly, the manufacturer ABB offers high-power chargers up to 350 kW [64].

### 9.5.2 SCENARIOS DESCRIPTION

*Scenario 1*: In this case, both parking lots and both D-PMUs are used as actuators and sensors for the control system, respectively (Figure 9.6). Simulation is performed with the MIMO LQG controller acting in closed-loop with the network in comparison with the system response in open loop.

*Scenario 2*: A comparison of the performance between PI and SISO LQG controllers closing the control loop with respect to the system response in open loop is devoted in this scenario. Parking lot 1 and D-PMU$_2$ represent the actuator and the sensor for the control system, respectively. Parking lot 2 and D-PMU$_1$ are not considered in this case (Figure 9.6).

*Scenario 3*: This case includes both parking lots and both D-PMUs. The simulation is carried out with the MIMO LQG controller, and it is contrasted with the open-loop system response.

*Scenario 4*: In this case, the resistive load connected at node 9 is increased linearly. The simulation is performed with the open-loop controller and in closed loop with the MIMO LQG controller.

In *Scenarios 2, 3*, and *4*, the load connected at bus 10 is increased 20 times to verify the efficiency of the proposed control system for regulating the network voltage in the presence of extreme load disturbances.

### 9.5.3 RESULTS

*Scenario 1*: The network voltage profile for the conditions in *Scenario 1* is presented in Figure 9.8. When the system operates in open loop, all buses except the source indicate lower voltage amplitudes than one suggested by the standard ANSI C84.1 [13], as exhibited by the dotted red line. When the closed-loop controller is enabled, the voltage profile improves significantly, settling its amplitude within the limits established by the standard ANSI C84.1 [13]. In fact, the voltage amplitude almost attains 1 pu at the smart measurement location (D-PMUs installed at nodes 5 and 13).

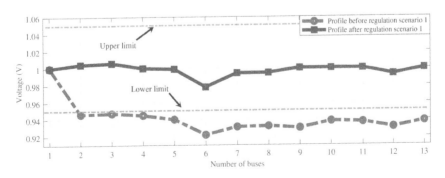

**FIGURE 9.8** Voltage profile in *Scenario 1* for the closed-loop with MIMO LQG controller and open-loop system responses.

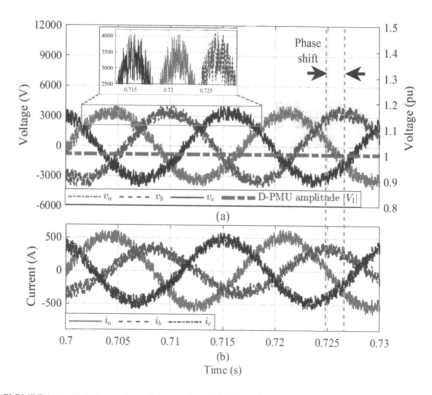

**FIGURE 9.9** Unbalanced conditions at bus 9. (a) Input/output D-PMU voltages. (b) Currents.

The proposed hierarchical controller can improve the voltage profile of the distribution network, even in the presence of unbalanced voltages/currents, as illustrated in Figure 9.9. Both measurement channels corresponding to the D-PMU₁ are polluted by adding a uniform noise of amplitude between ±10% to corroborate the noise tolerance in the proposed approach. The embedded finite impulse response filter in D-PMU₁ effectively rejects the electrical noise and the measurement signal used in the control system is noise free, as depicted the voltage amplitude by the dotted green line in Figure 9.9. The unbalance voltages and phase shift between current and voltage in phase C, are due to the unbalanced reactive loads connected along the network. However, the maximum variation of the voltage magnitudes is 0.93% (recalling that 3% is the upper limit reported in the ANSI C84.1 norm [13]).

Figure 9.10 points out the reactive power profile for all buses in the network. The reactive power is injected at nodes 4 and 9, as depicted in Figure 9.6, causing the reactive power in those buses to become greater than that of the other buses. Simultaneously, the controlled injection of reactive power increases the network capacity to distribute active power, improving the power factor, as indicated at buses 4 and 9 in Figures 9.11 and 9.12, and implying active power decongestion.

*Scenarios 2 and 3:* The dynamic response comparison for the PI and the SISO LQG controllers concerning the DS open-loop response, with a significant change in

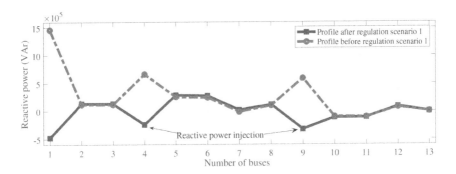

**FIGURE 9.10** Reactive power profile before and after regulation in *Scenario 1*.

load, is displayed in Figure 9.13. At 0.5 s, the breaker is closed changing the operating conditions of the network by adding capacity. Without a controller, the system operates outside the reliability enforcing limits established in the ANSI C84.1 standard (below 0.95 pu) [13], as shown by the black dotted line measuring at bus 13. At full load, a voltage drop of 14% is noticed. In contrast, the closed-loop system responses, when the PI and LQG controllers are incorporated, illustrated by the black and blue lines, respectively. With both controllers, the system returns to the desired operating value (1 pu). Nonetheless, the LQG controller exhibits a faster response than the PI controller. The LQG controller takes 205 ms to reach the steady state; in comparison, the PI controller takes 445 ms. Simultaneously, the controlled reactive power injection increases the capacity in the network to distribute active power, improving the power factor, as shown at buses 4 and 9 of Figure 9.14.

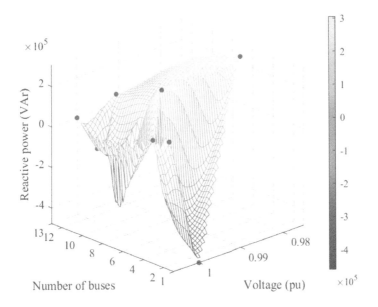

**FIGURE 9.11** Reactive power impacts on the voltage profile in *Scenario 1*.

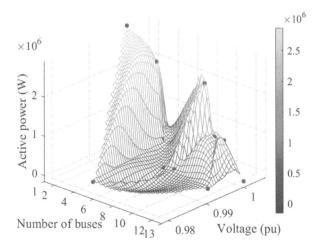

**FIGURE 9.12** Active power decongestion in *Scenario 1*.

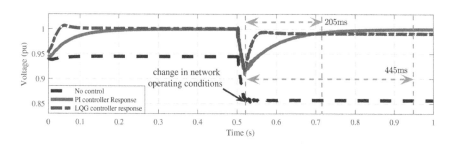

**FIGURE 9.13** Voltage profile dynamic response in *Scenario 2* for the DS equipped with PI and LQG controllers, and the open-loop system response.

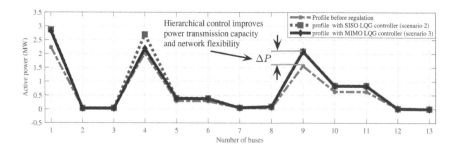

**FIGURE 9.14** Active power profile for *Scenarios 2* and *3* with respect to the open-loop system response.

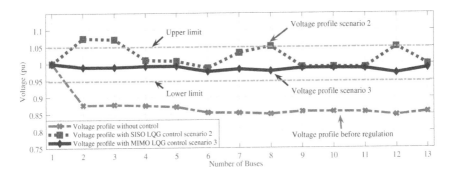

**FIGURE 9.15**  Voltage profile for *Scenarios* 2 and 3 with respect to the open-loop system response.

The voltage and reactive power at all buses during the highest load condition are depicted in Figures 9.15 and 9.16, respectively. Under open-loop system conditions (red dotted line), the voltage profile throughout the network is below the limits established by the ANSI C84.1 standard [13], exhibiting high levels of reactive power at bus 9. This is a consequence of the connected loads on the adjacent branches. Whereas, the voltage profile in *Scenarios* 2 and 3, when the system operates in closed loop with LQG controllers, is ostensibly improved by injecting reactive power from both parking lots. However, the LQG controller in case 2 only injects reactive power into the node 4 supporting the voltage magnitude. The MIMO control strategy used in *Scenario* 3 is more useful to improve the network voltage profile (about 1 pu), injecting reactive power at buses 4 and 9.

*Scenario* 4: In this case, the power absorbed by the resistive load connected to bus 9 is linearly increased from 1 kW to 7 MW to analyze grid flexibility via the network decongestion in open loop and closed loop with the MIMO LQG controller. Such variations along the transferred power between nodes 4 and 9, and the voltage level at node 9 with respect to the change in resistive load connected at node 9, are closely analyzed in Figure 9.17. In the open-loop behavior, the voltage level is reduced to 0.51 pu, whereas the load increases up to 6 MW, affecting the voltage profile of the entire network. Meanwhile, in closed loop, the voltage level remains

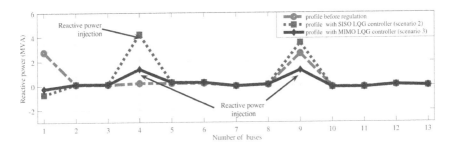

**FIGURE 9.16**  Reactive power behavior for *Scenarios* 2 and 3 with respect to the open-loop system response.

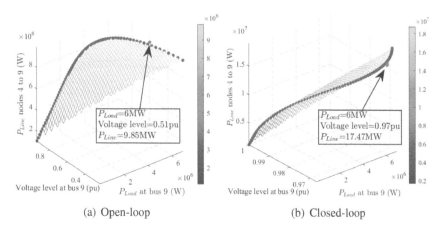

(a) Open-loop                                    (b) Closed-loop

**FIGURE 9.17** Bus 9 active power decongestion comparison in open loop vs. closed loop *Scenario 4.*

close to 1 pu and the active power transmission capacity is significantly increased, i.e., the active power transported by the line increases from 9.85 to 17.47 MW for the same load level, which in turn enables to quantitatively increasing the flexibility of the distribution grid.

Similar results are observed in the active power flowing between nodes 1 and 4, and the voltage at node 4 (Figure 9.18). In the open-loop condition, the voltage level is reduced to 0.77 pu, when the load increases up to 5.5 MW. The closed-loop system does that the voltage behaves close to 1 pu. In this test, the capacity to transport active power increases from 10.3 MW to 16.5 MW, when the load increases up to 5.5 MW.

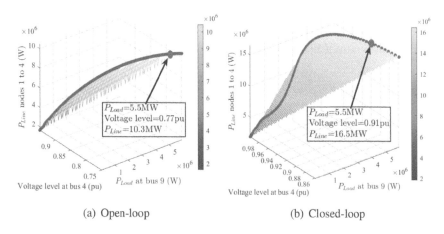

(a) Open-loop                                    (b) Closed-loop

**FIGURE 9.18** Bus 4 active power decongestion comparison in open loop vs. closed loop *Scenario 4.*

## 9.6 CONCLUSIONS

In this chapter, a new Volt/Var control framework for the operation of distribution systems is proposed. This framework implements a hierarchical control structure that coordinates the injection of reactive power from the EV chargers grouped in parking lots to improve the voltage profile and provide ancillary services to the distribution networks. Each charger is cooperatively controlled by an aggregator that coordinates the injection of reactive power, reducing the information flow that must be handled by the controller and the DSO. The control structure exploits the ER-based system identification, an optimal control structure hinged on the Bellman's principle of optimality and remote measurements from D-PMU devices. It is confirmed that the proposed technique contributes to voltage support of the power distribution system, even in the presence of significant disturbances of up to 14%, recovering the voltage profile in less than 205 ms, and protecting sensitive equipment connected to the power distribution network.

The proposed hierarchical controller provides the following benefits to the network: (i) it can regulate the voltage at all grid buses during extreme operating conditions; and (ii) it expands the capacity to transport active power, making the network more flexible. These features are achieved even when the control system is configured with measurements, the controller and parking lots located at different buses, and distant geographical locations.

## ACKNOWLEDGMENTS

Gabriel E. Mejia-Ruiz thanks to the graduated program in Engineering-Doctorate at the National Autonomous University of Mexico and Conacyt for the scholarship under grant (1044979). M.R.A. Paternina and A. Zamora acknowledge financial support from the Project Support Program for Research and Technological Innovation of UNAM (DGAPA, PAPIIT-2021) through the project TA101421. Also, they gratefully thank to the CEMIE-Redes, strategic project PE-A-04.

## REFERENCES

[1] S. Amamra and J. Marco, "Vehicle-to-grid aggregator to support power grid and reduce electric vehicle charging cost," *IEEE Access*, vol. 7, pp. 178528–178538, 2019.

[2] J. P. Cerdeira Bento and V. Moutinho, "$CO_2$ emissions, non-renewable and renewable electricity production, economic growth, and international trade in italy," *Renewable and Sustainable Energy Reviews*, vol. 55, pp. 142–155, 2016.

[3] N. Y. Amponsah, M. Troldborg, B. Kington, I. Aalders, and R. L. Hough, "Greenhouse gas emissions from renewable energy sources: A review of lifecycle considerations," *Renewable and Sustainable Energy Reviews*, vol. 39, pp. 461–475, 2014.

[4] Y. M. Atwa, E. F. El-Saadany, M. M. A. Salama, and R. Seethapathy, "Optimal renewable resources mix for distribution system energy loss minimization," *IEEE Transactions on Power Systems*, vol. 25, no. 1, pp. 360–370, 2010.

[5] A. K. Banhidarah and A. S. Al-Sumaiti, "Heuristic search algorithms for optimal locations and sizing of distributed generators in the grid: A brief recent review," in 2018 Advances in Science and Engineering Technology International Conferences (ASET), pp. 1–5, 2018.

[6] A. S. Al-Sumaiti, "The role of regulation in the economic evaluation of renewable energy investments in developing countries," in 2013 7th IEEE GCC Conference and Exhibition (GCC), pp. 39–43, 2013.

[7] A. S. Al-Sumaiti and M. M. A. Salama, "Review on issues related to electric energy demand in distribution system for developing countries," in 3rd IET International Conference on Clean Energy and Technology (CEAT) 2014, pp. 1–6, 2014.

[8] R. Kommalapati, A. Kadiyala, M. T. Shahriar, and Z. Huque, "Review of the life cycle greenhouse gas emissions from different photovoltaic and concentrating solar power electricity generation systems," *Energies*, vol. 10, no. 3, 2017.

[9] V. Calderaro, V. Galdi, F. Lamberti, and A. Piccolo, "A smart strategy for voltage control ancillary service in distribution networks," *IEEE Trans. Power Systems*, vol. 30, no. 1, pp. 494–502, 2015.

[10] J. Wang, G. R. Bharati, S. Paudyal, O. Ceylan, B. P. Bhattarai, and K. S. Myers, "Coordinated electric vehicle charging with reactive power support to distribution grids," *IEEE Trans. Ind. Inform.*, vol. 15, no. 1, pp. 54–63, 2019.

[11] A. Kulmala, S. Repo, and P. Jrventausta, "Coordinated voltage control in distribution networks including several distributed energy resources," *IEEE Trans. Smart Grid*, vol. 5, no. 4, pp. 2010–2020, 2014.

[12] I. Voltages, "Iec standard iec 60038," *Ed*, 2009.

[13] N. E. M. Association *et al.*, "American national standards institute (ansi) c84.1, voltage ratings for electric power systems and equipment," *Rosslyn, VA*, 2016.

[14] S. C. Dhulipala, R. V. A. Monteiro, R. F. d. Silva Teixeira, C. Ruben, A. S. Bretas, and G. C. Guimares, "Distributed model-predictive control strategy for distribution network volt/var control: A smart-building-based approach," *IEEE Trans. Indus. Applic.*, vol. 55, no. 6, pp. 7041–7051, 2019.

[15] G. Cavraro and R. Carli, "Local and distributed voltage control algorithms in distribution networks," *IEEE Trans. Power Systems*, vol. 33, no. 2, pp. 1420–1430, 2018.

[16] Y. M. Atwa and E. F. El-Saadany, "Probabilistic approach for optimal allocation of wind-based distributed generation in distribution systems," *IET Renewable Power Generation*, vol. 5, no. 1, pp. 79–88, 2011.

[17] A. S. Al-Sumaiti, M. H. Ahmed, S. Rivera, M. S. El Moursi, M. M. A. Salama, and T. Alsumaiti, "Stochastic pv model for power system planning applications," *IET Renewable Power Generation*, vol. 13, no. 16, pp. 3168–3179, 2019.

[18] A. S. Al-Sumaiti, M. M. Salama, and M. El-Moursi, "Enabling electricity access in developing countries: A probabilistic weather driven house based approach," *Applied Energy*, vol. 191, pp. 531–548, 2017.

[19] A. S. Al-Sumaiti, M. Salama, M. El-Moursi, T. S. Alsumaiti, and M. Marzband, "Enabling electricity access: revisiting load models for ac-grid operation - part i," *IET Generation, Transmission Distribution*, vol. 13, no. 12, pp. 2563–2571, 2019.

[20] Y. M. Atwa and E. F. El-Saadany, "Probabilistic approach for optimal allocation of wind-based distributed generation in distribution systems," *IET Renewable Power Generation*, vol. 5, no. 1, pp. 79–88, 2011.

[21] A. S. Al-Sumaiti, M. H. Ahmed, S. Rivera, M. S. El Moursi, M. M. A. Salama, and T. Alsumaiti, "Stochastic pv model for power system planning applications," *IET Renewable Power Generation*, vol. 13, no. 16, pp. 3168–3179, 2019.

[22] O. P. Mahela, B. Khan, H. H. Alhelou, and P. Siano, "Power quality assessment and event detection in distribution network with wind energy penetration using stockwell transform and fuzzy clustering," *IEEE Transactions on Industrial Informatics*, vol. 16, no. 11, pp. 6922–6932, 2020.

[23] T. Strasser, F. Andrn, J. Kathan, C. Cecati, C. Buccella, P. Siano, P. Leito, G. Zhabelova, V. Vyatkin, P. Vrba, and V. Mak, "A review of architectures and concepts for intelligence

in future electric energy systems," *IEEE Transactions on Industrial Electronics*, vol. 62, no. 4, pp. 2424–2438, 2015.

[24] L. Gan, U. Topcu, and S. H. Low, "Optimal decentralized protocol for electric vehicle charging," *IEEE Transactions on Power Systems*, vol. 28, no. 2, pp. 940–951, 2013.

[25] Y. M. Atwa and E. F. El-Saadany, "Optimal allocation of ess in distribution systems with a high penetration of wind energy," *IEEE Transactions on Power Systems*, vol. 25, no. 4, pp. 1815–1822, 2010.

[26] Y. M. Atwa and E. F. El-Saadany, "Optimal allocation of ess in distribution systems with a high penetration of wind energy," *IEEE Transactions on Power Systems*, vol. 25, no. 4, pp. 1815–1822, 2010.

[27] L. Gan, U. Topcu, and S. H. Low, "Optimal decentralized protocol for electric vehicle charging," *IEEE Transactions on Power Systems*, vol. 28, no. 2, pp. 940–951, 2013.

[28] G. E. Coria, A. M. Sanchez, A. S. Al-Sumaiti, G. A. Ratt, S. R. Rivera, and A. A. Romero, "A framework for determining a prediction-of-use tariff aimed at coordinating aggregators of plug-in electric vehicles," *Energies*, vol. 12, no. 23, 2019.

[29] M. Etezadi-Amoli, K. Choma, and J. Stefani, "Rapid-charge electric-vehicle stations," *IEEE Transactions on Power Delivery*, vol. 25, no. 3, pp. 1883–1887, 2010.

[30] X. Liu, "Research on flexibility evaluation method of distribution system based on renewable energy and electric vehicles," *IEEE Access*, vol. 8, pp. 109249–109265, 2020.

[31] E. Heydarian-Forushani, M. E. H. Golshan, and P. Siano, "Evaluating the operational flexibility of generation mixture with an innovative techno-economic measure," *IEEE Transactions on Power Systems*, vol. 33, no. 2, pp. 2205–2218, 2018.

[32] M. Di Somma, G. Graditi, and P. Siano, "Optimal bidding strategy for a der aggregator in the day-ahead market in the presence of demand flexibility," *IEEE Transactions on Industrial Electronics*, vol. 66, no. 2, pp. 1509–1519, 2019.

[33] S. Dalhues, Y. Zhou, O. Pohl, F. Rewald, F. Erlemeyer, D. Schmid, J. Zwartscholten, Z. Hagemann, C. Wagner, D. M. Gonzalez, H. Liu, M. Zhang, J. Liu, C. Rehtanz, Y. Li, and Y. Cao, "Research and practice of flexibility in distribution systems: A review," *CSEE Journal of Power and Energy Systems*, vol. 5, no. 3, pp. 285–294, 2019.

[34] I. Pavi, T. Capuder, and I. Kuzle, "A comprehensive approach for maximizing flexibility benefits of electric vehicles," *IEEE Systems Journal*, vol. 12, no. 3, pp. 2882–2893, 2018.

[35] C. Baron, A. S. Al-Sumaiti, and S. Rivera, "Impact of energy storage useful life on intelligent microgrid scheduling," *Energies*, vol. 13, no. 4, 2020.

[36] C. Eyisi, A. S. Al-Sumaiti, K. Turitsyn, and Q. Li, "Mathematical models for optimization of grid-integrated energy storage systems: a review," in 2019 North American Power Symposium (NAPS), pp. 1–5, 2019.

[37] J. Valinejad, M. Marzband, M. Korkali, Y. Xu, and A. S. Al-Sumaiti, "Coalition formation of microgrids with distributed energy resources and energy storage in energy market," *Journal of Modern Power Systems and Clean Energy*, vol. 8, no. 5, pp. 906–918, 2020.

[38] K. Rahbar, J. Xu, and R. Zhang, "Real-time energy storage management for renewable integration in microgrid: An off-line optimization approach," *IEEE Transactions on Smart Grid*, vol. 6, no. 1, pp. 124–134, 2015.

[39] B. Mohandes, S. Acharya, M. S. E. Moursi, A. S. Al-Sumaiti, H. Doukas, and S. Sgouridis, "Optimal design of an islanded microgrid with load shifting mechanism between electrical and thermal energy storage systems," *IEEE Transactions on Power Systems*, vol. 35, no. 4, pp. 2642–2657, 2020.

[40] L. A. Wong, V. K. Ramachandaramurthy, P. Taylor, J. Ekanayake, S. L. Walker, and S. Padmanaban, "Review on the optimal placement, sizing and control of an energy storage system in the distribution network," *Journal of Energy Storage*, vol. 21, pp. 489 – 504, 2019.

[41] Y. Yang, S. Bremner, C. Menictas, and M. Kay, "Battery energy storage system size determination in renewable energy systems: a review," *Renewable and Sustainable Energy Reviews*, vol. 91, pp. 109–125, 2018.

[42] Y. Wang, K. T. Tan, X. Y. Peng, and P. L. So, "Coordinated control of distributed energy-storage systems for voltage regulation in distribution networks," *IEEE Trans. Power Delivery*, vol. 31, no. 3, pp. 1132–1141, 2016.

[43] V. B. Pamshetti, S. Singh, and S. P. Singh, "Combined impact of network reconfiguration and volt-var control devices on energy savings in the presence of distributed generation," *IEEE Systems Journal*, vol. 14, no. 1, pp. 995–1006, 2020.

[44] D. Barros, W. L. A. Neves, and K. M. Dantas, "Controlled switching of series compensated transmission lines: challenges and solutions," *IEEE Trans. Power Delivery*, vol. 35, no. 1, pp. 47–57, 2020.

[45] J. Clairand, J. Rodrguez-Garca, and C. lvarez-Bel, "Smart charging for electric vehicle aggregators considering users preferences," *IEEE Access*, vol. 6, pp. 54624–54635, 2018.

[46] S. C. Dhulipala, R. V. A. Monteiro, R. F. d. Silva Teixeira, C. Ruben, A. S. Bretas, and G. C. Guimares, "Distributed model-predictive control strategy for distribution network volt/var control: a smart-building-based approach," *IEEE Trans. Indus. Applic.*, vol. 55, no. 6, pp. 7041–7051, 2019.

[47] M. Shafie-khah, E. Heydarian-Forushani, G. J. Osrio, F. A. S. Gil, J. Aghaei, M. Barani, and J. P. S. Catalo, "Optimal behavior of electric vehicle parking lots as demand response aggregation agents," *IEEE Transactions on Smart Grid*, vol. 7, no. 6, pp. 2654–2665, 2016.

[48] A. Majzoobi and A. Khodaei, "Application of microgrids in supporting distribution grid flexibility," *IEEE Transactions on Power Systems*, vol. 32, no. 5, pp. 3660–3669, 2017.

[49] E. Veldman and R. A. Verzijlbergh, "Distribution grid impacts of smart electric vehicle charging from different perspectives," *IEEE Transactions on Smart Grid*, vol. 6, no. 1, pp. 333–342, 2015.

[50] I. Kamwa, R. Grondin, D. J., and S. Fortin, "A minimal realization approach to reduced-order modeling and modal analysis for power sytems response signals," *IEEE Trans. Power Syst.*, vol. 8, no. 3, pp. 1020–1029, 1993.

[51] J. Dobrowolski, F. Segundo, M. Paternina, *et al.*, "Inter-area oscillation control based on eigensystem realization approach," in *IEEE ROPEC*, pp. 1–6, IEEE, 2018.

[52] Y. Liu, L. Wu, and J. Li, "D-pmu based applications for emerging active distribution systems: A review," *Electric Power Systems Research*, vol. 179, p. 106063, 2020.

[53] J. Sanchez-Gasca and D. Trudnowski, "Identification of electromechanical modes in power system," tech. rep., IEEE PES Task Force on Identification of Electromechanical Modes of the Power System Stability, June 2012.

[54] J. de la O Serna, J. Ramirez, A. Zamora Mendez, and M. Paternina, "Identification of electromechanical modes based on the digital Taylor-Fourier transform," *IEEE Trans. Power Syst.*, vol. 31, pp. 206–215, Jan 2016.

[55] N. Duan and E. M. Stewart, "Frequency event categorization in power distribution systems using micro pmu measurements," *IEEE Trans. Smart Grid*, vol. 11, no. 4, pp. 3043–3053, 2020.

[56] H. H. Mller, C. A. Castro, and D. Dotta, "Allocation of pmu channels at substations for topology processing and state estimation," *IET Generation, Transmission Distribution*, vol. 14, no. 11, pp. 2034–2045, 2020.

[57] H. Qin and J. W. Kimball, "Generalized average modeling of dual active bridge dcdc converter," *IEEE Trans. on Power Electron.*, vol. 27, no. 4, pp. 2078–2084, 2012.

[58] G. E. Meja-Ruiz, J. R. Rodrguez, M. R. A. Paternina, N. Muoz-Galeano, and A. Zamora, "Grid-connected three-phase inverter system with lcl filter: Model, control and experimental results," in *IEEE ISGT Latin America*, pp. 1–6, 2019.

[59] M. Kesler, M. C. Kisacikoglu, and L. M. Tolbert, "Vehicle-to-grid reactive power operation using plug-in electric vehicle bidirectional offboard charger," *IEEE Trans. Industrial Electronics*, vol. 61, no. 12, pp. 6778–6784, 2014.

[60] J. Bay, *Fundamentals of Linear State Space Systems. Electrical Engineering Series*, WCB/McGraw-Hill, 1999.

[61] F. Wilches-Bernal, R. H. Byrne, and J. Lian, "Damping of inter-area oscillations via modulation of aggregated loads," *IEEE Trans. Power Systems*, vol. 35, no. 3, pp. 2024–2036, 2019.

[62] I. Sharma, C. Caizares, and K. Bhattacharya, "Smart charging of pevs penetrating into residential distribution systems," *IEEE Trans. Smart Grid*, vol. 5, no. 3, pp. 1196–1209, 2014.

[63] Tesla Manufacture, "Introducing v3 supercharging," 2020.

[64] ABB Manufacture, "High power charging infrastructure," 2020.

# 10 Blockchain-based Decentralized, Flexible, and Transparent Energy Market

*Mohd Adil Sheikh[1], Vaishali Kamuni[1],*
*Mohit Fulpagare[1], Udaykumar Suryawanshi[1],*
*Sushama Wagh[1], and Navdeep M. Singh[1]*
[1]VJTI, Mumbai, India

## CONTENTS

## 10.1 INTRODUCTION

While exploiting the advantages of renewable energy sources (RESs) availability at any location in the grid, right from low-capacity distribution end (e.g., solar) to the high-capacity transmission or generation end, various injecting points are made accessible to the users to encourage bidirectional power flow [1, 2]. However, as side effects it has also provided open doors for mischievous/malicious users to enter in the system leading to the threat of data manipulation, serious privacy and confidentiality issues with worst situation as blackout. This kind of manipulated operating situations would result in not only financial loss but even a valuable human life loss because of failure of energy flow network. The rising awareness of green energy has

233

already pushed the efforts of renewable generation at all locations leading to installation of distributed generators (DGs) resulting in the development of microgrids at distribution end to the smart grids at larger network. As described in ref [3], these renewable-based DGs are small, can be located near the load side, as well as can be connected to the grid in a distributed form. In the future, the electricity market has to accommodate the change in the structure of the generating system as more number of RES will get added to the existing structure. However, in the traditional electricity market, some intermediate retailers [4, 5] will be employed for the communication between the RES suppliers and consumers.

Conversely, microgrids, being a small-scale distribution system [3], consist of various DGs, energy storage systems (ESS), and loads and prove to be most effective in utilizing RES. In case of microgrids, the consumers and DGs can communicate directly instead of involving an intermediate agency that benefits both the parties and is also considered to be conducive for the RES development [6]. However, the traditional system consists of a centralized organization for managing transaction in microgrids and leads to the following problems [7–9]:

1. The high cost imposed on DG suppliers for the regulation and operation of the transaction center.
2. The concern of fairness, information confidentiality, and transparency arises because of trust issues between traders and transaction centers.
3. The security of information, especially, the transaction details are at a greater stake because the trading mode is centralized.

Hence, a mechanism is required which can overcome the aforementioned issues and make the overall system transparent, secure, and reliable leading to flexibility in operation.

Meanwhile, the blockchain, being a distributed public ledger technology has gained quite momentum in the field of power system for overcoming the limitations of the existing centralized system. Blockchain utilizes peer-to-peer (P2P) transactions along with the help of digital signatures, consensus mechanism, encryption process, and other mechanisms [10]. Blockchain is a type of non-forgery and non-tampering system, which is in the form of a chain consisting of different data blocks at a particular timestamp and is assured by cryptography. Each block of the chain consists of the hashed messages obtained from actual data, the hash information of previous blocks, and the time at which it is stamped. As each block has a dependency on the previous block hash value, if anyone needs to modify any information of a particular block then it has to modify all the following blocks after the selected particular block which means that the complete structure of the blockchain needs to be rehashed. Thus the advantage of this structure is that the information cannot be hacked by any cyber-attacker, as it requires the manipulation of the complete chain which will require high computation power as well as computation cost and time. Different consensus mechanisms like proof of work (PoW) and proof of stake (PoS) are used in blockchain for ensuring that all the nodes of the P2P network reach a global truth and then only new blocks, if any, are added to the existing chain in the blockchain which enhances the security of the network. With the use of blockchain, the cost of the transaction is reduced as it eliminates the additional cost imposed by

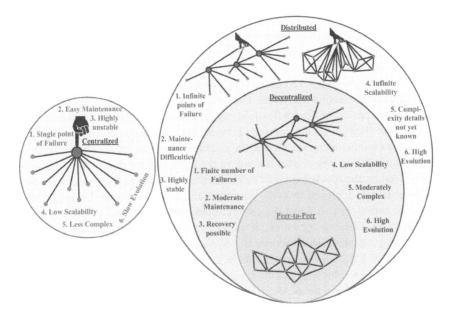

**FIGURE 10.1** Comparison of centralized, decentralized, and distributed system.

the transaction organization and also ensures the security of data. Figure 10.1 represents the difference between centralized, decentralized, and a distributed system.

Even though the research in the area of blockchain and its application in energy sources is gaining momentum, several preliminary results have been published in the literature. For implementing automation and security between microgrids, Kang [11] with the use of smart contracts, established an energy trading system that was completely decentralized. For avoiding leakage of information, ref. [12] proposed a consortium blockchain with a derivative method. Reference [13] presented an over-grid which is a fully decentralized microgrid and is a P2P representation of the actual grid. The energy trading process between the electric vehicle (EV) and distribution network was developed by ref. [14], which employed Byzantine based consensus mechanism for enhancing security. Similarly, the work described in refs. [15–17] concentrated on EV and charging station selection in the blockchain framework and also tested the proposed system for various attacks and the most well-known Sybil attack. For promoting P2P energy trading, ref. [18] with help of consortium blockchain proposed decentralized energy trading in a secure manner. A machine-to-machine electricity trading market is described in ref. [19] which investigates the communication technology between consumers and producers in a blockchain framework.

In this chapter, the dependency of the aggregator facilitating energy exchange between DGs and the traditional grid is eliminated by utilizing the blockchain framework. The data available from advanced metering infrastructure (AMI), such as electricity consumption and generation, is stored in a tamper-proof manner in the blockchain. Another important aspect of blockchain is the introduction of smart contracts, which programmatically help in examining the flexibility of expected energy from different DGs along with amalgamated penalties or rewards and the rules for matching the electrical energy demand with electrical energy generation at the grid level. The chapter further explores the effectiveness of blockchain in providing flexibility which is represented by considering different scenarios of energy and financial transaction exchange. The development of the smart grid had already opened many opportunities for flexibility in the operation of the grid; however, introducing blockchain in such a complex network not only increases the flexibility but also assures secure operation. Hence, the consumers/prosumers are more attracted to participate in the energy market because of the immutability, non-forgery in the system offered by the blockchain. For example, the flexibility in the process of energy exchange can be achieved from the consumer end as with the introduction of blockchain the consumer can select the energy suppliers on their own in a secured manner. The consumer can opt for clean sources of energy or communicate with some neighboring suppliers which reduce the cost of electricity, as well as the return on investment of RES, can be increased. In case of financial transactions, the smart contract helps in achieving flexibility among the participants of the P2P network as the participants can define electrical energy prices like the sales price, and also levy penalties or rewards can be awarded to participants based on energy management. Finally, the chapter concludes with various possible suggestions for utilizing blockchain in the electrical energy markets for enhancing the flexibility of the smart grid and also the overall security of the system. The main objectives of the chapter are as follows:

- A blockchain framework is proposed for the energy market eliminating the requirement of aggregator in the bidirectional energy trading system of the smart grid.
- The flexibility in the operation of smart grid was already existing because of the employment of bidirectional features, smart devices, facility for consumer participation, and interactions; however, there was a hindrance in this process due to security protocols. In view of this, the chapter highlights the usefulness of blockchain in enhancing the flexibility of the smart grid by assuring security with help of consensus mechanisms and smart contracts.
- The security of energy flow information is achieved by employing consensus mechanisms which help in eliminating malicious information getting added in the ledger of critical energy information and thus avoid manipulation of energy data which can result in various severe consequences on the operation of smart grid (e.g., demand–supply mismatch, tripping of essential loads, switching on non-essential loads, cascading events).
- By employing smart contracts, the financial aspect of the energy trading is secured as the financial data is programmatically written and added in the blockchain along with the associated penalties and rewards for the effective energy trading.
- The effectiveness of blockchain for enhancing the flexibility together with security is tested in a different scenario and the employment of smart contract and consensus mechanism confirms the secure operation of smart grid and thus leading to a flexible operation.

## 10.2 PRELIMINARIES OF BLOCKCHAIN

In prevailing computer structures, manipulation of data is possible as the existing data collection and storing mechanism endows a centralized environment that intensifies the likelihood of an attack. In contrast, the blockchain framework provides an essential structure for accumulating data from several units, sending plain text from different communication links, and also storing the information in form of a database [20]. The blockchain can be visualized as a data log consisting of records in form of a timestamped block and it forms a complete distributed data structure. The details of the formulation of blockchain and the storage mechanisms are explained in the following section.

### 10.2.1 DATA SIGNING, VERIFICATION, AND AUTHENTICATION

For maintaining the secrecy of critical data, the data signing mechanism in a cryptography framework is employed. The secure environment is established with help of the data signing mechanism and proves to be effective in assuring security. Cipher data is encrypted information which is impossible to be retrieved by an adversary without valid decryption. In the data signing mechanism, a public and private key is assigned to each node in the network. To decrypt any information, a secret key is required which is also known as the private key and it should have restricted access for assuring security against the adversary. For the different classes of algorithms, the key lengths are different [21]. In the network, a public key is accessible by all the existing nodes. Each node stored data consists of information about public keys of all nodes, individual-specific private keys, accumulated blocks, and consensus details, and the process of data signing and data verification and authentication is summarized in Table 10.1

### 10.2.2 DATA MINING AND GENERATION OF BLOCKS

As shown in Figure 10.2, the message stored in the chain is linked with cryptographically encoded data blocks. For data authentication and the signing of data, several types of hash functions are employed.

SHA-256 helps in the successful completion of the generation and mining of each block. In the blockchain framework, every block comprises the respective block number, timestamp, data content, hash value of the previous block, the value of current hash, and finally, the nonce value. The explanation of every element of the blocks in the blockchain is detailed in Table 10.2.

SHA-256 has different features like the block size of it is 512 bits, whereas the message size is less than $2^{64}$ bits and the 32 bits is the word size and as an output as a 256 bit digest. The compression function processes a message block of 512 bit and an intermediate hash value of 256 bit [20].

The algorithm is executed in two steps:

1. Pre-processing: In the pre-processing, the overall message $(S)$ [20] is formulated with help of the complete characteristic of block and is given as:

$$S = N + DC_N + TS_N + NV_N^K + \# P_{N-1} \tag{10.1}$$

**TABLE 10.1**

**Data Signing Versus Verification and Authentication**

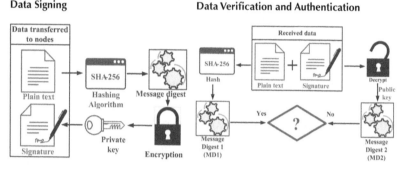

| Step | Data Signing | Data Verification and Authentication |
|---|---|---|
| 1 | The plaintext is collected and processed by a secure hash algorithm (SHA) to obtain a message digest (MD) | The encrypted data is received by each node, where, message digest 1 (MD1) verifies the plaintext which is hashed. Simultaneously, message digest 2 (MD2) decrypts the signature with help of the public senders key |
| 2 | MD is encrypted as a digital signature using a private key, where decryption requires the same node public key | A comparison of MD1 and MD2 is carried out and if both MDs are found consistent, then the data received is assumed to be authentic, else it is inferred to be corrupt data |
| 3 | The information is then broadcasted to the rest of the nodes through communication links | For data verification, each node should conform to a unique conclusion. In order to obtain consensus, the cumulative count of nodes reaching an agreement of a single conclusion should be higher than 51% |

**TABLE 10.2**

**Various Attributes of Block Content**

| Factors | Definition |
|---|---|
| $N$ | The title of the blocks or block number |
| $DC$ | Data content—block transaction details |
| $TS$ | Time stamp, the time instant of update in the block |
| $\#P$ | Hash value of previous block |
| $\#C$ | Hash value of present (or current) block |
| $NV$ | Nonce value is a number incremented by one every time a transaction is completed and it is a solution to the puzzle problem |
| $K$ | Nonce value of the next block |

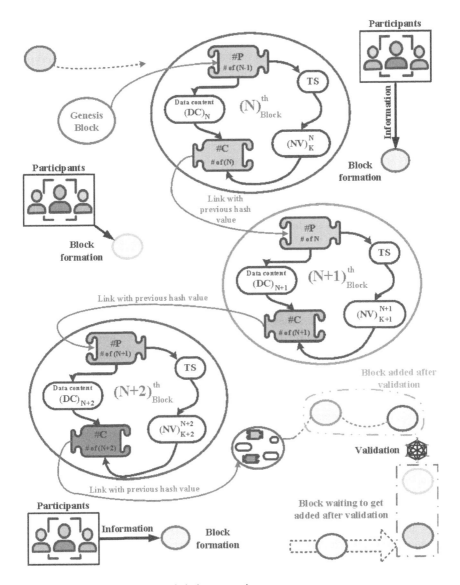

**FIGURE 10.2**   Block content and chain connections.

2. Hash computation: In the hash computation step, the message $S$ is again hashed with SHA-256 for generating MD. The target value is required to be greater than or equal to the final hash, as stated:

$$F_\# = \#(SHA - 256, \#(SHA - 256, S)) \qquad (10.2)$$

The computational complexity of the problem intensifies if the value of the target hash is small, resulting in the difficulty in finding a competent nonce. For validating the requirement of the target hash, the miner needs to find the nonce value and broadcast it to the rest of the miners or nodes. The accomplishment of consensus by more than 51% nodes, after verification, will result in an update of the resultant hash value in the block. After fulfillment of the previously mentioned step, the crypto-graphical linkage is allowed with the previous ledger.

### 10.2.3 Significance of Smart Contract

The smart contract is defined as "A computerized transaction protocol that executes the terms of a contract," which was introduced by Nick Szabo in 1994 [21]. Smart contracts are the codes executed to evince the logic of transactions in the blockchain, such as solidity which is a high-level language for developing smart contracts. The final state and the smart contracts are part of these Ethereum blocks, which is the result of executing the contracts. These contracts are preserved in the form of byte codes. Once the parties go through the contract and are satisfied with the terms, the linkage between the smart contract and the blockchain is executed as program code. Each node of the network receives the program code after validation and then is deployed to a particular block of the blockchain which allows it to observe the real-time status of smart contracts.

The consensus will not be achieved if there is any inconsistency between the results of different nodes. Thus, the deterministic nature of smart contracts is required. This feature ensures that the same output should be produced for a par-ticular input by a smart contract. Figure 10.3 presents the smart contract operation. These contracts are byte code instructions, which are stored in the blockchain for the Ethereum Virtual Machine (EVM) [22]. Writing such contracts usually involves the use of a higher-level language like Solidity. The contract (program code) is then appended to the blockchain after all the parties sign the contract [22].

On a similar line of traditional contracts, the smart contract involves basically three phases: generation, release, and execution as shown in Figure 10.4. In the case of contract generation, the multiple entities clinch over the contract, develop and organize the contract text, and finally, the program is verified for consistency with the text. The entities with a thorough discussion decide the rights and tasks of each entity in the initial draft of the contract. Then a person having preexisting knowl-edge about legal formalities will review and confirm the draft in terms of legal-ity, and in the last, the text is finalized on the paper. A technician now develops a

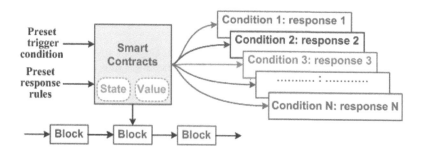

**FIGURE 10.3**  Smart contract configuration.

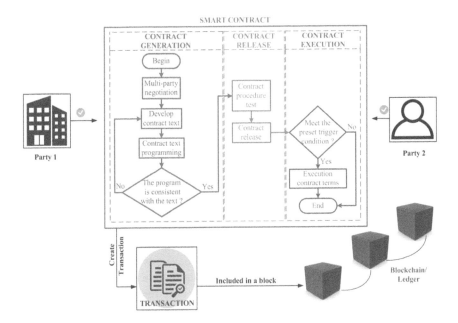

**FIGURE 10.4** Smart contract lifecycle.

program keeping the contract text as a reference and a trial test is conducted on a virtual machine for checking the consistency between the text and the program. If the program is found consistent, then the next phase is initiated otherwise the steps are repeated again.

The process of releasing the contract is relatively simple, once the program test run is passed, the contract can be released to all the nodes in the network. In case of execution, a preset trigger condition is verified, if all the conditions are satisfied then the contract is automatically executed, if the conditions are not satisfied then the program will be stopped instantly.

## 10.3 ARCHITECTURE OF THE ENERGY MARKET IN AGGREGATOR BASED SYSTEM AND A BLOCKCHAIN FRAMEWORK

### 10.3.1 AGGREGATOR-BASED ENERGY MARKET

The traditional grid consists of a centralized power plant that supplies power in one direction to different consumers. However, due to an increase in power demand, the current grid fails to maintain the demand–supply requirements. Furthermore, the traditional grid utilizes fossil fuels which are tremendously depleting and also lack the capability of self-healing, automatic operation, real-time information, capacity limitations, and one-way communication.

The modern electrical grid utilizes information and communication technology (ICTs) features to manage the exchange of information between consumers and utility in a completely automated manner. In addition, various sensors are deployed on the grid which helps in sensing various system data that helps to analyze the complete system. Moreover, the smart grid allows a bidirectional flow of power with full visibility. The advantage of having a sensor is that the consumer can have real-time information about their consumption profile and can opt for energy management activity. In the smart grid, the energy can be distributed and managed more efficiently with help of smart devices and automated systems, resulting in financial profits. Overall, high efficiency, high reliability, pricing methods based on real-time, flexibility, is achieved with the smart grids.

In smart grids, the consumer having RES units can also participate in the energy exchange process with the grid, such consumers with RES units are designated as DGs which can support the grid during peak demand with the power available from RES. An aggregator is a bridging element between the grid and the DGs, for the energy exchanging process. The DGs can have both role play i.e. they can switch the role between consumer and producer depending on the amount of energy available from the RES unit. The grid can coordinate for additional demand from DG units, in case, if they have a willingness to participate in the energy trading process by bidding their quotation for the energy supplied and the associated financial cost. The DGs also have the advantage of opting for energy from the grid and the neighboring DGs for individual demand–supply mismatch with help of an aggregator. The connection of aggregator is processed through an independent system operator which continuously monitors the line and load data of the grid and DG unit. The major role of the aggregator is to collect and store the information about energy available, stored, required, associated costs of all the participating DGs, and the grid as shown in Figure 10.5. The aggregator initiates the "virtual" trading process by selling the available energy from the DGs to the grid or in some cases between DGs. In an aggregator-based

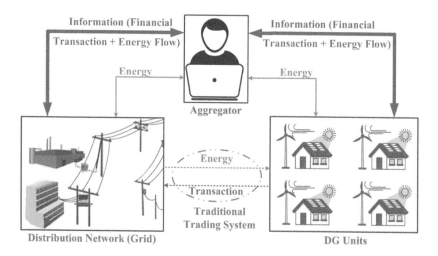

**FIGURE 10.5** Architecture of aggregator based energy market.

energy market, the grid quotes its requirement of energy and pricing to the aggregator which then announces price information to available DGs for selling their additional stored power to the grid. This process of continuous information exchange between aggregator and DGs can be considered as "virtual bargaining" and continues until DGs and aggregator reach a common point of the frame. Once the deal is finalized with the aggregator the DGs supply the required power to the grid for the decided time-frame and a corresponding reward is given to the DGs owner.

Even though aggregator-based trading is quite popular, it has limitations in respect of flexibility and security. The major concern is the authenticity of the aggregator, if the aggregator is suspicious it may lead to a complete mismatch of the demand and supply. There may be a possibility that even though there is a requirement of suppose $m$ $kWh$ at a price of $x$ \$ the aggregator if malicious may quote different energy $n$ $kWh$ and price $y$ \$ to the participating units and thus lead to trust issues between different parties. The issues of flexibility arise in the aggregator based system as the participants have to contact the aggregator for any communication between two participants for the requirement of energy and the associated prices. Hence, the response rate of the energy trading process depends on the aggregator which makes the complete system quite rigid.

## 10.3.2    ENERGY MARKET IN A BLOCKCHAIN FRAMEWORK

To overcome the issue of security and flexibility in the aggregator based system, a blockchain-based framework for enhancing energy market operation is proposed as shown in Figure 10.6. In the blockchain framework, the need for any intermediary is eliminated and the energy trading process is completely transparent between two parties. The energy trading process is initiated by the formation of a P2P network with different participating DGs and the traditional grid acting as a node of the network. The advantage of the P2P network is that each corresponding node has information of all the other nodes in the P2P network. The energy trading process is highly affected by the decision taken at every step by each node of the P2P network. The two major information which are required in such an energy trading process is the information about energy requirement and the associated financial records. This information can be easily accessed by attackers which can manipulate the information through the open ports provided with the aim of encouraging interconnection at the consumer ends.

With the introduction of blockchain an immutable data ledger is formulated which ensures that the malicious activity is minimized. The data of energy and information from the participating unit is stored in a hash format in blocks with a proper timestamp and is linked to form a chain of blocks i.e. blockchain. The new block if any is added to the existing chain after verifying it with help of existing nodes of the P2P network agreeing on a consensus. The requirement of energy by a DG or traditional grid is broadcasted in the P2P network and the corresponding nodes (DGs or Grid) having the capacity to satisfy requirements respond accordingly. The participating units bid their amount of contribution in energy requirement by hashing the amount

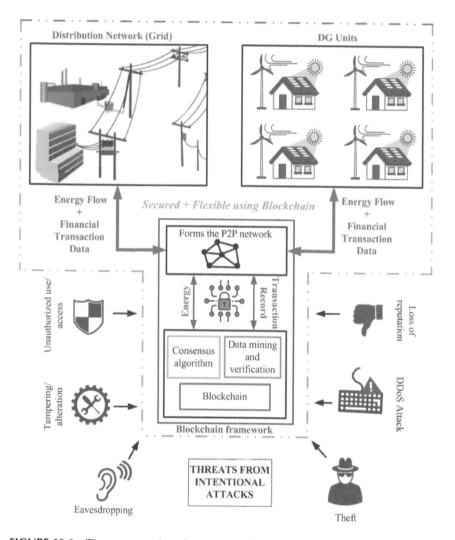

**FIGURE 10.6** The representation of communication between DGs and the traditional grid in a blockchain framework.

of energy committed and associated financial charges. To access the information of bid the consumer unit has to perform the decryption process using a hash function and public key. The verification is carried out to check MD1 and MD2 and the process is continued if the resultant of MDs is the same as described in Table 10.1. The data mining and verification process then validates the hash value with help of a consensus algorithm and then the new block is added to the chain. The smart contract is used to check the condition between the participating unit and consumer unit based on which validation process is again carried out with help of the consensus mechanism.

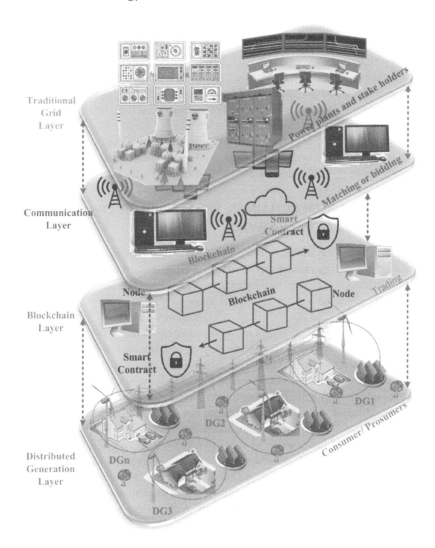

**FIGURE 10.7** The layered architecture of blockchain-based energy market.

The blockchain-based energy market can be realized as a four-layer architecture as shown in Figure 10.7. The details of each layer are:

- *Energy layer*: This layer consists of the players which can either generate energy or help in distributing it. The power plants and stakeholders are part of the energy layer.
- *Communication layer*: The pre-bargain process between the participating unit and consumer unit is executed in this layer. Here, the consumer units initiate a request for buying energy and the participating units respond with the help of two protocols, i.e., matching or bidding. The consumer unit is

matched with a participating unit according to the minimum selling price and then the participating unit submits their bids and the consumer units commit to one.

- *Blockchain layer*: The layer is responsible for all the transaction exchange and utilizes a smart contract for assuring transaction between two parties. With the use of ledgers, the process of buying and selling between two parties is more robust, scalable, and efficient.
- *Consumer/Prosumer layer:* This layer consists of different consumers and prosumers which is decided based on the amount of generation available with the help of DGs installed in the household.

## 10.4   ROLE OF BLOCKCHAIN IN THE FLEXIBILITY OF DISTRIBUTION NETWORK

It can be seen that the introduction of blockchain has assured security along with flexibility in the energy market. In the energy market, exchange of financial data and energy information is extensively carried out. The details of blockchain providing flexibility are given below:

- *Access to financial assets*: In a traditional energy market the financial services are not easily accessible from the user end, however, with the introduction of blockchain users can access all the financial services. The individual owner can set up a blockchain-based digital wallet that can instantaneously send, receive, or as well as store money anywhere around the globe without any additional fees. Thus the individual has access to all the financial information related to his own credits, loans, and insurances.
- *Availability of information at every point of energy trading*: In the blockchain, a P2P network is formulated with all the players participating in the energy trading process. The nodes of the P2P network has access to all the information of other participating units. Thus the users or participants can track any information immediately without any additional financial requirement. The participants not only have access to the information but also verify the information in case if the process is found malicious at any point of time.
- *The privacy and security of the individual identity*: With the introduction of blockchain, the credentials are verified in a decentralized manner which allows users to create a digital identity. This type of identity can be carried to any platform in a secure manner and it can be accessed only when needed and also decide the parts of information to be shared. Thus the dignity and privacy of all the participating individuals are maintained in the process.
- *Decentralized operation*: The key advantage of employing the blockchain is the decentralization of the complete process where each party can communicate with each other without any intermediary. Thus the process is more secure as well as less complex and also financially more feasible. The participants can directly communicate their requirements with the suppliers and agree on a common framework.

**TABLE 10.3**

**Comparison between Aggreagtor and Blockchain-Based Energy Market**

| Parameters | Aggregator Based | Blockchain Based |
|---|---|---|
| Network type | Centralized | Distributed |
| Third party involvement | Yes | No |
| Control | Full control stays with aggregator | Control stays with all participants |
| Single point of failure | Yes | No |
| Hackable | More prone to hack and data leaks | Less prone to hacks and data leaks |
| Easy to use | Intuitive and easy to use | Not easy to use |
| Maintenance | Easy | Moderate |
| Power consumption | High power is consumed | Power is consumed efficiently |
| Exchange fees | Higher fees | Less fees |
| Cost | Expensive | Inexpensive |

Thus, it can be seen that with the inclusion of the blockchain, the energy market is more transparent, non-forged, and non-tampered. The summary of comparison between aggregator and blockchain based energy market is given in Table 10.3.

## 10.5 FLEXIBILITY IN ENERGY TRADING WITH THE HELP OF BLOCKCHAIN

As discussed in Section 10.4, the role of the blockchain in enhancing system flexibility is vital. The energy trading between the different participating unit is described:

- System initialization
  The blockchain network consists of authorized nodes with all information to each other, making them a P2P network. The smart meters of the distribution network (DN) and DGs act as nodes and forms the P2P network. Every smart meter needs to be approved as a legal node to participate in the network. To be the authorized node, each participant has its unique key and signature. The new node to be involved in the network generates it's own public and private key for encryption of sensed data. Once the new node is legally authorized, the new smart meter downloads the other block's data for storage. After this, the data synchronization occurs with the available nodes and forms the P2P network. Let's say a P2P network community 'C' consists of nodes of DN and DGs:

$$C : nodes_n = DN_1, DN_2, ....., DN_j, DG_1, DG_2, ....DG_i$$
$$C : nodes_n = node_1, node_2, ....., node_n$$

(10.3)

where, $n = i + j$, is the total number of nodes in the community. Each node has a unique signature and public key to encrypt the sensed data of smart meters.

$$Node_n = PK_n, Signature_n \qquad (10.4)$$

Thus each node is initialized accordingly with distinct signatures and keys.

- Data requirement upload to the network
  Once the nodes are authorized in the network, the cryptography process begins. The sensed data from the smart meter is utilized to calculate the cost parameters. The predefined parameters like how much is the demand, available energy supply, peak time, and load variation are considered to evaluate the cost function and these values are set by users. After this, the smart meter of each participating unit packs the data and encrypts the signature, which is passed to other authorized nodes in the network. The encrypting is done using the public keys of each node. The equation can be expressed as follows:

$$Node_n = encrypt_{PK} \{data_n, signature_n, timestamp\} \qquad (10.5)$$

The encrypted data is shared with the other nodes for further validation of block data.

- Block validation
  When the data is received by other nodes (especially the consumer units), it starts decrypting the data using its own unique private key. Once the data is decrypted the process of consensus is executed between the nodes for adding the valid block to the chain. When the block is validated it is added to the chain or else the block is discarded as represented in Figure 10.8.
- Proof-of-stake (PoS) consensus
  The consensus is one of the important parts of blockchain cryptography. The process of PoS is similar to PoW but the method of reaching the valid block is different. In POS, instead of miners, some validators bid for some

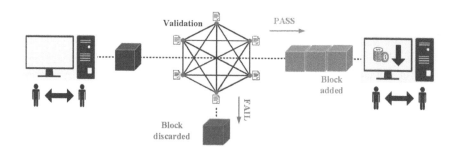

**FIGURE 10.8** Validation of the block for adding in the blockchain.

of their coins as a stake in the system. Following that, the validators bet on the blocks that they feel will be added next to the chain. When the block gets added, the validators get a block reward in proportion to their stake. With this, the concept of coindays can be taken into the picture. Coindays is the number of holding days multiplied by currency. When the node is having high coindays means there is no consumption of any computational power but it will add to block generation data. Thus, whenever the PoS block is found by nodes, it will clear the coindays by giving the participates a reward from it. When the consensus is achieved by authorized nodes the block is appended to the block.

* Final reply
The data received by nodes are decrypted and verified through PoS consensus. After which the respective node is signed and sent to the master node or validator. If the data is synchronized with the validator's guess, then the rewards are acquired to the nodes. Finally, the blocks are appended to the chain.

With the introduction of PoS, for an attacker to forge into the system, it needs to acquire 51% of the cryptocurrency which is difficult as well as expensive. Another drawback is that since the attacker has 51% share in a network, i.e., it holds a majority of shares, and if the value of the stakes falls, then the value of holdings by the attacker will also fall. Thus for an attacker to acquire 51% of the stakes is quite difficult as well as expensive and even though if it successfully acquires stakes then it is not feasible financially to attack such a system where the attacker is the major player on his own.

## 10.6  FLEXIBILITY IN A FINANCIAL TRANSACTION WITH HELP OF SMART CONTRACTS

In a smart contract-based energy trading the participating units (consumer as well as production units) are free to get in and out of the trading market, the confidentiality of the price quoted by each production unit is maintained until it is cleared, the terms and conditions in the contract should be executed automatically.

The energy trading process is initiated by the consumption unit broadcasting its energy requirements in the P2P network by depositing a token amount (virtual) to the address of the smart contract. Thus the deposit of such a virtual amount for initiating the request avoids any false request from the intruders trying to enter into the system. Based on the broadcasted energy requirements the production units which are part of the P2P network respond by quoting its prices and the amount of energy it can contribute. This information is collected from all the production units which are willing to participate in the energy trading process. The quotation is sealed in a hash format using the process of data signing described in Section 10.2.1. A token amount is taken as a deposit during the submission of the hash quotation from all the production units and hence helps in avoiding any malicious units getting added into the process. The smart contract keeps this quotation of the production unit completely unique and confidential.

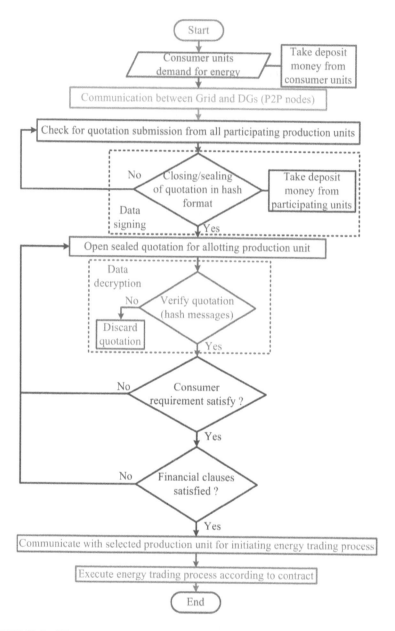

**FIGURE 10.9** The smart contract based energy trading.

The seal quotation is opened by the consumer unit for allotting the production unit. The smart contract then checks the value of a hash function with the sealed hash function, i.e., it compares the MD1 and MD2 described in the data decryption process of Section 10.2. If any mismatch is found in the hash value then the quotation will be discarded. Once all the quotations are verified the consumer requirement for the energy and the associated financial charges are checked. Finally, after checking the financial and energy requirement the contact between the selected production unit and consumer unit is established and the energy trading process is executed as per the terms defined in the contract.

For making the energy trading process more realistic, the following conditions are written in a smart contract. The conditions are divided into three parts:

- *Supply-demand balance*: This situation is the most favorable situation for the energy trading process. In this case, the energy trading process is executed as per the defined normas and the token amount taken from all the parties is returned.
- *Penalty to production unit*: In case the production unit agreed to allot a certain amount of energy to the consumption unit, however, after the opening of the quotation if the unit backed out at the last moment due to certain unavoidable circumstances, then a penalty is levied to such unit. In such cases, the production unit token amount is not returned and an equivalent charge based on the deferred energy requirement is charged which is given by (9.6):

$$Penalty = \beta \cdot \Delta M \qquad (10.6)$$

where, $\beta$ represents the penalty factor whose value changes according to the situation and is generally varies from 0 to 1. The term $\Delta M$ represents the difference in the amount of energy that the production unit failed to supply to the consumption unit after commitment.

- Reward to production unit: In this case, if the consumer unit denies the energy requirement from few production units or wants to terminate the energy trading process in between, in such a scenario the production units will be eligible for rewards. Similar to the previous case the token amount taken from the consumption unit is retained and a charge represented in (10.7) is levied based on the energy requirement:

$$Reward = \alpha \cdot \Delta N \qquad (10.7)$$

where, $\alpha$ represents the reward factor whose value changes according to the situation and is generally varies from 0 to 1. The term $\Delta N$ represents the difference in the amount of energy that the consumer unit denied from the production unit.

Thus with the use of a smart contract, the final settlement between the production and consumption unit is achieved which is highly flexible as well as secure.

## 10.7  CONCLUSION

The proposed energy market in a blockchain framework have proved to be more secure and flexible as compared to the aggregator-based energy market. The key advantage achieved by blockchain in an energy market is the elimination of untrusted intermediaries making the system more transparent, decentralized, and tamper-proof. The flexibility in terms of energy information in addition to financial transactions is achieved with help of blockchain consensus mechanisms and smart contracts respectively. With the use of PoS consensus, the probability of an attack on the system decreases, and the loss of energy information is minimized. The smart contract with help of program helps to secure financial transaction by applying penalties and rewards to the participating unit in a bidding framework and also build the process expensive in initial stage so that any malicious units are restricted from entering into the system. In the future, the energy trading process between different entities in a power system may be considered and various consensus mechanisms for achieving flexibility in the different scenarios will be considered.

## REFERENCES

[1] B. Shen, Y. Han, L. Price, H. Lu, and M. Liu, "Techno-economic evaluation of strategies for addressing energy and environmental challenges of industrial boilers in china," *Energy*, vol. 118, pp. 526–533, 2017.

[2] T. Lv and Q. Ai, "Interactive energy management of networked microgrids-based active distribution system considering large-scale integration of renewable energy resources," *Applied Energy*, vol. 163, pp. 408–422, 2016.

[3] W. Lee, L. Xiang, R. Schober, and V. W. Wong, "Direct electricity trading in smart grid: A coalitional game analysis," *IEEE Journal on Selected Areas in Communications*, vol. 32, no. 7, pp. 1398–1411, 2014.

[4] M. Vasirani and S. Ossowski, "Smart consumer load balancing: state of the art and an empirical evaluation in the spanish electricity market," *Artificial Intelligence Review*, vol. 39, no. 1, pp. 81–95, 2013.

[5] B.-G. Kim, S. Ren, M. Van Der Schaar, and J.-W. Lee, "Bidirectional energy trading and residential load scheduling with electric vehicles in the smart grid," *IEEE Journal on Selected Areas in Communications*, vol. 31, no. 7, pp. 1219–1234, 2013.

[6] T. Cui, Y. Wang, S. Nazarian, and M. Pedram, "An electricity trade model for microgrid communities in smart grid," in *ISGT 2014*. IEEE, 2014, pp. 1–5.

[7] E. Mengelkamp, J. Gärttner, K. Rock, S. Kessler, L. Orsini, and C. Weinhardt, "Designing microgrid energy markets: A case study: The brooklyn microgrid," *Applied Energy*, vol. 210, pp. 870–880, 2018.

[8] S. Hesmondhalgh, "Is neta the blueprint for wholesale electricity trading arrangements of the future?" *IEEE Transactions on Power Systems*, vol. 18, no. 2, pp. 548–554, 2003.

[9] M. M. Roggenkamp, R. L. Hendriks, B. C. Ummels, and W. L. Kling, "Market and regulatory aspects of trans-national offshore electricity networks for wind power interconnection," *Wind Energy*, vol. 13, no. 5, pp. 483–491, 2010.

[10] S. Nakamoto, "Bitcoin: A peer-to-peer electronic cash system," *Manubot, Tech. Rep.*, 2019.

[11] E. S. Kang, S. J. Pee, J. G. Song, and J. W. Jang, "A blockchain-based energy trading platform for smart homes in a microgrid," in *2018 3rd international conference on computer and communication systems (ICCCS)*. IEEE, 2018, pp. 472–476.

[12] K. Gai, Y. Wu, L. Zhu, M. Qiu, and M. Shen, "Privacy-preserving energy trading using consortium blockchain in smart grid," *IEEE Transactions on Industrial Informatics*, vol. 15, no. 6, pp. 3548–3558, 2019.

[13] D. Croce, F. Giuliano, I. Tinnirello, A. Galatioto, M. Bonomolo, M. Beccali, and G. Zizzo, "Overgrid: A fully distributed demand response architecture based on overlay networks," *IEEE transactions on automation science and engineering*, vol. 14, no. 2, pp. 471–481, 2016.

[14] A. Sheikh, V. Kamuni, A. Urooj, S. Wagh, N. Singh, and D. Patel, "Secured energy trading using byzantine-based blockchain consensus," *IEEE Access*, vol. 8, pp. 8554–8571, 2019.

[15] U. Asfia, V. Kamuni, S. Sutavani, A. Sheikh, S. Wagh, and N. Singh, "A blockchain construct for energy trading against sybil attacks," in 2019 27th Mediterranean Conference on Control and Automation (MED). IEEE, 2019, pp. 422–427.

[16] U. Asfia, V. Kamuni, A. Sheikh, S. Wagh, and D. Patel, "Energy trading of electric vehicles using blockchain and smart contracts," in 2019 18th European Control Conference (ECC). IEEE, 2019, pp. 3958–3963.

[17] V. Kamuni, U. Asfia, S. Sutavani, A. Sheikh, and D. Patel, "Secure energy market against cyber attacks using blockchain," in 2019 6th International Conference on Control, Decision and Information Technologies (CoDIT). IEEE, 2019, pp. 1792–1797.

[18] Z. Li, J. Kang, R. Yu, D. Ye, Q. Deng, and Y. Zhang, "Consortium blockchain for secure energy trading in industrial internet of things," *IEEE transactions on industrial informatics*, vol. 14, no. 8, pp. 3690–3700, 2017.

[19] J. J. Sikorski, J. Haughton, and M. Kraft, "Blockchain technology in the chemical industry: Machine-to-machine electricity market," *Applied energy*, vol. 195, pp. 234–246, 2017.

[20] G. Liang, S. R. Weller, F. Luo, J. Zhao, and Z. Y. Dong, "Distributed blockchain-based data protection framework for modern power systems against cyber attacks," *IEEE Transactions on Smart Grid*, vol. 10, no. 3, pp. 3162–3173, 2018.

[21] I. Bashir and M. Blockchain, "Packt publishing," Birmingham, UK, 2017.

[22] S. Hua, E. Zhou, B. Pi, J. Sun, Y. Nomura, and H. Kurihara, "Apply blockchain technology to electric vehicle battery refueling," in *Proceedings of the 51st Hawaii International Conference on System Sciences*, 2018.

# 11 Integrated Operation and Planning Model of Renewable Energy Sources with Flexible Devices in Active Distribution Networks

*Vijay Babu Pamshetti[1] and Shiv Pujan Singh[2]*
[1]B V Raju Institute of Technology, Narsapur, Telangana, India
[2]Indian Institute of Technology (BHU) Varanasi,
Varanasi, India

## CONTENTS

## NOMENCLATURE

### Indices

| | |
|---|---|
| $i/j$, $s$ | Index for bus, scenario |
| $t$, $T$ | Index for time, Total time duration |

### Sets

| | |
|---|---|
| $\Omega_{bus}$ | set of buses |
| $\Omega_{bess}$ / $\Omega_{sop}$ | set of buses installed with BESS/SOP |
| $\Omega_{PV}$ / $\Omega_{Wd}$ | set of buses installed with PV/wind generation |
| $\Omega_{DR}$ | set of buses participate in DR scheme |

### Parameters

| | |
|---|---|
| $N_L/N_{pV}$ | Total number of loads, installed PV |
| $N_s$ /$N_{rs}$ | Total number of scenarios, reduced scenarios |
| $Prob(s)$ | Probability of $s^{th}$ scenario |
| $\alpha_{l,s}^{L}, \alpha_{l,s}^{PV}, \alpha_{l,s}^{Wd}$ | Probability of $s^{th}$ scenario in load, PV generation and wind generation |
| $P_{nom,i,s}^{D,t}$ / $Q_{nom,i,s}^{D,t}$ | Nominal active/reactive power demand at $i^{th}$ bus for $s^{th}$ scenario at $t^{th}$ hour |
| $d, r$ | Rate of interest, rate of growth |
| $C_{P,bess}^{inv}$ / $C_{E,bess}^{inv}$ | Investment cost of power and energy capacity of BESS |
| $C_{sop}^{inv}$ | Investment cost of SOP |
| $C_{sub}^{t}$ | Active power purchased cost from grid at $t^{th}$ hour |
| $C_{emis}$ | Carbon (CO2) emission cost |
| $C_{ens}$ | Cost of energy not served |
| $C_{bess}^{OM}$ / $C_{sop}^{OM}$ | O & M cost of BESS/SOP |
| $V^{min}$ / $V^{max}$ | Minimum and maximum bus voltage |
| $Q_{sop,i}^{min}$ / $Q_{sop,i}^{max}$ | Minimum and Maximum reactive power limits of SOP at $i^{th}$ bus |
| $I_{br}^{rated}$ | Rated current flow in branch $br$ |
| $tap_{oltc}^{min}$ / $tap_{oltc}^{max}$ | Minimum and maximum OLTC transformer taps |
| $step_{cb}^{max}$ | Capacitor bank maximum step change |
| $\bar{\Lambda}$ | Maximum number of DR customers |
| $\gamma_i^{min}$ / $\gamma_i^{max}$ | Minimum and maximum limits of demand flexibility at $i^{th}$ bus |
| $\Delta t$ | Change in time |

## *Variables*

| | |
|---|---|
| $P^{rated}_{bess,i}$ / $S^{rated}_{sop,i}$ | Rated power of BESS/SOP at $i^{th}$ bus |
| $\pi_s$ | Probability of scenario $s$ |
| $SOC^t_i$ | State of charge of BESS at $i^{th}$ bus at $t^{th}$ hour |
| $V^t_{i,s}$ | Bus voltage at $i^{th}$ bus for $s^{th}$ scenario at $t^{th}$ hour |
| $P^{ch,t}_i$ / $P^{dch,t}_i$ | Active power charge/discharge of BESS at $i^{th}$ bus at $t^{th}$ hour |
| $P^t_{bess,i}$ / $Q^t_{bess,i}$ | Active/reactive power of BESS at $i^{th}$ bus at $t^{th}$ hour |
| $S^{rated}_{bess,i}$ | Rated MVA capacity of BESS at $i^{th}$ bus |
| $P^t_{sop,i}$ / $Q^t_{sop,i}$ | Active/reactive power of SOP at $i^{th}$ bus at $t^{th}$ hour |
| $P^{t,loss}_{sop,i}$ | Active power loss of SOP at $i^{th}$ bus at $t^{th}$ hour |
| $S_{sop,ij}$ | MVA capacity of SOP at $ij^{th}$ bus |
| $P^t_{sub,s}$ | Active power taken from substation for $s^{th}$ scenario at $t^{th}$ hour |
| $P^t_{ens,s}$ | Active power not served for $s^{th}$ scenario at $t^{th}$ hour |
| $P^t_{i,s}$ / $Q^t_{i,s}$ | Active and reactive power injection at $i^{th}$ bus for $s^{th}$ scenario at $t^{th}$ hour |
| $I^t_{br,s}$ | Current flowing through branch for $s^{th}$ scenario at $t^{th}$ hour |
| $tap^t_{oltc}$ | OLTC transformer tap position at $t^{th}$ hour |
| $step^t_{cb}$ | Capacitor bank step position at $t^{th}$ hour |
| $\gamma^t_i$ | Demand flexibility at $i^{th}$ bus at $t^{th}$ hour |
| $P^{D,t}_{i,s}$ / $Q^{D,t}_{i,s}$ | Active/reactive power demand at $i^{th}$ bus for $s^{th}$ scenario at $t^{th}$ hour after DR program |
| $P^{D,t}_{i,s,0}$ / $Q^{D,t}_{i,s,0}$ | Active/reactive power demand at $i^{th}$ bus for $s^{th}$ scenario at $t^{th}$ hour before DR program |

## 11.1 OVERVIEW

In current scenario, the integration of renewable energy sources (RESs) especially, wind and photovoltaic (PV) generation, has been tremendously increasing due to the three key factors, namely environmental concerns, technological innovation, and new government policies. However, distribution system operators (DSOs) are experiencing significant challenges that are impeding a reliable, resilient, and economically advantageous deployment of RESs in large scale. In order to cope with the intermittency and uncertainty associated with RES, it is indispensable to introduce additional operational flexibility into the system. Though, operation of soft open point (SOP), battery energy storage systems (BESS), demand response (DR), and dynamic network reconfiguration (DNR) has emerged as potential solution for the increases operational flexibility in the system, an efficient operation and planning strategy is much needed for proper coordination of multiple aforementioned devices. In this context, a new two-stage coordinated optimization model has been developed for integrated operation and planning of RES, BESS, and SOP devices. In stage 1, optimal allocation of RES, BESS, and SOP devices has been carried out simultaneously. Whereas, stage 2 performs optimal active and reactive power scheduling of RES, BESS, and SOP devices considering on-load tap changer (OLTC) transformer tap, capacitor bank (CB) steps, and status of switches. The objective of the proposed methodology is to minimize the total investment and operating cost of RES, BESS, and SOP devices. Besides, it includes cost of purchased power from substation, cost of energy not served (ENS), and cost of $CO_2$ emission. Meanwhile, to address high-level uncertainties related to renewable generation and load demands, a stochastic module has been adopted which produces several scenarios by Monte Carlo simulation (MCS) and Roulette wheel mechanism (RWM) at each hour. Further, K-means clustering technique has been employed to determine the reduced scenarios with high quality and diversity. The proposed framework has been implemented on IEEE 33-bus distribution system for different cases and is solved by using proposed hybrid optimization solver, which is performed in MATLAB-GAMS environment. The test results demonstrate the effectiveness of proposed framework in improving the system efficiency, enhancing the reliability, flexibility and reducing carbon emission footprint of distribution systems, when compared with the traditional planning schemes.

## 11.2 INTRODUCTION: BACKGROUND

According to the report of global trends in renewable energy investment published in the year 2020 [1], RESs across the world targets about 721 gigawatts of new capacity in wind, solar, and other non-hydro renewable power technologies over the next decade. Figure 11.1 depicts the different types of RES and their share to meet government targets with deadlines between 2020 and 2030. Among all RESs, the share of solar PV and wind generations dominates other RES (e.g., Biomass & waste and geothermal), which is about 64 and 33%, respectively as shown in Figure 11.1.

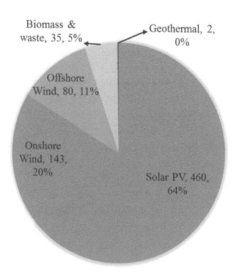

**FIGURE 11.1**  Pie chart of RES additions required to meet government targets by 2030, GW.

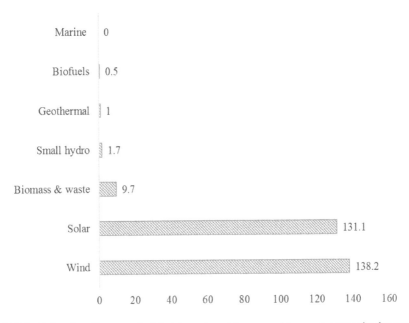

**FIGURE 11.2**  Bar diagram of global investment in renewable energy capacity by sector in 2019.

Figure 11.2 depicts the global investment in RES capacity by sector in 2019. Again the investments in solar and wind generations are much higher than other RES, which are about $131.1 billion and $138.2 billion, respectively. There are two main reasons for these changes, one is downward trend in costs per megawatt for solar PV, and second one is further rise in activity in offshore wind, both off the coasts of Europe and in the sea off mainland China and Taiwan [1].

## 11.2.1  MOTIVATION AND AIM

As demonstrated in the background, RESs make a crucial part of the solution for environmental sustainability; hence, they will play an important role in power systems. However, large-scale integration of RES often poses a number of technical challenges in the system from the stability, reliability, and power quality perspectives. This is because integrating RESs introduces significant operational variability and uncertainty to the distribution system, making operation, planning, and control rather complicated. Besides, large-scale penetration of RESs will necessarily involve a process of adapting and changing the existing infrastructure because of their intrinsic characteristics, such as intermittency and variability.

The arbitrary allocation of RES units in the system may lead to an uncertain increase in the feeder's power flows, resulting in network congestion and increased losses in the network. Therefore, RES allocation should be carefully planned to maximize the system efficiency. The high proliferation of RES and their integration to distribution systems can adversely impact on feeder's voltage profile, due to reverse power. Besides, the intermittent nature of RESs, such as wind and solar, and uncertain behavior of the load can cause voltage variation in distribution systems. The conventional voltage and reactive power control strategies in distribution systems are expected to face numerous challenges. When RES units are interconnected to a distribution system, they can significantly change the system voltage profile and interfere with the conventional local control strategies of OLTC transformer, voltage regulator (VR), and capacitor banks (CBs). This interference leads to overvoltage, under voltage, increasing in system losses and excessive wear and tear of voltage control devices. In order to mitigate voltage violations, reactive power control issues and facilitate a seamless integration for large penetration of RES, which are needed to coordinate with advanced distribution management schemes (ADMS) such as volt/VAR control, network reconfiguration and demand response.

On the other hand, the installation of flexible devices, such as BESS and SOP, along with RESs, has become one of the most viable solutions to facilitate the increased penetration of RES as well as increased controllability and flexibility to the operation of distribution systems. Energy storage system (ESS) has a capability to level the mismatch between variable power generation and demand by store energy during periods of low electricity demand (price) or high RES power production, and then release it during periods of peak demand and low RES production. However, SOP has ability to provide active power flow control between adjacent feeders, reactive power compensation, and voltage regulation in a distribution system. Despite the high capital costs, their deployment in distribution systems is in the upward trend. If storage and SOP can be optimized for the right location, size, and duration along with being produced at a very low cost, then load can be scheduled to match the generation profile so as to reduce the overall cost of energy. Hence, proper planning and operation of RES along with flexible devices in active distribution network is crucial for enhancing the performance of distribution system.

### 11.2.2 RELATED WORK

Table 11.1 summarizes the recent survey on planning and operation of RES along with flexible devices (i.e., BESS and SOP) considering ADMS schemes. In order to improve the reliability and flexibility of the operation of ADN, authors proposed various methods [2–16] for allocation of RES along with flexible devices in distribution system. Sizing of BESS has been performed in ref. [2] for eliminating system constraint violations and for reducing short-duration uncertainty of PV. Reference [3] reveals the significance of conservation voltage reduction (CVR) operation while allocation of BESS in high RES penetrated distribution network. Back-to-back voltage source converters (B2B VSC) flexible device in distribution network named as SOP, which can provide active power flow control between adjacent feeders, reactive power compensation, and voltage regulation in a distribution system. Normally, SOP devices are installed in place of tie-line switches in a distribution network [4]. In ref. [5] optimal planning of storage in active distribution network has been performed with the minimization of installation cost, operation cost, and reliability cost. A two-stage iterative process has been adopted in ref. [6] to determine the storage size at multiple locations in distribution network. Further, congestion and voltages are

**TABLE 11.1**

**Literature Survey**

| Ref. | RES | Un** | Flexible Devices | | | | ADMS Scheme | | | Objectives | | |
|---|---|---|---|---|---|---|---|---|---|---|---|---|
| | | | BESS | | SOP | | | | | | | |
| | | | P* | O# | P* | O# | DR | VVC/CVR | DNR | EC | RC | EMC |
| [2] | Y | Y | Y | Y | N | N | N | N | N | Y | N | N |
| [3] | Y | Y | Y | Y | N | N | N | Y | N | Y | N | N |
| [5] | Y | Y | Y | Y | N | N | N | N | N | Y | Y | N |
| [6] | Y | N | Y | Y | N | N | N | Y | N | Y | N | N |
| [9] | Y | Y | N | Y | N | N | Y | Y | N | Y | N | N |
| [12] | Y | N | N | N | N | Y | N | Y | N | Y | N | N |
| [14] | Y | N | Y | Y | N | Y | N | N | N | Y | N | N |
| [15] | Y | Y | N | N | Y | Y | N | N | N | Y | N | N |
| [17] | Y | Y | Y | Y | N | N | N | N | N | Y | Y | N |
| [18] | Y | Y | Y | Y | N | N | N | N | N | Y | Y | N |
| [19] | Y | Y | N | Y | N | Y | N | Y | N | Y | N | N |
| [20] | Y | Y | N | N | N | N | N | Y | Y | Y | N | N |
| [21] | Y | Y | N | Y | N | N | Y | N | N | Y | N | N |

N, not considered; Y, considered; RES*, renewable energy sources; P*, planning; O#, operation; un**, uncertainty; DR, demand response; VVC, volt/VAR control; CVR, conversation voltage reduction; DNR, distribution network reconfiguration; EC, economic cost; RC, reliability cost; EMC, emission cost.

managed through the optimal control of storage (active and reactive power), OLTCs, DG power factor. In ref. [7] coordinated scheduling of BESS and DR has been performed to minimize the active losses payments. A coordinated BESS and DG planning framework has been presented in ref. [8] for maximization of DISCO profit. In ref. [9], authors were formulated a framework of integrated CVR and DR scheme for ADMS. In ref. [10] a coordinated control scheme has been proposed for BESS and solar PV operation along with OLTC for voltage regulation. Reference [11] reveals that Volt–VAR technique (e.g., CVR) can enhance the available transfer capability (ATC) of the system. A coordinated volt–VAR control (VVC) method based on SOP has been present in ref. [12] for minimization of total operation costs and improvement in the voltage profile of distribution network. Similarly, two-stage robust optimization method has been presented in ref. [13] for effective utilization of SOPs in high RES-penetrated distribution network. Impact of BESS allocation in SOP based active distribution network has been presented in ref. [14]. Reference [15] reveals the benefits of coordinated allocation of DG, CBs, and SOPs. Similarly, in [16] reveals the benefits of coordinated allocation of DG Units and SOPs in active distribution networks. However, these previous studies [2–16] did not investigate the benefits of coordination allocation BESS and SOP in distribution network. Further, impact of integrated operation of CVR and DR has not been studied. Moreover, the benefits of reactive power support from BESS and SOP did not explored. In view of this, present chapter proposes a new coordinated planning of RES along with BESS and SOP devices in distribution network that incorporates ADMS schemes. The contributions of the present work are as follows:

- Proposed a new two-stage coordinated optimization framework for cooperative planning model of RES along with flexible devices in distribution network.
- Proposed an integrated planning model for RES that addresses the economic, operational, and environmental issues. The uncertainties of solar irradiance and load demand are entirely taken into account by using a stochastic module.
- Impact of integrated operation of ADMS schemes on RES allocation has been studied.
- Introduced a hybrid optimization solver to solve the large scale non-convex mixed-integer nonlinear programming (MINLP) problem without linearization or relaxation.

The proposed algorithm has been validated on well-known IEEE 33-bus distribution system.

## 11.3  UNCERTAINTIES MODELING OF RENEWABLE ENERGY SOURCES AND LOAD

### 11.3.1  UNCERTAINTY OF PHOTOVOLTAIC GENERATION

In this chapter, beta distribution has been employed to exhibits the uncertainty of solar irradiance [22] and probability density function (PDF) is given in (11.1):

$$PDF_s(s) = \begin{cases} \dfrac{\Gamma(\alpha+\beta)}{\Gamma(\alpha)\Gamma(\beta)} s^{(\alpha-1)}(1-s)^{(\beta-1)}; 0 \leq s \leq 1, \alpha, \beta \geq 0 \\ 0 \qquad\qquad\qquad\qquad\qquad\qquad \text{other wise} \end{cases} \tag{11.1}$$

where, $s$ is the solar irradiance, $\Gamma(\cdot)$ is the Gamma function, $\alpha$ and $\beta$ are function parameters and determined from the mean ($\mu$) and standard deviation ($\sigma$) of solar irradiance $s$ as given below:

$$\beta = (1-\mu)\left(\frac{\mu(1+\mu)}{\sigma^2} - 1\right); \quad \alpha = \frac{\mu \times \beta}{1-\mu} \tag{11.2}$$

Power outputs of each PV module are functions of solar irradiance, ambient temperature, and physical characteristics of each module. The output power in PV module can be determined by the power characteristic model as shown in (11.3):

$$P_{pv} = P_{STD}\left\{\frac{s}{1000}[1+\delta(T_{cell}-25)]\right\} \tag{11.3}$$

$$T_{cell} = T_{amb} + \left(\frac{NOCT-20}{800}\right)s \tag{11.4}$$

### 11.3.2 Uncertainty of Wind Generation

Weibull distribution has been employed to exhibits the uncertainty of wind speed [22] and PDF is given in (11.5):

$$PDF_w(V) = \left(\frac{k}{c}\right)\left(\frac{v}{c}\right)^{(k-1)} \exp\left(-\left(\frac{v}{c}\right)^k\right) \tag{11.5}$$

where, $k$, $c$ are Weibull parameters, called shape index and scale index, respectively. $v$ is the wind speed.

Power outputs of wind generation (WG) connected to wind turbine is a function of wind speed and physical characteristics of turbine. The output power in wind turbine can be determined by using (11.6) and (11.7), respectively

$$P_{WG} = \begin{cases} 0; & 0 \le v < v_{ci} \\ a + b \times v^3; & v_{ci} \le v < v_{rated} \\ P^n_{WG,rated}; & v_{rated} \le v < v_{co} \\ 0; & v_{co} \le v \end{cases} \tag{11.6}$$

$$\left.\begin{array}{l} a = \left(P^n_{WG,rated} \times v_{ci}^3\right) \Big/ \left(v_{ci}^3 - v_{rated}^3\right) \\[2mm] b = P^n_{WG,rated} \Big/ \left(v_{rated}^3 - v_{ci}^3\right) \end{array}\right\} \tag{11.7}$$

### 11.3.3 Voltage-Dependent Load Modeling

In medium-/low-voltage distribution network, most of the loads exhibit the voltage-dependant behavior. These loads are highly dependent on the voltage magnitude. There are two VDLMs namely exponential and polynomial loads. In this chapter, exponential load voltage dependents have been chosen to represent the load-to-voltage sensitivities in this study. The active and reactive power load are determined by using (11.8) and (11.9), respectively

$$P^D = P^D_{nom}\left(\frac{V}{V^{nom}}\right)^{k^p} \tag{11.8}$$

$$Q^D = Q^D_{nom}\left(\frac{V}{V^{nom}}\right)^{k^q} \tag{11.9}$$

Normal distribution function has been employed to exhibits the uncertainty of loads [22] and active and reactive power load probability density function is given in equations (11.10) and (11.11), respectively

$$PDF_{pd}(P^D) = \frac{1}{\sqrt{2\pi}\sigma_p} \exp\left[-\frac{(P^D_{nom} - \mu_p)^2}{2\sigma_p^2}\right] \tag{11.10}$$

$$PDF_{qd}(Q^D) = \frac{1}{\sqrt{2\pi}\sigma_q} \exp\left[-\frac{(Q^D_{nom} - \mu_q)^2}{2\sigma_q^2}\right] \tag{11.11}$$

## 11.4   PROPOSED STOCHASTIC VARIABLE MODULE (SVM)

Stochastic Variable Module (SVM) has been introduced in order to model the uncertainties of variables such as solar irradiance and electrical load. In this paper, Normal probability distribution, Beta probability distribution and Weibull probability distribution have been employed to exhibit the uncertainties of load, solar irradiance and wind speed respectively [22].

As shown in Figure 11.3(a), the PDF of each random variable (e.g., load, solar irradiance) is divided into seven intervals having standard deviation ($\sigma$) and a probability ($\alpha_l^L$, $\alpha_l^{PV}$ and $\alpha_l^{Wd}$) of load, solar irradiance and wind speed in the interval ($l$), respectively. An RWM has been employed to model each probability level with a specific stochastic variable (e.g., load, solar irradiance, and wind speed) as shown in Figure 11.3(b). Generate a random number in the range of [0,1]. The generated random number value will fall in one of the intervals of Roulette wheel, which indicates the forecast error and its probability associated with random variable (e.g. load, solar irradiance and wind speed). Similarly, the same procedure is implemented generate the required number of scenarios ($N_s$). Then the solar irradiance has converted into PV power generation based on the PV module characteristics. Similarly, the wind speed has converted into wind

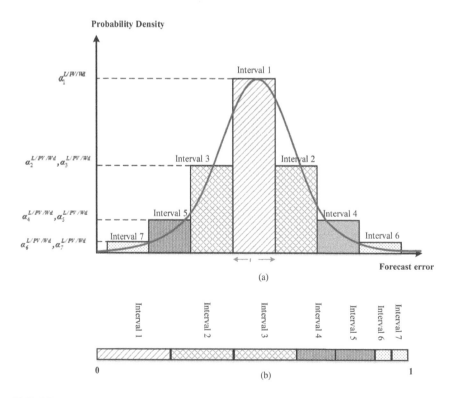

**FIGURE 11.3**   (a) Typical discretization of the PDF of the load/PV generation forecast error; (b) the Roulette Wheel Mechanism.

power generation based on the wind turbine characteristics. The structure of the $s^{th}$ scenario and corresponding normalized probability have been given in (11.2) and (11.13), respectively.

$$s = \left[ P_{nom,i,s}^{D}, P_{j,s}^{PV}, P_{k,s}^{Wd} \right]; \forall \begin{cases} i = 1,2,...N_L; \\ j = 1,2,...N_L; \\ k = 1,2,...N_{Wd}; \\ s = 1,2,...N_s \end{cases} \tag{11.12}$$

$$\Pr ob(s) = \frac{\left( \prod\limits_{L=1}^{N_L} (\alpha_{l,s}^{L}) \right) \left( \prod\limits_{PV=1}^{N_{PV}} (\alpha_{l,s}^{PV}) \right) \left( \prod\limits_{Wd=1}^{N_{Wd}} (\alpha_{l,s}^{Wd}) \right)}{\sum\limits_{s=1}^{N_s} \left( \left( \prod\limits_{L=1}^{N_L} (\alpha_{l,s}^{L}) \right) \left( \prod\limits_{PV=1}^{N_{PV}} (\alpha_{l,s}^{PV}) \right) \left( \prod\limits_{Wd=1}^{N_{Wd}} (\alpha_{l,s}^{Wd}) \right) \right)} \tag{11.13}$$

It is noteworthy to mention that, as the number of scenarios increases; the computational time will also increase. The DSO must decide the settings of control variables as fast as possible to accommodate the hourly operation schedule of a day. Therefore, an efficient scenario reduction method is required to decrease the computation time. For this purpose, K-mean clustering algorithm [23] has been employed to reduce the number of scenarios. The aim of this algorithm is to arrange original scenarios of solar irradiance and loads into clusters according to their likenesses. The centroid of each cluster is defined as the mean value of solar irradiance and loads allocated to each cluster. After implementing the scenario reduction technique, the probability of the achieved reduced number of scenarios ($N_{rs}$) is normalized using expression (11.14)

$$\pi_s = \frac{\Pr ob(s)}{\sum\limits_{s=1}^{N_{rs}} \Pr ob(s)}; \forall s = 1,2,...N_{rs} \tag{11.14}$$

## 11.5   FLEXIBLE DEVICES

### 11.5.1   BATTERY ENERGY STORAGE SYSTEM

BESSs can produce active and reactive powers locally. Figure 11.4 shows the direction of active and reactive powers of the BESS connected to the bus, which means BESS can produce and consume both the active and reactive powers simultaneously [24].

### 11.5.2   SOFT OPEN POINTS

SOP devices are installed between the adjacent feeders in place of normally open points (NOP) in distribution system [12] as shown in Figure 11.5.

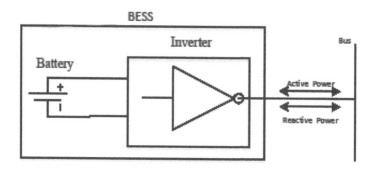

**FIGURE 11.4**   Direction of active and reactive power flow of BESS connected to bus.

There are two VSCs that are installed between the feeder endpoints and connected through a common dc link. One of which adopts PQ control mode and the other adopts Vdc-Q control mode. The VSC with PQ control is used to control the transmitted active power of the SOP and the reactive power injected by the converter, while the VSC with Vdc-Q control is used to maintain the voltage constant of the DC bus and inject reactive power. The active and reactive power can be controlled independently by the VSC controller in a active distribution network.

## 11.6   ACTIVE MANAGEMENT SCHEMES

### 11.6.1   Volt/VAR Control

VVC is a control strategy to manage the voltage magnitude within the statutory limits and reactive power flow throughout the distribution network in order to achieve the efficient operation of distribution network [25].

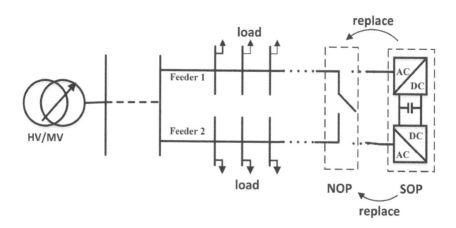

**FIGURE 11.5**   A typical soft open point connected to adjacent feeder.

## 11.6.2 Distribution Network Reconfiguration

The DNR is a well-established technique for the reduction of network losses, load balancing, and restoring the power supply of the outage area in distribution networks. It is the process of changing the open/close status of normally closed (sectionalized) and normally opened (tie) switches dynamically to obtain a new radial topology of the distribution network on the premise of the switching time limits according to the dynamic load variation over a specific period of time [26].

## 11.7 FRAMEWORK OF PROPOSED TWO-STAGE COORDINATED OPTIMIZATION MODEL

The proposed integrated planning model not only determines the location and capacity of RES, BESS and SOP, but also simulates the optimal active and reactive management of RES, BESS and SOP devices. The formulation of the proposed model in two stages corresponding to the integrated planning and operation model respectively has been shown in Figure 11.6.

In stage 1, (i.e., integrated planning level) the data provider unit is fed to the integrated planning optimization model, which involves the installation of RES, BESS, and SOP devices. In this stage, decision variables are the locations and capacities

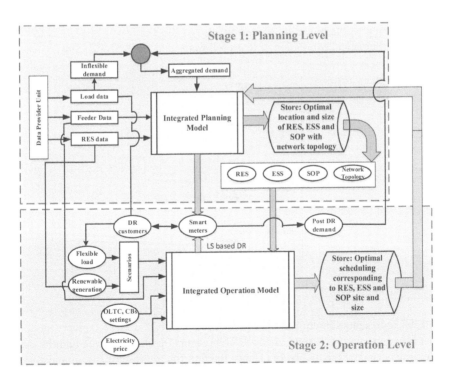

**FIGURE 11.6** Framework of proposed two-stage coordinated optimization.

of RES, BESS, and SOP devices to be installed. For coordination purpose, all the candidate planning proposals will be transferred to the integrated operation optimization model in stage 2 (i.e., integrated operation level) where optimal scheduling of RES, BESS, and SOP considering OLTC tap, CBs steps and branch switch operations would be performed considering uncertainties of renewable energy generation and loads. After this step, the posted scheduling of aforementioned devices corresponding to their capacities would be stored and fed back to the planning model in stage 1, in order to revise the previous planning scheme. This process is continued until the given time period. Finally, the output file presents the optimized site and size of RES, BESS, and SOP in an active distribution network.

## 11.8 OBJECTIVE FUNCTION

The objective function (OF) of the proposed methodology is to minimize the total economic cost, which is expressed as given in (11.15):

$$\min.OF = \overset{A}{\overbrace{IC}} + \overset{B}{\overbrace{OC + EPC + EC + RC}} \tag{11.15}$$

where, part $A$ in (11.15) is the investment cost (IC) of PV, Wind based DG, BESS and SOP units, which are derived from stage 1 (i.e., integrated planning level); part $B$ stands for total expected variable costs (e.g., operating costs (OC) of BESS and SOP units, energy purchased cost (EPC), emission cost (EC) and reliability cost (RC)) under all the possible operating scenarios.

The mathematical expression of IC as follows:

$$IC = \begin{bmatrix} EAC^{PV} \sum_{ij \in \Omega_{PV}} \left( C_{PV}^{inv} P_{PV,i}^{rated} \right) \\[2mm] +EAC^{Wd} \sum_{ij \in \Omega_{Wd}} \left( C_{Wd}^{inv} P_{Wd,i}^{rated} \right) \\[2mm] +EAC^{bess} \sum_{i \in \Omega_{bess}} \left( \begin{array}{c} C_{P,bess}^{inv} P_{bess,i}^{rated} \\ +C_{E,bess}^{inv} E_{bess,i}^{rated} \end{array} \right) \\[2mm] +EAC^{SOP} \sum_{ij \in \Omega_{sop}} \left( C_{sop}^{inv} S_{bess,ij}^{rated} \right) \end{bmatrix} \tag{11.16}$$

where, $EAC^x = \dfrac{d(1+d)^{LT_x}}{(1+d)^{LT_x}-1}; x \in \{PV, wind, bess, sop\}$

$EAC$ represents the equivalent annual cost, which is used to convert the total cost into annual cost. Where, $d$ and $LT$ represents the discount rate and the asset life time period, respectively. In (11.16), the first line represents the annual cost of power and energy capacity of BESS, whereas second line represents the annual cost of SOP.

As stage 2 represents the integrated operation optimization through repeated simulations for minimizing the total expected variable costs (B), which is given in (11.15), they can be determined by using following mathematical expressions:

$$OC = \left[ \sum_{t=1}^{T} \left( \sum_{s=1}^{N_{rs}} \pi_s \times \left( \begin{array}{c} \sum_{i \in \Omega_{PV}} C_{PV}^{OM,t} P_{PV,i,s}^t + \sum_{i \in \Omega_{Wd}} C_{Wd}^{OM,t} P_{Wd,i,s}^t \\ + \sum_{i \in \Omega_{bess}} C_{bess}^{OM,t} P_{bess,i,s}^t + \sum_{i \in \Omega_{sop}} C_{sop}^{OM,t} S_{sop,i,s}^t \end{array} \right) \right) \right] \times D_y \quad (11.17)$$

$$EPC = \left[ \sum_{t=1}^{T} \left( \sum_{s=1}^{N_{rs}} \pi_s \times \left( C_{sub}^t P_{sub,s}^t \right) \right) \right] \times D_y \quad (11.18)$$

$$EC = \left[ \sum_{t=1}^{T} \left( \sum_{s=1}^{N_{rs}} \pi_s \times \left( C_{emis} P_{sub,s}^t \right) \right) \right] \times D_y \quad (11.19)$$

$$RC = \left[ \sum_{t=1}^{T} \left( \sum_{s=1}^{N_{rs}} \pi_s \times \left( C_{ens} P_{ens,i,s}^t \right) \right) \right] \times D_y \quad (11.20)$$

Equation (11.17) represents the operation and maintenance cost of BESS and SOP. Equation (11.18) represents the EPC, usually results in power loss reduction. Equation (11.19) is helpful to reduce the carbon emission emitted from the substation. The cost of ENS is considered as the reliability cost (RC) of the distribution network as given in (11.20). Here $D_y$ is the number of days in a year, which is used to convert the daily cost to the annual cost.

## 11.8.1 CONSTRAINTS

The following planning and operational model constraints should be satisfied:
- Maximum installation of RES, BESS, and SOP at each bus

$$\left. \begin{array}{l} 0 \leq P_{PV,i}^{rated} \leq P_{PV,i}^{max} \\ 0 \leq P_{Wd,i}^{rated} \leq P_{Wd,i}^{max} \\ 0 \leq P_{bess,i}^{rated} \leq P_{bess,i}^{max} \\ 0 \leq S_{sop,ij}^{rated} \leq S_{sop,ij}^{max} \end{array} \right\} \quad (11.21)$$

- power flow equations for each scenario

$$\left. \begin{array}{l} P_{i,s}^t = V_{i,s}^t \sum_{i \in \Omega_{bus}} V_{j,s}^t \left( G_{ij} \cos \theta_{ij,s}^t + B_{ij} \sin \theta_{ij,s}^t \right) \\ Q_{i,s}^t = V_{i,s}^t \sum_{i \in \Omega_{bus}} V_{j,s}^t \left( G_{ij} \sin \theta_{ij,s}^t - B_{ij} \cos \theta_{ij,s}^t \right) \end{array} \right\} \quad (11.22)$$

- Nodal voltage constraint for each scenario

$$V_i^{\min} \leq V_{i,s}^t \leq V_i^{\max} \qquad (11.23)$$

- Branch capacity constraint for each scenario

$$0 \leq I_{l,s}^t \leq I_l^{rated} \qquad (11.24)$$

- The OLTC transformer tap constraints

$$tap_{oltc}^{\min} \leq tap_{oltc,y}^t \leq tap_{oltc}^{\max} \qquad (11.25)$$

- CB step constraints

$$0 \leq step_{cb}^t \leq step_{cb}^{\max} \qquad (11.26)$$

- Load modeling:

$$P_{i,s,0}^D = P_{nom,i,s}^D \left( \frac{V_{i,s}^t}{V_{nom,i,s}^t} \right)^{k^p} \qquad (11.27)$$

$$Q_{i,s,0}^D = Q_{nom,i,s}^D \left( \frac{V_{i,s}^t}{V_{nom,i,s}^t} \right)^{k^q} \qquad (11.28)$$

In the present scenario, most of the feeder loads exhibit voltage-sensitive behavior [27]. Hence, savings achieved via CVR is greatly influenced by voltage dependent load type. The exponential voltage dependent load model has been chosen to represent the load-to-voltage sensitivities in this study. Equation (11.27) and (11.28) represents the active and power voltage dependent demand (i.e., exponential demand) respectively.

- DR constraints:

$$\sum_{t=1}^T P_{i,s}^{D,t} \Delta t = \sum_{t=1}^T P_{i,s,0}^{D,t} \Delta t \qquad (11.29)$$

$$\sum_{t=1}^T Q_{i,s}^{D,t} \Delta t = \sum_{t=1}^T Q_{i,s,0}^{D,t} \Delta t \qquad (11.30)$$

$$P_{i,s}^{D,t} = P_{i,s,0}^{D,t} \times \gamma_i^t \qquad (11.31)$$

$$Q_{i,s}^{D,t} = Q_{i,s,0}^{D,t} \times \gamma_i^t \qquad (11.32)$$

$$\left(1 - \gamma_i^{\min} \Lambda_i \right) \leq \gamma_i^t \leq \left(1 - \gamma_i^{\max} \Lambda_i \right) \qquad (11.33)$$

$$\sum_{i \in \Omega_{DR}} \Lambda_i \leq \overline{\Lambda} \qquad (11.34)$$

In this paper, price-based DR scheme [7] has been implemented, where it aims to shift demand rather than curtailing it, equations (11.29) and (11.30) represent the scheduling constraints of active and power demand flexibility that satisfy overall energy requirement over the complete day must be same. Equations (11.31) and (11.32) represent the active/reactive power demand flexibility by factor $\gamma_i^t$ at each node i at each hour t. Constraint (11.33) represents the limitation of scheduling demand flexibility at each node i and also noted that the energy of node i is redistributed in different time periods but not transferred to other nodes, where $\Lambda_i$ is a binary variable. If $\Lambda_i = 0$ then the node i does not participate in a DR program and vice versa. In equation (11.34), $\bar{\Lambda}$ represents the limit of maximum number of nodes which can participate in a DR program.

- SOP operation constraints:

$$P_{sop,i}^t + P_{sop,j}^t + P_{sop,i}^{t,loss} + P_{sop,j}^{t,loss} = 0 \tag{11.35}$$

$$\left.\begin{array}{l} P_{sop,i}^{t,loss} = A_{sop,i}\sqrt{\left(P_{sop,i}^t\right)^2 + \left(Q_{sop,i}^t\right)^2} \\[2mm] P_{sop,j}^{t,loss} = A_{sop,j}\sqrt{\left(P_{sop,j}^t\right)^2 + \left(Q_{sop,j}^t\right)^2} \end{array}\right\} \tag{11.36}$$

$$\left.\begin{array}{l} Q_{sop,i}^{\min} \leq Q_{sop,i}^t \leq Q_{sop,i}^{\max} \\[2mm] Q_{sop,j}^{\min} \leq Q_{sop,j}^t \leq Q_{sop,j}^{\max} \end{array}\right\} \tag{11.37}$$

$$\left.\begin{array}{l} \sqrt{\left(P_{sop,i}^t\right)^2 + \left(Q_{sop,i}^t\right)^2} \leq S_{sop,ij} \\[2mm] \sqrt{\left(P_{sop,j}^t\right)^2 + \left(Q_{sop,j}^t\right)^2} \leq S_{sop,ij} \end{array}\right\} \tag{11.38}$$

Equation (11.35) represents the active power constraints of SOP and equation (11.36) represents the active power loss in the SOP converters. Constraint (11.37) represents the limits of reactive power. Similarly, constraint (11.38) represents the limits of SOP capacity.

- BESS operation constraints:

$$SOC_i^t = SOC_i^{t-1} + \left(P_{bess,i}^t\right) \times \Delta t \tag{11.39}$$

$$SOC^{\min} \leq SOC_i^t \leq SOC^{\max} \tag{11.40}$$

$$SOC_i^{t=24} = SOC_i^{t=1} \tag{11.41}$$

$$\left.\begin{array}{l} P_i^{ch,\min} \leq P_i^{ch,t} \leq P_i^{ch,\max} \\[2mm] P_i^{dch,\min} \leq P_i^{dch,t} \leq P_i^{dch,\max} \end{array}\right\} \tag{11.42}$$

$$P_i^{ch,t} \times P_i^{dch,t} = 0 \qquad (11.43)$$

$$P_{bess,i}^t = \eta^{ch} P_i^{ch,t} - P_i^{dch,t} / \eta^{dch} \qquad (11.44)$$

$$Q_{bess,i}^t \leq \sqrt{\left(S_{bess,i}^{rated}\right)^2 - \left(P_{bess,i}^t\right)^2} \qquad (11.45)$$

Constraints (11.39) and (11.40) represent the status of state of charge (SOC) at a particular time and node, and SOC limits of BESS, respectively. Equation (11.41) regulates the energy remaining in the BESSs at the last time interval equal to the initial energy. Constraint (11.42) represents the limits of active power charging and discharging of BESS. Equation (11.43) ensures that only state (i.e., charging, discharging, or no action) to be active at each time interval. Equation (11.44) defines the active power of BESS power based on charging and discharging power with consideration of efficiency. Constraint (11.45) represents the reactive power injection/absorption constraint of BESSs converter.

## 11.9 PROPOSED SOLUTION METHODOLOGY

The problem formulated in this chapter is a large-scale non-convex MINLP problem, which is difficult to solve by conventional optimization techniques. To overcome this problem, the problem has been viewed as two models, namely, the planning and operation models. These models have been solved using the proposed two-stage framework as explained earlier in Section 11.7. The stage 1 and stage 2 are controlled by the decision variable sets, D1 and D2, respectively. The set D1 consists of candidate solutions for installations of RES, BESS, and SOP corresponding to the network topology, as illustrated in Figure 11.7. Similarly, set D2 consists of operational scheduling of RES, ESS, DR, SOP devices considering OLTC tap, CBs steps and status of switches as illustrated in Figure 11.8. In this chapter, stage 1 and stage 2 have been solved concurrently with the help of the evolutionary algorithms and GAMS optimization tool, respectively.

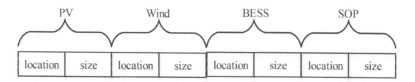

**FIGURE 11.7** Decision variable in planning stage (D1).

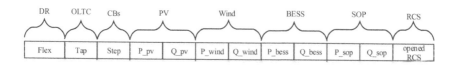

**FIGURE 11.8** Decision variable in operation stage (D2).

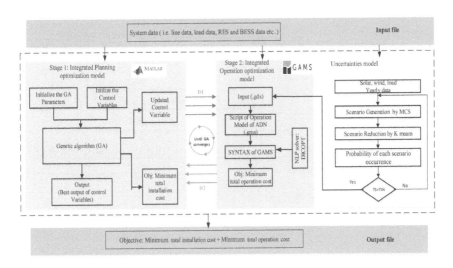

**FIGURE 11.9** Two stage coordinated optimization model solved by hybrid optimization solver.

Figure 11.9 illustrates the implementation of the proposed two-stage coordinated optimization model using the MATLAB-GAMS platform. In this chapter, a widely accepted Genetic algorithm (GA) [28] evolutionary algorithm has been employed to generate the candidate solutions D1 under the MATLAB environment. Each candidate solution (D1) obtained by stage 1, is transferred to the GAMS environment through GAMS data exchange (GDX) files. For each candidate solution D1, at stage 2, the problem relating the operation scheduling of RES, SOP, and BESS considering OLTC taps, CBs steps, and status of switches has been solved dynamically for each hour for each possible scenarios. The formulated MINLP problem in stage 2 has been solved by using DICOPT, MINLP solver under the GAMS environment [29]. Here, the objective is to recursively minimize the daily operation costs at the candidate installation variables provided by the GA solver. The solution of stage 2 yields the set of decision variables for the operation scheduling, i.e., D2 for each scenario *s* at hour *t*. This process is repeated until the GA solver converges.

## 11.10 SIMULATION RESULTS AND DISCUSSIONS

### 11.10.1 SIMULATION PLATFORM

The proposed two-stage coordinated optimization model has been implemented on MATLAB-GAMS environment. Detailed explanation of interfacing GAMS and MATLAB has been given in ref. [30] run on a computer with a Core i3 2.5-GHz processor and 4-GB RAM.

### 11.10.2 DATA AND ASSUMPTIONS

The performance of the proposed methodology has been tested on standard IEEE 33-bus distribution system. The detailed line data and load data of 33-bus system

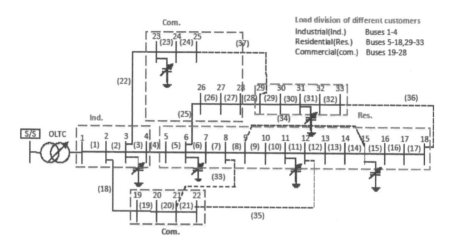

**FIGURE 11.10**  Modified 33 bus distribution system.

has been taken from [31] and also given in Appendix. In modified test system, it is assumed that OLTC transformer is connected between substation and node 1. The OLTC transformer can vary the substation secondary side voltage in the range of ±5% with 16 tap positions change in each step would be 0.625%. Six CBs are installed at buses no. 2, 6, 11, 15, 23, and 30, having capacity of 150 kVAR each. It is assumed that the lower and upper permissible voltage limits are 0.95 pu to 1.05 pu, respectively. The single line diagram of modified 33-bus distribution system has been shown in Figure 11.10. Typical load profile, power price purchased from grid, and PV generation output over 24 hours have been plotted in Figure 11.11. Exponential load model coefficients of different customers (industrial, residential, and commercial) have been taken from [32]. The wind speed and solar irradiance data have been taken from [33]. In this chapter, 500 number of scenarios (Ns) have been generated based on MCS and RWM as described in Section 11.4. Further, K-means clustering technique has been employed to reduce the number of scenarios to ten. The cost parameters of SOP and

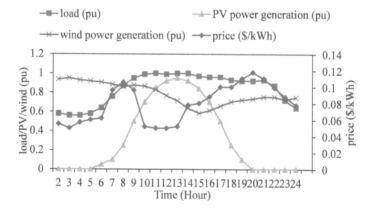

**FIGURE 11.11**  Typical PV, wind power generation, load, and its price profile.

BESS have been taken from [15] and [22], respectively. The cost of unserved energy is taken as 2000 $/MWh. The emission rate of purchased power from grid is taken as 0.4 tCO2e/MWh and the emission price is set as 60 $/tCO2e. It is assumed that load demand at all node participates to a DR program. The range of demand flexibility is assumed as ±20%. Maximum number of installed PV, wind, BESS and SOP in distribution network is assumed as two units each. The candidate locations for SOP units installation are the tie-line switches, which are connected between the buses (i.e., 8–21, 9–15, 12–22, 18–33, 25–29) as shown in Figure 11.10. The Parametric values of wind turbine generation, PV Generator, Soft Open Point Device, Battery Energy Storage System, Sodium Sulfur (NaS) have been given in Appendix.

### 11.10.3 Numerical Results and Discussions

The developed optimization problem has been solved considering five different cases (i.e. namely case 1 to case 5) as given in Table 11.2. Case 1 is corresponding to base case scenario, where investments on RES (i.e., PV and wind) are made. Case 2 and case 3 corresponds to scenarios, where investments on RES along with SOP and BESS are made respectively. Whereas in case 4, both BESS and SOP along with RES investments are made. Case 2, 3, and 4 are performed to show the importance of coordinated operation of SOP and BESS compared with acting alone. Case 5 corresponds to allocation of RES along with BESS and SOP considering ADMS scheme (i.e., VVC, DR, DNR).

Table 11.3 shows the most relevant variables such as investment cost, operation & maintenance (O & M) cost of BESS and SOPs, EPC from the grid, ENS cost and $CO_2$ emission cost of different cases. The results of case 2 reveals the coordinated allocation of RES and SOP yields savings about $256,618 (i.e., 13.721%) compared to that of case 1. Similarly, case 3 reveals the coordinated allocation of RES and BESS yields savings about $357,407 (i.e., 19.11%) compared to that of case 1. However, in case 4 coordinated allocation of RES along with SOP and BESS yields savings about $516,546 (i.e., 27.619%) compared to that of case 1. It reveals that allocation of RES along with SOP and BESS is more beneficial than that of SOP and BESS alone. The results of case 5 reveals the coordinated allocation of RES along with BESS and SOP incorporating ADMS scheme more beneficial and yields savings about $584,289 (i.e., 31.241%) compared to that of case 1. Table 11.4 summarizes the optimal location and size of installed RES, BESS, and SOP in different cases. It can be observed from

**TABLE 11.2**
**Different Cases Studied**

| Cases | RES | SOP | BESS | CVR | DR | DNR |
|---|---|---|---|---|---|---|
| Case 1 | ✓ | × | × | × | × | × |
| Case 2 | ✓ | ✓ | × | × | × | × |
| Case 3 | ✓ | × | ✓ | × | × | × |
| Case 4 | ✓ | ✓ | ✓ | × | × | × |
| Case 5 | ✓ | ✓ | ✓ | ✓ | ✓ | ✓ |

×-indicates not considered, ✓-indicates considered.

**TABLE 11.3**

**Results of 33 Bus Distribution System under Different Cases**

| Parameters | Case 1 | Case 2 | Case 3 | Case 4 | Case 5 |
|---|---|---|---|---|---|
| Investment cost ($\times 10^3$ \$) | 252.715 | 279.311 | 518.249 | 543.483 | 508.057 |
| O & M cost ($\times 10^3$ \$) | 1.756 | 4.844 | 4.558 | 7.17 | 6.9 |
| Energy purchased from grid ($\times 10^3$ \$) | 695.251 | 704.111 | 565.12 | 445.18 | 446.916 |
| Reliability cost ($\times 10^3$ \$) | 714.853 | 418.786 | 274.743 | 213.598 | 177.339 |
| Emission cost ($\times 10^3$ \$) | 205.714 | 206.620 | 150.212 | 144.313 | 146.789 |
| Total cost ($\times 10^3$ \$) | 1870.289 | 1613.671 | 1512.882 | 1353.743 | 1286.001 |
| Savings in total cost ($\times 10^3$ \$) | – | 256.618 | 357.407 | 516.546 | 584.289 |
| Total cost reduction (%) | – | 13.721 | 19.110 | 27.619 | 31.241 |

Table 11.4 that the requirement of BESS and SOP capacity substantially reduce in case 4, and case 5 compared with case 2, 3 which is occurred due to the coordinated allocation of RES along with SOP and BESS incorporating ADMS scheme.

## 11.10.4 EFFECT OF ENERGY CONSUMPTION, LOSSES, AND VOLTAGE PROFILE

Table 11.5 depicts the energy losses ($E_{loss}$) and energy consumption ($E_{cons}$) and their percentage reduction with compared to base case (i.e., case 1) for five different cases. The %$E_{loss}$ reduction yield goes up to 41.416, 61.567, 70.12, and 75.192 in case 2, case 3, case 4, and case 5 respectively. However, %$E_{cons}$ of the system has been increased

**TABLE 11.4**

**Optimal Location and Size of RES, BESS, and SOP in 33 Bus Distribution under Different Cases**

| Cases | RES/Flexible Devices | (Location, Size in MVA) | Total Capacity (MWh; MVA) |
|---|---|---|---|
| Case 1 | PV | (9, 0.33) (13,0.5) (21,0.105) | 0.935 |
| | Wind | (25,0.49) (27,0.53) (29,0.57) | 1.590 |
| Case 2 | PV | (9, 0.615) (13,0.240) (21,0.08) | 0.935 |
| | Wind | (25,0.53) (27,0.47) (29,0.59) | 1.590 |
| | SOP[#] | (12-22,0.5)(9-15,0.5) | 1.0 |
| Case 3 | PV | (9, 0.275) (13,0.66) (21,0.215) | 1.15 |
| | Wind | (25,0.616) (27,0.66) (29,0.68) | 1.956 |
| | BESS[*] | (17,1900;475) (33,2900;725) | (4.800;1.200) |
| Case 4 | PV | (9, 0.79) (13,0.19) (21,0.26) | 1.240 |
| | Wind | (25,0.64) (27,0.59) (29,0.69) | 1.920 |
| | SOP[#] | (12-22,0.41)(9-15,0.5) | 0.91 |
| | BESS[*] | (17,1925;480) (33,2782;690) | (4.707;1.170) |
| Case 5 | PV | (9, 0.72) (13,0.11) (21,0.24) | 1.070 |
| | Wind | (25,0.67) (27,0.476) (29,0.71) | 1.856 |
| | SOP[#] | (12-22,0.4)(9-15,0.5) | 0.90 |
| | BESS[*] | (17,1700;420) (33,2729;680) | (4.429;1.100) |

[#] (SOP rating in MVA)

[*] (BESS rating in MWh; MVA)

**TABLE 11.5**

**Energy Losses and Energy Consumption over a Year for Different Cases**

| Cases | Case 1 | Case 2 | Case 3 | Case 4 | Case 5 |
|---|---|---|---|---|---|
| $E_{loss}$ | 357.426 | 209.393 | 137.371 | 106.799 | 88.669 |
| $\Delta E_{loss}$ | – | 41.416 | 61.567 | 70.12 | 75.192 |
| $E_{cons.}$ | 2185.383 | 2203.432 | 2179.415 | 2217.74 | 2142.915 |
| $\Delta E_{cons.}$ | – | –0.826 | 0.273 | –1.481 | 1.943 |

by 0.826 and 1.481%, respectively. It happens because of improvement in voltage profile as seen Figure 11.12. Compared with case 1, case 5 achieved 1.943% reduction in $E_{cons}$. This happens due to lower the substation voltage to perform CVR operation, which results in lower voltage profile as seen in Figure 11.12.

Figure 11.12 shows the cumulative distribution of the average voltage values for different cases. In Figure 11.12 and Figure 11.13, it can be observed that there is a significant improvement in voltages due to the installation of BESS and SOP along with RES as seen in case 2 to case 4. Further improved voltage profile has been achieved as seen in case 5 with the participation of ADMS scheme. However, voltage profile in case 5 are flatten and maintained at lower portion of permissible limit. This happens due to the participation of DR, DNR program along with CVR operation.

## 11.10.5 EFFECT OF ACTIVE AND REACTIVE ENERGY DEMAND TAKEN FROM GRID

Table 11.6 depicts the active energy ($E^P_{grid}$) and reactive energy ($E^Q_{grid}$) taken from grid and their percentage reduction with compared to base case (i.e., case 1). With coordinated allocation of RES along with BESS and SOP, there is about 29.848%

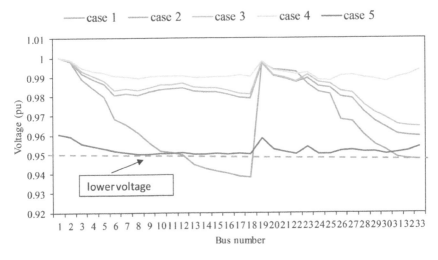

**FIGURE 11.12** Average voltage magnitude of each node for different cases.

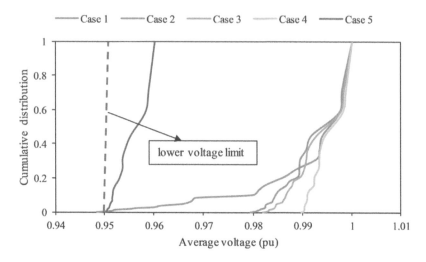

**FIGURE 11.13**  Cumulative probability distribution of average nodal voltages for different cases.

reduction in $E^P_{grid}$ as obtained in case 4. Further 1.204% more reduction in $E^P_{grid}$ has been achieved with the deployment of ADMS scheme as seen in case 5. Similarly, from Table 11.6, it can be observed that reactive power taken from grid $(E^Q_{grid})$ is drastically decreases. Case 4 achieves 64.515% reduction in $E^Q_{grid}$ achieved by coordinated allocation of RES along with BESS and SOP. Further 28.377% more reduction in $E^Q_{grid}$ integrated operation of ADMS scheme can be seen in case 5. This is achieved by the reactive compensation from SOPs and BESSs.

### 11.10.6  ERROR ANALYSIS UNDER DIFFERENT CASES

As mentioned earlier in Section 11.10.2, 500 number of scenarios (Ns) have been generated based on Monte Carlo simulation (MCS) and RWM as described in Section 11.5. The K-means clustering technique has been employed to reduce the number of scenarios to ten. The deterministic analysis has been performed for all ten scenarios. The error bar diagrams have been plotted in Figure 11.14 to depict graphical representation of variability of obtained results. Figure 11.14(a, b, and c) depicts

**TABLE 11.6**

**Active and Reactive Energy Demand Taken from Grid over a Year for Different Cases**

| Cases | Case 1 | Case 2 | Case 3 | Case 4 | Case 5 |
|---|---|---|---|---|---|
| $E^P_{grid}$(MWh) | 8571.407 | 8609.151 | 6808.710 | 6013.024 | 6116.195 |
| $\Delta E^P_{grid}$(%) | – | −0.440 | 20.565 | 29.848 | 28.644 |
| $E^Q_{grid}$(MVARh) | 12984.244 | 6762.186 | 5936.294 | 4607.496 | 922.959 |
| $\Delta E^Q_{grid}$(%) | – | 47.920 | 54.281 | 64.515 | 92.892 |

*Note:* $E^P_{grid}, E^Q_{grid}$ are active and reactive energy taken from grid, respectively.

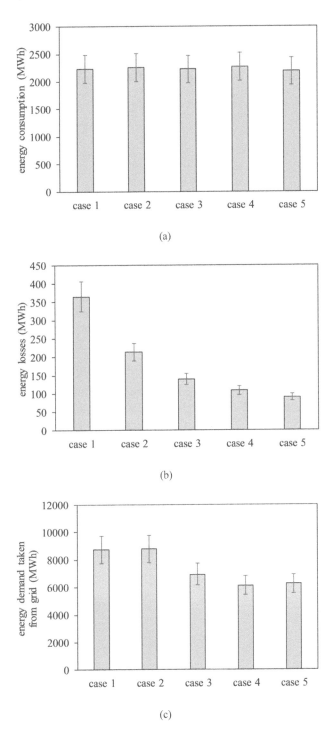

**FIGURE 11.14** Error bar diagram of (a) energy losses, (b) energy consumption, and (c) energy demand taken from grid over a year.

**TABLE 11.7**

**Energy Loss, Energy Consumption, and Energy Demand Taken from Grid over a Year for Different Cases**

| Cases | | Case 1 | Case 2 | Case 3 | Case 4 | Case 5 |
|---|---|---|---|---|---|---|
| $E_{loss}$ (MWh) | Avg. | 364.575 | 213.581 | 140.119 | 108.935 | 90.443 |
| | Std. | 41.143 | 24.103 | 15.813 | 12.293 | 10.207 |
| $E_{cons.}$ (MWh) | Avg. | 2229.090 | 2247.501 | 2223.003 | 2262.095 | 2185.773 |
| | Std. | 251.556 | 253.634 | 250.870 | 255.281 | 246.668 |
| $E_{grid}^{P}$ (MWh) | Avg. | 8742.835 | 8781.334 | 6944.884 | 6133.284 | 6238.519 |
| | Std. | 986.643 | 990.988 | 783.741 | 692.151 | 704.027 |

*Abbreviations*: $E_{loss}$, energy losses; $E_{cons.}$, energy consumption; $E_{grid}^{P}$, active energy taken from grid; Avg., average; Std., standard deviation.

error bar diagram of the active power loss, active power consumption, and active power taken grid under different cases, respectively. Further, Table 11.7 depicts the average and standard deviation of results under different cases. Among all cases, case 5 dominates the all cases in respect of reduction of $E_{loss}$, $E_{cons}$ and $E_{grid}^{P}$.

### 11.10.7 PERFORMANCE OF PROPOSED HYBRID OPTIMIZATION SOLVER

In this section, the convergence performance and robustness of the hybrid optimization solver has been tested and also compared with the conventional genetic algorithm (CGA) solver for case 7. For this purpose, the simulation has been performed 20 times with random initial data. For both solvers, same number of initial population (i.e., 30), maximum iterations (i.e., 200) have been chosen. Figure 11.15 shows

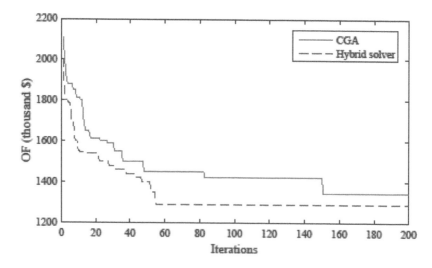

**FIGURE 11.15**   Convergence curve of different algorithms for case 5.

**TABLE 11.8**

**Performance of Proposed Hybrid Solver and CGA for Case 5**

| Solver | Best Value ($\times 10^3$ $) | Worst Value ($\times 10^3$ $) | Average Value ($\times 10^3$ $) | Standard Deviation ($\times 10^3$ $) | Time (Seconds) |
|---|---|---|---|---|---|
| CGA | 1345.65 | 2110.2 | 1460.01 | 132.55 | 310 |
| Hybrid solver | 1286 | 2002.6 | 1348.25 | 122.233 | 125 |

*Abbreviation:* CGA, conventional genetic algorithm.

the convergence pattern of both solvers. Further, Table 11.8 depicted the capability parameters such as best, worst, average, and standard deviation of fitness value of both solvers. With the comparison of CGA, proposed hybrid optimization solver converges faster, as seen in Figure 11.15. Besides, it has achieved good capability parameters as seen in Table 11.8. It is noteworthy to mention that almost 60% of computational time has been reduced using proposed hybrid solver compared with CGA. Furthermore, an attempt has been made for solving the present large complex MINLP problem using the DICOPT solver in GAMS [26], but the solution process did not converge even after 1 hour of run, therefore it was terminated.

## 11.11 CONCLUSION

This chapter investigated the coordinated allocation of RES along with flexible devices in distribution network considering ADMS schemes. A two-stage coordinated optimization framework for integrated planning model of RES along with flexible devices has been presented. Simulation results verify the efficiency and feasibility of the proposed planning model and hybrid optimization solver. The numerical simulations allow following conclusions to be drawn:

- Deployment of DNR, VVC, and/or DR scheme can reduce the requirement of RES, BESS, and SOP capacity substantially. Moreover, these schemes also result in reduction of network power loss, power consumption, electricity purchase cost, and reduction in $CO_2$ emission.
- Reactive power compensation through BESS and SOP devices is a potential candidate for VAR support for voltage regulation.
- Integrated operation of VVC, DNR, and DR schemes can improve and flatten the voltage profile without violating the permissible voltage limits.
- With hybrid optimization solver, almost 60% of computational time has been reduced compared with CGA solver.

It can be concluded that the deployment of ADMS schemes, while planning of RES along with BESS and SOP devices can simultaneously improve the economic benefits and enhance the reliability, flexibility, and environmental performance of distribution network.

# APPENDIX

### TABLE A.1
### Parametric Values of Wind Turbine Generation

| Parameters | Value (unit) |
| --- | --- |
| Rated capacity | 100 (kW) |
| Life time | 20 (years) |
| Per unit investment cost | 1219 ($/kWh) |
| Per unit operation and management cost | 0.047 ($/kWh) |
| Rated wind speed | 7.7 (m/s) |
| Cut-in wind speed | 4.3 (m/s) |
| Cut-out wind speed | 17.9 (m/s) |

### TABLE A.2
### Parametric Values of PV Generator

| Parameters | Value (unit) |
| --- | --- |
| Rated capacity | 100 (kW) |
| Life time | 20 (years) |
| Per unit investment cost | 1281 ($/kWh) |
| Per unit operation and management cost | 0.047 ($/kWh) |
| Rated illumination intensity | 500 (W/m$^2$) |

### TABLE A.3
### Parametric Values of Soft Open Point Device

| Parameters | Value (unit) |
| --- | --- |
| SOP economical service life | 20 (years) |
| SOP unit capacity investment | 308.8 ($/kVA) |
| Installation of SOP capacity | 1000 (kVA) |
| Coefficient of annual maintenance cost | 0.01 |

### TABLE A.4
### Parametric Values of Battery Energy Storage System, Sodium Sulfur (NaS)

| Parameters | Value (unit) |
| --- | --- |
| Power conversion system (PCS) | 371 ($/kW) |
| Life time | 15 (years) |
| Storage section | 327 ($/kWh) |
| operation and management cost | 0.35 ($/MWh) |
| Range of state of charge (SOC) | 10~90 |
| Efficiency | 85 % |

## TABLE A.5
## Line and Load Data of IEEE 33 Bus Distribution System

| Branch No. | Sending End | Receiving End | R (Ω) | X (Ω) | Real Power Load at Receiving End Node (kW) | Reactive Power Load at Receiving End Node (kVAr) |
|---|---|---|---|---|---|---|
| 1 | 1 | 2 | 0.0922 | 0.047 | 100 | 60 |
| 2 | 2 | 3 | 0.493 | 0.2511 | 90 | 40 |
| 3 | 3 | 4 | 0.366 | 0.1864 | 120 | 80 |
| 4 | 4 | 5 | 0.3811 | 0.1941 | 60 | 30 |
| 5 | 5 | 6 | 0.819 | 0.707 | 60 | 20 |
| 6 | 6 | 7 | 0.1872 | 0.6188 | 200 | 100 |
| 7 | 7 | 8 | 1.7114 | 1.2351 | 200 | 100 |
| 8 | 8 | 9 | 1.03 | 0.74 | 60 | 20 |
| 9 | 9 | 10 | 1.044 | 0.74 | 60 | 20 |
| 10 | 10 | 11 | 0.1966 | 0.065 | 45 | 30 |
| 11 | 11 | 12 | 0.3744 | 0.1238 | 60 | 35 |
| 12 | 12 | 13 | 1.468 | 1.155 | 60 | 35 |
| 13 | 13 | 14 | 0.5416 | 0.7129 | 120 | 80 |
| 14 | 14 | 15 | 0.591 | 0.526 | 60 | 10 |
| 15 | 15 | 16 | 0.7463 | 0.545 | 60 | 20 |
| 16 | 16 | 17 | 1.289 | 1.721 | 60 | 20 |
| 17 | 17 | 18 | 0.732 | 0.574 | 90 | 40 |
| 18 | 2 | 19 | 0.164 | 0.1565 | 90 | 40 |
| 19 | 19 | 20 | 1.5042 | 1.3554 | 90 | 40 |
| 20 | 20 | 21 | 0.4095 | 0.4784 | 90 | 40 |
| 21 | 21 | 22 | 0.7089 | 0.9373 | 90 | 40 |
| 22 | 3 | 23 | 0.4512 | 0.3083 | 90 | 50 |
| 23 | 23 | 24 | 0.898 | 0.7091 | 420 | 200 |
| 24 | 24 | 25 | 0.896 | 0.7011 | 420 | 200 |
| 25 | 6 | 26 | 0.203 | 0.1034 | 60 | 25 |
| 26 | 26 | 27 | 0.2842 | 0.1447 | 60 | 25 |
| 27 | 27 | 28 | 1.059 | 0.9337 | 60 | 20 |
| 28 | 28 | 29 | 0.8042 | 0.7006 | 120 | 70 |
| 29 | 29 | 30 | 0.5075 | 0.2585 | 200 | 600 |
| 30 | 30 | 31 | 0.9744 | 0.963 | 150 | 70 |
| 31 | 31 | 32 | 0.3105 | 0.3619 | 210 | 100 |
| 32 | 32 | 33 | 0.341 | 0.5302 | 60 | 40 |
| 33* | 21 | 8 | 2 | 2 | - | - |
| 34* | 9 | 15 | 2 | 2 | - | - |
| 35* | 12 | 22 | 2 | 2 | - | - |
| 36* | 18 | 33 | 0.5 | 0.5 | - | - |
| 37* | 25 | 29 | 0.5 | 0.5 | - | - |

* indicates tie lines of meshed distribution system.

# REFERENCES

[1] Global trends in renewable energy investment, 2020, https://www.fs-unep-centre.org/wp-content/uploads/2020/06/GTR_2020.pdf.

[2] Q. Li, R. Ayyanar, and V. Vittal, "Convex optimization for des planning and operation in radial distribution systems with high penetration of photovoltaic resources," IEEE Transactions on Sustainable Energy, vol. 7, no. 3, pp. 985–995, 2016.

[3] Y. Zhang, S. Ren, Z. Y. Dong, Y. Xu, K. Meng, and Y. Zheng, "Optimal placement of battery energy storage in distribution networks considering conservation voltage reduction and stochastic load composition," IET Generation, Transmission & Distribution, vol. 11, no. 15, pp. 3862–3870, 2017.

[4] W. Cao, J. Wu, N. Jenkins, C. Wang, and T. Green, "Benefits analysis of soft open points for electrical distribution network operation," Applied Energy, vol. 165, pp. 36–47, 2016.

[5] M. Sedghi, A. Ahmadian, and M. Aliakbar-Golkar, "Optimal storage planning in active distribution network considering uncertainty of wind power distributed generation," IEEE Transactions on Power Systems, vol. 31, no. 1, pp. 304–316, 2015.

[6] S. W. Alnaser and L. F. Ochoa, "Optimal sizing and control of energy storage in wind power-rich distribution networks," IEEE Transactions on Power Systems, vol. 31, no. 3, pp. 2004–2013, 2015.

[7] A. Soroudi, P. Siano, and A. Keane, "Optimal DR and ESS scheduling for distribution losses payments minimization under electricity price uncertainty," IEEE Transactions on Smart Grid, vol. 7, no. 1, pp. 261–272, 2015.

[8] H. Saboori and R. Hemmati, "Maximizing disco profit in active distribution networks by optimal planning of energy storage systems and distributed generators," Renewable and Sustainable Energy Reviews, vol. 71, pp. 365–372, 2017.

[9] M. S. Hossan and B. Chowdhury, "Integrated CVR and demand response framework for advanced distribution management systems," IEEE Transactions on Sustainable Energy, vol. 11, no. 1, pp. 534–544, 2019.

[10] T. Tewari, A. Mohapatra, and S. Anand, "Coordinated control of oltc and energy storage for voltage regulation in distribution network with high pv penetration," IEEE Transactions on Sustainable Energy, 2020.

[11] D. Shukla and S. P. Singh, "Aggregated effect of active distribution system on available transfer capability using multi-agent system based itd framework," IEEE Systems Journal, 2020.

[12] P. Li, H. Ji, C. Wang, J. Zhao, G. Song, F. Ding, and J. Wu, "Coordinated control method of voltage and reactive power for active distribution networks based on soft open point," IEEE Transactions on Sustainable Energy, vol. 8, no. 4, pp. 1430–1442, 2017.

[13] H. Ji, C. Wang, P. Li, F. Ding, and J. Wu, "Robust operation of soft open points in active distribution networks with high penetration of photovoltaic integration," IEEE Transactions on Sustainable Energy, vol. 10, no. 1, pp. 280–289, 2018.

[14] L. Bai, T. Jiang, F. Li, H. Chen, and X. Li, "Distributed energy storage planning in soft open point based active distribution networks incorporating network reconfiguration and DG reactive power capability," Applied Energy, vol. 210, pp. 1082–1091, 2018.

[15] L. Zhang, C. Shen, Y. Chen, S. Huang, and W. Tang, "Coordinated allocation of distributed generation, capacitor banks and soft open points in active distribution networks considering dispatching results," Applied energy, vol. 231, pp. 1122–1131, 2018.

[16] J. Wang, N. Zhou, C. Chung, and Q. Wang, "Coordinated planning of converter-based DG units and soft open points incorporating active management in unbalanced distribution networks," IEEE Transactions on Sustainable Energy, 2019.

[17] A. M. A. El Motaleb, S. K. Bekdache, and L. A. Barrios, "Optimal sizing for a hybrid power system with wind/energy storage based in stochastic environment," Renewable and Sustainable Energy Reviews, vol. 59, pp. 1149–1158, 2016.

[18] A. Nagarajan and R. Ayyanar, "Design and strategy for the deployment of energy storage systems in a distribution feeder with penetration of renewable resources," IEEE Transactions on Sustainable Energy, vol. 6, no. 3, pp. 1085–1092, 2014.

[19] P. Li, H. Ji, C. Wang, J. Zhao, G. Song, F. Ding, and J. Wu, "Optimal operation of soft open points in active distribution networks under three-phase unbalanced conditions," IEEE Transactions on Smart Grid, vol. 10, no. 1, pp. 380–391, 2017.

[20] V. B. Pamshetti, S. Singh, and S. P. Singh, "Combined impact of network reconfiguration and volt-var control devices on energy savings in the presence of distributed generation," IEEE Systems Journal, vol. 14, no. 1, pp. 995–1006, 2020.

[21] H. Geramifar, M. Shahabi, and T. Barforoshi, "Coordination of energy storage systems and DR resources for optimal scheduling of microgrids under uncertainties," IET Renewable Power Generation, vol. 11, no. 2, pp. 378–388, 2016.

[22] R. Li, W. Wang, X. Wu, F. Tang, and Z. Chen, "Cooperative planning model of renewable energy sources and energy storage units in active distribution systems: A bi-level model and pareto analysis," Energy, vol. 168, pp. 30–42, 2019.

[23] F. Scarlatache, G. Grigoras, G. Chicco, and G. Cˆart,inˇa, "Using k-means clustering method in determination of the optimal placement of distributed generation sources in electrical distribution systems," in 2012 13th International Conference on Optimization of Electrical and Electronic Equipment (OPTIM). IEEE, 2012, pp. 953–958.

[24] S. Singh, S. P. Singh, and V. B. Pamshetti, "Energy efficiency and peak load management via CVR and distributed energy storage in active distribution grid," International Transactions on Electrical Energy Systems, vol. 30, no. 3, pp. e12224, 2020.

[25] V. B. Pamshetti, and S. P. Singh, "Optimal coordination of PV smart inverter and traditional volt-VAR control devices for energy cost savings and voltage regulation," International Transactions on Electrical Energy Systems, vol. 29, no. 7, pp. e12042, 2019.

[26] V. B. Pamshetti, S. Singh, and S. P. Singh, "Reduction of energy demand via conservation voltage reduction considering network reconfiguration and soft open point," International Transactions on Electrical Energy Systems, vol. 30, no. 1, pp. e12147, 2020.

[27] Z. Wang and J. Wang, "Review on implementation and assessment of conservation voltage reduction," IEEE Transactions on Power Systems, vol. 29, no. 3, pp. 1306–1315, 2013.

[28] M.-R. Haghifam and O. Malik, "Genetic algorithm-based approach for fixed and switchable capacitors placement in distribution systems with uncertainty and time varying loads," IET Generation, Transmission & Distribution, vol. 1, no. 2, pp. 244–252, 2007.

[29] I. E. Grossmann, J. Viswanathan, A. Vecchietti, R. Raman, E. Kalvelagen et al., "Gams/dicopt: A discrete continuous optimization package," GAMS Corporation Inc, vol. 37, p. 55, 2002.

[30] P. V. Babu and S. Singh, "Optimal placement of dg in distribution network for power loss minimization using nlp & pls technique," Energy Procedia, vol. 90, pp. 441–454, 2016.

[31] M. E. Baran and F. F. Wu, "Network reconfiguration in distribution systems for loss reduction and load balancing," IEEE Power Engineering Review, vol. 9, no. 4, pp. 101–102, 1989.

[32] A. Padilha-Feltrin, D. A. Q. Rodezno, and J. R. S. Mantovani, "Volt-var multiobjective optimization to peak-load relief and energy efficiency in distribution networks," IEEE Trans. Power Del., vol. 30, no. 2, pp. 618–626, Apr. 2015.

[33] Renewable energy sources data https://maps.nrel.gov/nsrdb-viewer.

# 12 Network-Driven Flexibility Planning of Active Distribution Network

*Mahnaz Moradijoz*
Tarbiat Modares University, Tehran, Iran

## CONTENTS

## 12.1 NETWORK-DRIVEN FLEXIBILITY PROVIDERS

In Chapter 6, it has been shown that considering islanding mode operation of distributed energy resources (DERs) in active distribution networks (ADNs) would result in planning scheme having more structural flexibility. In that chapter, FMG boundaries have been determined based on the locations of protective devices before solving the expansion planning problem. It is obvious that optimal determination of these boundaries will increase network flexibility compared to the case in which the boundaries are predefined. These boundaries can be optimally designed by appropriate allocation of sectionalizing switches (SSs). In other words, existence and proper arrangement of SSs, which can be considered as network-driven flexibility providers, are other requirements to design more flexible ADNs.

In this chapter, in order to better quantify the flexibility resulting from the DER integration in the distribution network, a model for the optimal planning of SSs,

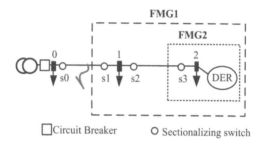

FIGURE 12.1   A simplified ADN with FGM formation capability.

which are considered as devices determining the boundaries of flexible microgrids (FMGs) in ADNs, is presented.

## 12.2   PROBLEM DESCRIPTION

A simplified ADN is given in Figure 12.1. Consider a case in which a fault occurs in line 0-1. After the fault, the restoration process is as follows:

Step 1: The circuit breaker operates and de-energizes all of the downstream load points.

Step 2: If any SSs exist, the faulty area is isolated by opening the adjacent upstream and downstream SSs.

Step 3: Based on the available DER power output and the SS locations, a microgrid is formed to restore power to some downstream customers.

Step 4: The breaker is closed in order to restore power to as many disrupted upstream customers as possible.

The restoration process in steps 1–2 and 4 are like that of in conventional distribution networks. The third step is devoted to the ADNs. Under the aforesaid fault in Figure 12.1, FMGs 1 and 2 shown with dashed and dotted rectangles respectively, could be formed. Formation of each of these FMGs depends on the power output of the DER in bus 2, and the existence or inexistence of switches s0, s1, s2, and s3. Thus, it can be stated that ADNs play two important roles:

• Fault isolation.
• FMG boundary determination.

As mentioned, the first role is considered in both passive and active distribution networks, whereas the second role is devoted to ADNs. In other words, in active networks depending on the location of SSs, various FMGs can be formed to restore loads.

## 12.3   SECTIONALIZING SWITCH PLACEMENT MODEL

As stated, it is assumed that ADN can be operated as a set of FMGs after a fault. As a result, in the long-term optimal FMG boundaries should be defined. Each FMG can restore power to its load points after a corresponding fault. However, due to the power generation limit of DERs in an FMG, some loads may be curtailed. In other words, in the short term, DNO should make a decision on the curtailment status of load points in an

FMG. Therefore, it can be stated that there are two sets of variables in the model including planning phase variables, i.e., optimal switch allocation variables, and operational phase ones, i.e., curtailment status of load points in an FMG. Since the set of the *candidate* FMGs that can restore power to a load point during a specific component outage, i.e., $MG_{ksb}$, is determined based on the *candidate* locations for installation of SSs and location of DERs before solving the optimization problem; a two-stage optimization problem can be defined for modeling SS allocation problem in ADN. As it can be seen in the following subsections, such decomposing technique simplifies the solution method of the problem.

### 12.3.1 First Stage: Load Points Curtailment Status Determination in Candidate FMGs

As mentioned before, the determination of the optimal load shedding procedure, i.e., the optimal values of $\delta_{shed}$ is a wait-and-see decision. The objective function of the first stage depends on the load shedding policy. There are different polices such as ENS cost minimization and optimal dispatch method of [1]. Since the policy used here is to curtail the loads with the low ENS cost in each FMG, the objective function of the first stage can be formulated by Equation (12.1).

For each outage event leading to the formation of a specific FMG, the operational constraints of DER units should be satisfied as shown in Equations (12.2)–(12.4). In addition to these general constraints, specific constraints related to DER type should be considered in this stage. For example, energy limited-based constraints of ESSs should be considered in this stage.

$$\text{Min}\left\{\text{OF}^{\text{FMG}}_{\omega,t,m}\right\} = \sum_{b \in \text{BMG}_m} \delta^{\text{shed}}_{\omega,b,t,m} P^d_{b,t} \text{CD}_b \tag{12.1}$$

$$P^{\text{Gen}}_{\omega,t,m} \geq (1+\xi^{res}) \sum_{b \in \text{BMG}_m} (1-\delta^{\text{shed}}_{\omega,b,t,m}) P^d_{b,t} \tag{12.2}$$

$$P^{\text{Gen}}_{\omega,t,m} = \sum_{b \in \text{BMG}_m} P^{\text{der}}_{\omega,b,t} \tag{12.3}$$

$$P^{\text{der}}_{\omega,b,t} \leq P^{\text{der,max}}_{\omega,b,t} \tag{12.4}$$

### 12.3.2 Second Stage: SS Optimal Allocation

The optimal locations of the SSs, i.e., the optimal values of $X(s)$ are here-and-now decisions, which are determined in the second stage.

#### 12.3.2.1 Objective Function

Mathematically, the aim of the SS placement problem is to choose the number and location of the switches in order to optimize the following conflicting objectives (12.5):

- Minimizing the cost of the SSs.
- Improving network reliability.

$$\text{OF}^{AS} = \text{OF}^{IS} + \text{OF}^{\text{ENS}} \tag{12.5}$$

The SS cost includes capital investment, installation, and operation and maintenance costs, which can be calculated using Equation (12.6) in which $X(s)=1$ if a switch is installed in location $s$, and otherwise, $X(s)=0$. It should be noted that in evaluating a project, the planning horizon should be compatible with the lifetime of utilized expansion equipment. In order to ensure compliance with this requirement, there are various methods for economic appraisal of a project, including the present worth method, rate of return method, and annual cost method [2]. In this study, the last method, i.e., annual cost method, is used and all cash flows of solutions are converted to a series of uniform annual cash flows.

$$OF^{IS} = \sum_s IC^{SW} X(s) \tag{12.6}$$

Commonly, the second objective is quantified using the expected energy not supplied (ENS) cost. Generally, in the SS placement problem in the passive distribution networks, the ENS cost is calculated using Equation (12.7). The interruption times of the load points can be calculated using the conditional statement given in Equation (12.8) in which $t^{int}$ represents the interruption time of a specific load point, $S_k$ denotes the set of load points whose electrical demand can be supplied by switching actions from the main grid, i.e., the load points that experience outage time equal to the switching time due to the restoration by the main grid, and $R_k$ represents the set of the load points that are being interrupted during the clearance time of the fault. $R_k$ and $S_k$ depend on the locations of the SSs and vice versa, consequently Equation (12.8) is a non-linear term.

$$OF^{ENS} = \sum_k \lambda_k^f \times \left( \sum_b P_b^d \times CD_b \times t_b^{int} \right) \tag{12.7}$$

$$t_b^{int} = \begin{cases} t_k^{sw} & if \ b \in S_k \\ t_k^{rep} & if \ b \in R_k \\ 0 & otherwise \end{cases} \tag{12.8}$$

The restoration of the loads by FMGs complicates the conditional statement required for the calculation of $t^{int}$ in ADNs. In such environments, the flowchart given in Figure 12.2 can be used to calculate the interruption time of the load points affected by a specific fault. On one hand, the flowchart given in Figure 12.2 indicates that the conditional statement required for the calculation of the ENS cost in ADNs is very complicated. On the other hand, similar to Equation (12.8), the conditional statement in ADNs is a non-linear term. In order to deal with the mentioned issues, another approach for the calculation of the ENS cost is proposed. The proposed method uses two main points as follows:

1. Utilizing continuous non-negative variables instead of discrete variable in the process of ENS calculation.
2. Modeling conditional statement related to the flowchart depicted in Figure 12.2 by a set of constraints.

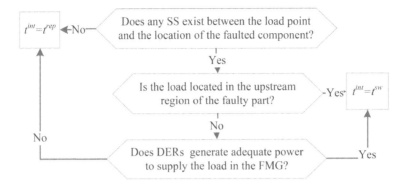

**FIGURE 12.2** Flowchart of $t^{int}$ calculation in ADN.

As given in Equation (12.9), ENS cost can be quantified by aggregating the load points' interruption costs resulted from the outages in the network components. It should be noted that in this chapter, it is assumed that the uncertain behavior of DERs can be modeled by a set of scenarios.

$$\text{OF}^{\text{ENS}} = \sum_{\omega}\sum_{b}\sum_{k}\sum_{t} \rho_{\omega}\text{CENS}_{\omega,b,k,t} \tag{12.9}$$

where CENS is a continuous non-negative variable that shows the ENS cost of load point $b$ arising from a fault in component $l$ in scenario $\omega$, and time period $t$. In other words, instead of using discrete term $P^d \times \text{CDF} \times t^{int}$ given in Equation (12.7), a continuous positive variable, i.e., CENS is used to calculate the ENS cost. Moreover, the conditional statement related to the flowchart depicted in Figure 12.2 is modeled as a set of constraints that determines the lower bound of CENS as explained in the next section.

### 12.3.2.2 Constraints

According to the flowchart given in Figure 12.2, it is clear that the outage duration of the load points affected by a specific fault is at least equal to the switching time of the circuit breakers and SSs acting to isolate the fault location. Consequently, the load point interruption cost related to the mentioned outage duration can be calculated using Equations (12.10) and (12.11). In Equation (12.11), $\tau_t$ is determined based on the number of the stages used for modeling load and DER generation variability.

$$\text{CENS}_{\omega,b,k,t} \geq \text{CDF}^{sw}_{b,k,t} \tag{12.10}$$

$$\text{CDF}^{sw}_{b,k,t} = \frac{\lambda^f_k}{8760} t^{sw}_{k,b} P^d_{b,t} \text{CD}_b \tau_t \tag{12.11}$$

If there is no switch between the load point and the location of the faulted component, the load point will experience an outage time equal to the clearance time of the fault. Therefore, if ‖(.) is intended to perform the Boolean OR operator on components of a vector, Equations (12.12) and (12.13) can be used to restrict the load point interruption cost based on the location, number, and the switching time of the protective devices, the customer damage function (CDF), and the clearance time of the fault. Clearance time includes repair duration of a fault along with the localization and isolation times of the fault. It is worth mentioning that since the optimization approach seeks to find the solution that minimizes Equation (12.9), CENS is fixed in the lower bound, so there is no need for equality operators used in Figure 12.2.

$$CENS_{\omega,b,k,t} \geq CDF_{b,k,t}^{rep}\left(1- \underset{s\in A_{k,b}}{||} X(s)\right) \tag{12.12}$$

$$CDF_{b,k,t}^{rep} = \frac{\lambda_k^f}{8760} t_{k,b}^{rep} P_{b,t}^d\, CD_b\, \tau_t \tag{12.13}$$

If a switch exists between the load point and the faulted component, the outage time of the load point will depend on the location of the load. If the load is located in the upstream region of the faulty part, the outage time of the load point will be limited to the switching time of the protective devices. Therefore, the mentioned loads will be restored by the main grid. The outage time of the affected downstream load points depends on the distribution system structure. In case of passive distribution systems, the mentioned load points are remained unsupplied for a time equal to the clearance time of the fault. However, if the system is an active one, the embedded DERs bring in new opportunities to supply the customers fed through the load points located in the downstream region of the faulty part, and consequently, the load points can be restored at a time less than the clearance time of the fault. In other words, based on the existence or inexistence of the adjacent upstream and downstream SSs opened to isolate the faulty part, different FMGs with different service areas can be formed to restore power to a specific load point. Therefore, it can be stated that for a certain fault and certain affected load point, there is a set of FMGs denoted by $MG_{kb}$ which can restore power to the load point. The formation of each FMG in this set depends on the following conditions:

1. At least one switch should exist to disconnect the FMG from the faulty part.
2. All switches that form an FMG with bigger service area should not exist.

Considering the aforementioned conditions, constraint (12.14) can be utilized to restrict the interruption cost of the affected load points located in the downstream region of the faulted component.

$$CENS_{\omega,b,k,t} \geq CDF_{b,k,t}^{rep}\, \delta_{\omega,b,t,m}^{shed}\left(1- \underset{s\in M_{k,m}}{||} X(s)\right)\underset{s\in M_{k,m}}{||} X(s) \tag{12.14}$$

$$\forall k \in BF_b,\ \forall m \in MG_{k,b}$$

where $M_{lk}$ denotes the set of switches and at least one of them should exist for the formation of the $k$th FMG under a faulty condition in component $l$, and $M'_{lk}$ represents the set of switches and all of them should not exist for the formation of the $k_{th}$

FMG under the faulty condition in component $l$. For more clarification, suppose that a fault occurs in line 1 (line 0–1) of the network given in Figure 12.1. The existence of s1 leads to the formation of the first microgrid, i.e., FMG1. However, the inexistence of s1 and existence of s2 or s3 result in the formation of the second microgrid, i.e., FMG2. It is clear that the load point 2 can be supplied both in FMG1 and FMG2. Therefore, $M_{11}=\{s1\}$, $M'_{11}=\{\}$. $M_{12}=\{s2, s3\}$, $M'_{12}=\{s1\}$, $MG_{12}=\{FMG1, FMG2\}$.

## 12.4  SOLUTION METHOD

Mathematically, the SS placement model is a two-stage MINLP problem. The first stage of the model is a simple linear optimization problem and can be solved easily. However, the second stage of the problem is an MINLP optimization problem. Such MINLP problem cannot be straightforwardly solved by conventional optimization methods. Consequently, in this chapter, an MILP-based method is employed to solve the second stage of the proposed SS placement problem as described in the following.

The first nonlinear term of the model is the right member in (12.12). Bearing in mind that the output of the Boolean OR operator is a binary number, and CENS is a non-negative variable, (12.12) can be simply linearized as follows:

$$\text{CENS}_{\omega,b,k,t} \geq \text{CDF}_{b,k,t}^{\text{rep}}\left(1 - \sum_{s \in A_{k,b}} X(s)\right) \tag{12.15}$$

Considering the definitions given in (12.16) and (12.17), (12.12) can be rewritten as represented in (12.18). According to the definitions given in (12.16) and (12.17), if all $X(s)$ belonging to $M_{lk}$ are zero, then $Y_{lk}$ will be zero, otherwise, $Y_{lk}$ will be 1. Similar philosophy holds about $Y'_{lk}$ belonging to $M'_{lk}$. These constraints can be formulated as a linear term as given in (12.19) and (12.20).

$$Y_{k,m} \triangleq \underset{s \in M_{k,m}}{\|} X(s) \tag{12.16}$$

$$Y'_{k,m} \triangleq \underset{s \in M'_{k,m}}{\|} X(s) \tag{12.17}$$

$$\text{CENS}_{\omega,b,k,t} \geq \text{CDF}_{b,k,t}^{\text{rep}}\, \delta_{\omega,b,t,m}^{\text{shed}}\left(Y_{k,m} - Y_{k,m} Y'_{k,m}\right),\ \forall k \in BF_b,\ \forall m \in MG_{k,b} \tag{12.18}$$

$$\frac{\sum_{s \in M_{k,m}} X(s)}{\sum_{s \in M_{k,m}} 1} \leq Y_{k,m} \leq \sum_{s \in M_{k,m}} X(s) \tag{12.19}$$

$$\frac{\sum_{s \in M'_{k,m}} X(s)}{\sum_{s \in M'_{k,m}} 1} \leq Y'_{k,m} \leq \sum_{s \in M'_{k,m}} X(s) \tag{12.20}$$

**TABLE 12.1**

**Truth Table**

| $Y'_{k,m}$ | $Y_{k,m}$ | $Y'_{k,m}Y_{k,m}$ | $Y'_{k,m}+Y_{k,m}$ |
|---|---|---|---|
| 0 | 0 | 0 | 0 |
| 0 | 1 | 0 | 1 |
| 1 | 0 | 0 | 1 |
| 1 | 1 | 1 | 2 |

Since $\delta^{\text{shed}}$ is independent decision variable of the first stage and it can be used as an input parameter in the second stage, $Y \times Y'$ is the only nonlinear term in the model. In order to linearize this term, the Truth table of the term is constructed as represented in the non-shaded part of Table 12.1. Moreover, the results of using the summation operator between the $Y$ and $Y'$ are shown in the shaded part of Table 12.1. Considering this table and using the definition given in (12.21), the linear form of (12.18) can be extracted as given in (12.22) and (12.23). Therefore, the second stage of the model is converted to a linear model.

$$\chi_{k,m} \triangleq Y'_{k,m}Y_{k,m} \tag{12.21}$$

$$\text{CENS}_{\omega,b,k,t} \geq \text{CDF}^{\text{rep}}_{b,k,t}\,\delta^{\text{shed}}_{\omega,b,t,m}\left(Y_{k,m}-\chi_{k,m}\right), \forall k \in BF_b,\ \forall m \in MG_{k,b} \tag{12.22}$$

$$Y'_{k,m}+Y_{k,m}-1 \leq \chi_{k,m} \leq \frac{Y'_{k,m}+Y_{k,m}}{2} \tag{12.23}$$

## 12.5  NUMERICAL RESULTS

In this chapter, numerical results for the passive and active distribution systems are examined. For this purpose, the simulations are performed on the 33 bus distribution network of Chapter 6 which is depicted in Figure 12.3. The electrical energy injected by gridable parking lots (GPLs) depends on the amount of energy stored in the batteries of vehicles. In addition to the constraints given in problem formulation section, the following constraint should be taken into account in the first stage of the model:

$$P^{\text{der}}_{\omega,b,t'} \leq \frac{\eta^{\text{dis}}\,SOC_{\omega,b,t'}}{t_k^{\text{rep}}} \tag{12.24}$$

The capital investment and installation cost of the SS is considered to be US $4700 [2]. The life period of the SS is assumed to be 15 years. The overall switching time of the switches and breakers is considered to be 10 minutes. Also, the annual interest rate is assumed to be 12.5%. Both sides of each line are considered to be candidate for installation of SSs.

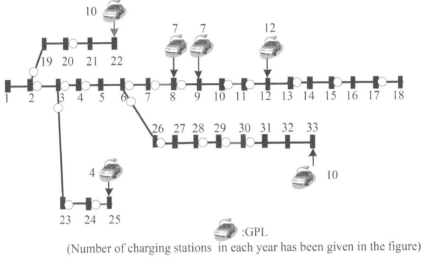

(Number of charging stations in each year has been given in the figure)

○:SS

**FIGURE 12.3** Optimal placement of SSs in the first case.

## 12.5.1 FIRST CASE RESULTS

In this case, the SS placement problem is solved in the passive distribution system. In other words, this case ignores the power injection capability of the GPLs under outage events. Therefore, only the second stage of the problem is solved.

The optimal number of the SSs which should be installed is 21. The total cost including the system ENS and SS costs is 296930 ($). If there is no SS in the system, the total cost will be 1183300 ($).Therefore, compared to the passive distribution system lacking SSs, the total cost in the first case is decreased by 74%. The optimum locations for the SSs are given in Figure 12.3. As can be seen, all switches are installed in the left side of the lines. The reason is that in this case, switches are used only for fault isolation. Therefore, in this case, switches can only reduce the time of the power interruption for the load points located in the upstream region of the faulty part. Hence, the downstream load points experience an interruption time that is equal to the clearance time of the fault. Consequently, if a switch is installed in the left side of a specific line, its switching actions can reduce the interruption time of the upstream load points under faulty condition in that line.

## 12.5.2 SECOND CASE RESULTS

In the second case, the GPLs with the capacities given in Figure 12.3 are considered as alternative supply sources under outage events. Therefore, the proposed two-stage problem is solved.

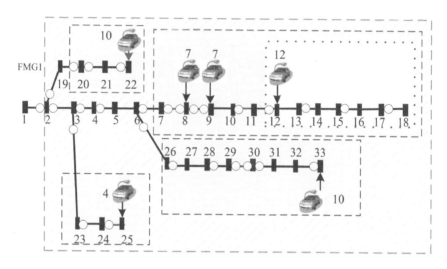

**FIGURE 12.4**  Optimal placement of SSs in the second case.

The optimal locations of SSs obtained by solving the proposed placement prob-lem are given in Figure 12.4. As can be seen in this case, the optimal number of SSs is increased to 28. Moreover, in this case, SSs are installed in both sides of several lines, whereas in the first case SSs are installed only in one side of the lines. The reason is that in this case, SSs play two roles, namely, isolation of faults and forma-tion of FMGs, some of which are shown with dashed and dotted-rectangles in Figure 12.4. Consequently, the installation of the SSs becomes more economical than the first case in which the SSs only isolate the fault. In addition, GPL located in bus 12 has led to placement of an SS in the right side of line 11-12, whereas in the previous case, the optimal location of this SS was on the left side of that line. By appropriate switching actions of SS in line 11-12, an FMG whose boundaries are shown with dotted-rectangle in Figure 12.4, can be formed under fault occurrence in line 10-11 or 11-12.

The total cost including the system ENS and SS costs is 241750 ($), which decreases by 18.58% compared to the first case. Moreover, the expected value of the ENS cost in this case is decreased by 21.32% compared to the previous case. This means that the flexibility of the ADN for supplying demands under outage events is increased by 21.32% by enabling FMG formation through appropriate placement of SSs.

## 12.6  SUMMARY AND CONCLUSION

In this chapter, an MILP-based two-stage approach for network-driven flexibility providers planning, i.e., SS placement, in ADNs is proposed to address the con-tribution of FMGs in reliability enhancement of the ADNs. This framework can accommodate the complicated SS placement problem in ADNs in which formation of FMGs is considered as a power restoration tool. Simulation results prove that the optimal placement of SSs in ADNs can enable restoration of the power to the interrupted load points by FMGs, enhance the network flexibility, and consequently can bring clear economic and technical benefits to DNO.

## REFERENCES

[1] S. Lei, C. Chen, H. Zhou, and Y. Hou. Routing and scheduling of mobile power sources for distribution system resilience enhancement. IEEE Transactions on Smart Grid, 2018.

[2] H. Seifi and M. S. Sepasian. Electric power system planning: issues, algorithms and solutions, Springer Science & Business Media, 2011.

# 13 Optimization Techniques for Reconfiguration of Energy Distribution Systems

*Meisam Mahdavi[1], Pierluigi Siano[2],
and Hassan Haes Alhelou[3]*
[1]São Paulo State University, São Paulo, Brazil
[2]University of Salerno, Fisciano (SA), Italy.
[3]University College Dublin, Dublin 4, Ireland

## CONTENTS

## 13.1 INTRODUCTION

Distribution network is an important part of the power system infrastructure that links high-voltage transmission system [1–10] to end users of electric energy [11–13] in order to deliver the electricity produced by generation system [14–16] (or generation companies (GENCOs) in deregulated environments [17]) to individual customers [18, 19]. The purpose of a power system [20] is providing reliable and safe electricity for its consumers in an economical way [21–23].

Power distribution systems in urban areas are constructed as a meshed structure and usually operated in a radial topology by opening sectional switches (normally closed) and closing tie-lines (normally open), because many operation and control activities (e.g., voltage control and protection) are based on radial structure. In simple terms, operation of distribution networks in radial configuration is cheaper because of easier coordination of protective instruments (simple protection and coordination schemes), lower fault and short circuit currents, and better voltage and power flow control. Tie-lines interconnect ends of radial feeders and/or provide connections to alternative supply points. Some of tie and sectionalizing switches are controlled manually (manual switches) and some others may be controlled automatically (remote controlled switches) [24–29].

In modern power distribution networks [30], a number of remote controlled switches are allocated to provide emergency connections and for reliability improvement [31]. In these networks, switches along feeders are automated and can be controlled using communication systems to improve network flexibility [32]. The main reason for distribution utilities (or distribution companies (DISCOs) in deregulated power systems) to invest in switching devices is to increase the network flexibility in order to prevent prolonged failures and reduce the number of customers isolated by faults [33].

The main objectives of distribution system reconfiguration (DSR) are power loss minimization [34], power quality maximization, network adequacy improvement [35–38], system stability increment [39], reliability enhancement [40], lines loading optimization [41], reduction of line maintenance costs [42–44], supply capacity expansion, load balancing, distribution generation (DG) [45, 46] penetration increment, service restoration, quick fault isolation, and implementation of preventive maintenance plans [19, 32, 47–49]. In this process, the configuration of the system changes with the status of the switches while the network has to maintain its radiality and energized loads. In general, flexibility of distribution system increases by network reconfiguration when load changes and faults happen [50, 51].

DSR can be classified into two types: static and dynamic. In static (single-stage) DSR, distribution system topology is considered to be fixed for yearly, seasonal, monthly, or shorter period. In simple terms, load is considered to be constant during these specific timeframes and switching operations are not executed frequently; while in dynamic (multi-stage) reconfiguration, the topology permanently changes with the real-time operational conditions and requirements using automatic switches. In multi-stage DSR, the distribution network has more flexibility compared to static one [24, 31, 52].

In order to simplify the dynamic reconfiguration problem, most studies determine the operation time first and then solve the problem for each sub-period (multi-period reconfiguration). In this way, network topologies remain unchanged for a given time period [24].

DSR is a large-scale combinational optimization problem including decision variables, an objective function, and a set of constraints that can often contain nonlinearities. The feasible search space in DSR is typically large, non-convex, and hard to explore. Hence, determining good-quality solutions for the DSR problem is always a challenging task. In order to cope with this issue, distribution system researchers have dedicated their efforts to develop efficient methodologies to find the best possible solution for the DSR. In this regard, classical optimization methods, heuristics, and metaheuristics have played prominent roles in the DSR solution. Since the DSR problem was first proposed in 1975 [29], classical optimization has been presented as an important tool in order to find good-quality solutions for this problem. Later, heuristic methods were adopted in DSR as a solution strategy to avoid limitations presented by classical optimization [53], for example, there is no complete mathematical model, high nonlinearities, or extremely high computational effort. Finally, by improving heuristic performances in the DSR, metaheuristics were introduced.

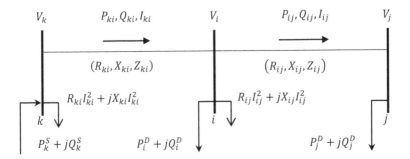

**FIGURE 13.1**    Single-line diagram of a three-node system.

In this chapter, first, mathematical model of DSR is presented and then the most important optimization techniques related to DSR problem is described and classified.

## 13.2   DSR MODELING

Figure 13.1 illustrates a radial distribution system with two lines (branches) and three buses (nodes).

According to this figure, the following set of equations can be used to calculate power flow [54, 55].

$$P_k^S = P_{ki} + R_{ki}|I_{ki}|^2 \tag{13.1}$$

$$Q_k^S = Q_{ki} + X_{ki}|I_{ki}|^2 \tag{13.2}$$

$$P_{ki} = P_{ij} + P_i^D + R_{ij}|I_{ij}|^2 \tag{13.3}$$

$$Q_{ki} = Q_{ij} + Q_i^D + X_{ij}|I_{ij}|^2 \tag{13.4}$$

$$|V_k|^2 = |V_i|^2 - 2(R_{ki}P_{ki} + X_{ki}Q_{ki}) + |Z_{ij}|^2|I_{ki}|^2 \tag{13.5}$$

$$|V_i|^2 = |V_j|^2 - 2(R_{ij}P_{ij} + X_{ij}Q_{ij}) + |Z_{ij}|^2|I_{ij}|^2 \tag{13.6}$$

$$P_{ij} = P_i^D \tag{13.7}$$

$$Q_{ij} = Q_j^D \tag{13.8}$$

where,

$$|I_{ij}| = \frac{|S_{ij}|}{|V_j|} \tag{13.9}$$

$$|S_{ij}| = \sqrt{P_{ij}^2 + Q_{ij}^2} \tag{13.10}$$

$$|Z_{ij}| = \sqrt{R_{ij}^2 + X_{ij}^2} \tag{13.11}$$

In above equation:

$P_k^S, Q_k^S$: Active and reactive power provided by substation at bus $k$.

$P_{ij}, Q_{ij}$: Active and reactive power flow from bus $i$ to $j$ (branch $i$-$j$).

$|S_{ij}|, |Z_{ij}|$: Amplitudes of apparent (complex) power and impedance of branch $i$-$j$.

$P_i^D, Q_i^D$: Active and reactive power demand at bus $i$.

$R_{ij}, X_{ij}$: Resistance and reactance of branch $i$-$j$.

$|I_{ij}|$: Amplitude of current flow from bus $i$ to $j$.

$|V_j|$: Voltage magnitude of bus $j$.

DSR is a complex and combinatorial optimization problem including integer and continuous variables with nonlinear and linear constrains. An efficient solution for this problem requires the selection of the most appropriate topology among all feasible configurations [56].

The DSR problem with objective of loss ($P_{Loss}$) reduction considering equality and inequality constraints can be formulated as follows. The goal is to reconfigure the distribution network in order to obtain minimal power loss considering operational and technical restrictions.

## 13.2.1  OBJECTIVE FUNCTION

The objective function for loss minimization can be represented by (13.12) [57, 58].

$$\text{Min } P_{\text{Loss}} = \sum_{ij \in \Omega^l} R_{ij} |I_{ij}|^2 \tag{13.12}$$

where, $\Omega^l$ is set of all branches (including normal and tie lines).

## 13.2.2  CONSTRAINTS

Following limits need to be considered during optimization of equation (13.12) without any violation:

- *Nodal active power balance*
  For any electrical network, the Kirchhoff's first law has to be established. For real power, this condition can be described as follows by general representation of equation (13.3).

$$P_i^S + \sum_{k=1}^{NB} P_{ki} = \sum_{j=1}^{NB} P_{ij} + \sum_{j=1}^{NB} R_{ij} |I_{ij}|^2 + P_i^D \qquad \forall \; ij, \; ki \in \Omega^l, \; i = 1, 2, \ldots, NB \tag{13.13}$$

  where, $NB$ is the total number of buses.

- *Nodal reactive power balance*
  Kirchhoff's first law for reactive power can be written as (13.14) by general representation of equation (13.4).

$$Q_i^S + \sum_{k=1}^{NB} Q_{ki} = \sum_{j=1}^{NB} Q_{ij} + \sum_{j=1}^{NB} X_{ij} |I_{ij}|^2 + Q_i^D \; \forall \; ij, \; ki \in \Omega^l, \; i = 1, 2, \ldots, NB \tag{13.14}$$

- *Voltage magnitude limit*
Change in voltage magnitude of buses is not desirable in distribution systems because of stability and technical concerns. Therefore, the voltage magnitude at each bus must be maintained within its limits (acceptable operating bounds) during the optimization process in the following manner [59]:

$$V_{min} \leq |V_i| \leq V_{max} \qquad \forall\ i = 1,\ 2,\dots,NB \qquad (13.15)$$

where, $V_{min}$ and $V_{max}$ are the lower and upper bound of bus voltage limits, respectively.
- *Voltage drop constraint*
Generally, voltage drop of each branch based on the Kirchhoff's second law (Kirchhoff's voltage law) can be exhibited by (13.6). Kirchhoff's voltage law (KVL) indicates that net sum of voltage drops of all branches (links) in a planar loop has to be equal to zero. Therefore, this equation does not include open branches while in reconfiguration problem some of branches need to be opened and others (tie-switches) must be closed. For resolve this problem, equation (13.6) need to be revised as (13.16) [60].

$$|V_i|^2 = |V_j|^2 - 2\left(R_{ij}P_{ij} + X_{ij}Q_{ij}\right) + |Z_{ij}|^2|I_{ij}|^2 + b_{ij}\ \forall\ ij \in \Omega^l,\ i = 1,\ \dots,NB$$
$$(13.16)$$

In (13.16), $b_{ij}$ is a slack variable that needs to be zero when branch $i\text{-}j$ is closed (i.e., the equation (13.6) is satisfied) and be a real number when the corresponding branch is open. Let us consider what happens when a switch becomes open. In this situation, $|I_{ij}| = P_{ij} = Q_{ij} = 0$ and therefore, $b_{ij} = |V_i|^2 - |V_j|^2$. According to equation (13.15), the maximum value of this continuous variable ($b_{ij}$) will be $V_{max}^2 - V_{min}^2$, and its minimum value is $V_{min}^2 - V_{max}^2$. Therefore, the following constraint can be considered for the slack variable to provide $b_{ij} = 0$ when the branch is being operated and $b_{ij} \neq 0$ when the branch is out of service.
- *Slack variable limit*
As it mentioned earlier, the slack variable constraint need to be defined as follows [61]:

$$|b_{ij}| \leq \left(V_{max}^2 - V_{min}^2\right)\left(1 - y_{ij}\right) \qquad \forall\ ij \in \Omega^l \qquad (13.17)$$

In above equation, $|b_{ij}|$ is absolute value of $b_{ij}$. Also, $y_{ij}$ is a binary variable that represents the status of each switch (0 and 1 for open and closed switches, respectively). It can be observed that $b_{ij} = 0$ for closed branches ($y_{ij} = 1$) and for open ones it will be $V_{min}^2 - V_{max}^2 \leq b_{ij} \leq V_{max}^2 - V_{min}^2$.
- *Permissible line current capacity*
The current magnitude in each branch must satisfy the branch's capacity, because of thermal and security problems. This limitation can be presented as follows [62, 63]:

$$|I_{ij}| \leq I_{ij}^{max} \qquad \forall\ ij \in \Omega^l \qquad (13.18)$$

$I_{ij}^{\max}$ is current rating limit of branch $i$-$j$. This rating value is determined according to the thermal rating limit and permissible operating capacity. Thermal limit that is defined by manufactory shows maximum current capacity of a branch without any line damage. The permissible operating capacity is a margin that a line can carry current without any security problems (a part of lines' capacity should be allocated for maintaining stability and reliability of the system in contingency states, e.g., line outages). In order to match constraint (13.18) to the proposed DSR model, (13.19) is defined [64]:

$$0 \le \left| I_{ij} \right| \le I_{ij}^{\max} y_{ij} \qquad \forall \; ij \in \Omega^l \tag{13.19}$$

This revision helps to satisfy constrains (13.16) and (13.17) for open switches, because the amount of $\left| I_{ij} \right|$ will be zero in this case. Equation (13.18) can be used instead of (13.19) by including $y_{ij}$ in the objective function and constraints (13.13), (13.14), and (13.16), but it increases the computational burden.

- *Radiality constraint*
This constraint indicates that the weekly meshed distribution network has to be operated in radial structure because of advantages were mentioned in Section 13.1. Distribution systems may be operated as meshed structure during the switching actions because of operation of some tie-lines. This is an important constrain, because non-radial structure found after reconfiguration is not practicable. In addition, large part of computational time and effort is related to find feasible radial topologies. In addition to radial solutions, the connectivity of the network has to be maintained during switching operation (i.e., none of buses must be separated from the network except for transfer nodes). An appropriate definition of this condition can help us to find more accurate solutions in shorter times. Most of specialized literature was used following equation to find feasible solutions in distribution networks with one substation.

$$\sum_{ij \in \Omega_l} y_{ij} = NB - 1 \tag{13.20}$$

It means sum of branches under operation (closed switches) has to be one unit less than total nodes. This condition is based on spanning trees in graph theory and its implementation in reconfiguration problem is easy. However, there are several substation nodes in real distribution systems. Therefore, (13.20) in this form cannot be employed to find the radial configurations. Equation (13.21) is an extension of this condition to include multi-substation distribution networks [65].

$$\sum_{ij \in \Omega_l} y_{ij} = NB - NB_s \tag{13.21}$$

where, $NB_s$ is total number of substation buses. This constraint can be applied for reconfiguration of both single- and multi-substation distribution networks. Nevertheless, (13.21) cannot guarantee network connectivity in distribution systems with transfer nodes. Power balance equations (13.13) and (13.14) can meet connectivity condition if there are no transfer buses in the network. Thus, a way to resolve this shortcoming is considering small demand values for transferring nodes. However, these small values may reduce the accuracy of solutions. Another way for solving this problem is to defined following constraint in addition to (13.21).

$$\sum_{k=1}^{NB} y_{ki} + \sum_{j=1}^{NB} y_{ij} \geq 1 \qquad \forall \ ij, ki \in \Omega^l \qquad (13.22)$$

This inequality constraint represents that at least one branch has to be connected to every node. In this way, no isolated transfer buses appeared in proposed configurations. Nevertheless, considering (13.22) in the problem constraints leads to increase the computing time. Although, these two mentioned schemes can help constraint (13.21) to find feasible radial topologies in distribution networks with several substations and transfer nodes, these modifications cannot guarantee radial configuration when size or complexity of the network increases. Therefore, following minimal spanning tree problem can be used instead of (13.21) [66].

$$\text{Min} \sum_{ij \in \Omega_l} w_{ij} \beta_{ij} \qquad (13.23)$$

subject to:

$$\sum_{j \in N_i} \beta_{ij} = 1 \forall \qquad i = 1, 2, \ldots, NB - 1 \qquad (13.24)$$

$$\sum_{l \in N_m} x_{ml,e} = 1 \qquad \forall \ m = 1, 2, \ldots, f - 1 \qquad (13.25)$$

$$\beta_{ij} + \beta_{ji} + x_{ml,e} + x_{lm,e} = 1 \qquad \forall \ ij \in \Omega^l \qquad (13.26)$$

where,

$w_{ij}$: Weight of the branch $i$-$j$.
$N_i$: Set of buses directly connected to bus $i$ in meshed network.
$N_m$: Set of nodes directly connected to node $m$ in dual graph of meshed network.
$f$: Total number of nodes in dual graph.
$\beta_{ij}$: Orientation variable of branch $i$-$j$.
$x_{ml,e}$: Status of the branch connecting node $m$ to $l$ in dual graph. Index $e$ is used to distinguish between multiple branches connecting the same two nodes.

Although, solving the equations (13.23–13.26) simultaneously can provide the radial configuration, but solution of two optimization problems (DSR and minimum spanning tree) increases the computational burden. Moreover, solving minimal spanning tree problem for large distribution systems is not easy because determination of dual graph and finding its branches which connect different nodes is boring and time-consuming. In simple words, we need to draw the dual graph correctly and find the relationship between its nodes and main graph nodes for each case study. Also, this method can only applied to meshed distribution systems with planar graph. In addition, determination of appropriate values for $w_{ij}$ is important and needs additional calculations. Therefore, set of equations (13.27–13.32) can be defined to satisfy the radiality constraint instead of (13.23–13.26) without checking the network loops and solving an additional optimization problem [67].

$$\beta_{ij} + \beta_{ji} = y_{ij} \qquad \forall \ ij \in \Omega^l \tag{13.27}$$

$$y_{ij} = y_{ji} \forall \ ij \in \Omega^l \tag{13.28}$$

$$\sum_{j \in N_i} \beta_{ij} = 1 \qquad \forall \ i = 2, 3, \ldots, NB \tag{13.29}$$

$$\beta_{1j} = 0 \qquad \forall j \in N_1 \tag{13.30}$$

$$0 \leq y_{ij} \leq 1 \qquad \forall \ ij \in \Omega^l \tag{13.31}$$

$$\beta_{ij} \in \{0, 1\} \forall \ i = 2, 3, \ldots, NB, \ j \in N_i \tag{13.32}$$

Constraints (13.27–13.32) can guarantee the radial structure of the distribution network with or without transfer nodes. However, these constraints were defined for distribution networks with one substation bus. Also, (13.28) is not necessary and it just increases computation time. Therefore, set of equations (13.27) and (13.31–13.34) can be considered in order to meet radiality condition for multi-substation distribution networks [68].

$$\beta_{ij} + \beta_{ji} = y_{ij} \qquad \forall \ ij \in \Omega^l \tag{13.27}$$

$$\sum_{j \in N_i} \beta_{ij} \leq 1 \qquad \forall \ i \in \Omega^b \tag{13.33}$$

$$\sum_{j \in N_i} \beta_{ij} = 0 \forall \ i \in \Omega^s \tag{13.34}$$

$$0 \leq y_{ij} \leq 1 \qquad \forall \ ij \in \Omega^l \tag{13.31}$$

$$\beta_{ij} \in \{0, 1\} \forall \ i \in \Omega^b, \ j \in N_i \tag{13.32}$$

where, $\Omega^b$ is set of buses without substation and $\Omega^s$ is set of buses including substation. Following equations define abovementioned radiality constraints in simpler terms [25]:

$$\beta_{ij} + \beta_{ji} = y_{ij} \qquad \forall \ ij \in \Omega^l \tag{13.27}$$

$$\sum_{j \in N_i} \beta_{ij} = 1 \qquad \forall \ i \in \Omega^b \tag{13.35}$$

$$\beta_{ij} = 0 \qquad \forall \ i \in \Omega^s, \ j \in N_i \tag{13.36}$$

$$0 \le y_{ij} \le 1 \quad \forall \ ij \in \Omega^l \qquad (13.31)$$

$$\beta_{ij} \in \{0, 1\} \forall \ i \in \Omega^b, \ j \in N_i \qquad (13.32)$$

Constraint (13.31) is not mandatory because $y_{ij}$ is a binary variable and this form of definition increases the computing time of the optimization algorithm. In addition, determination of sets $N_i$ for each distribution network takes time and needs more calculations. For these reasons following equations can be presented [69].

$$\beta_{ij} + \beta_{ji} = y_{ij} \quad \forall \ ij \in \Omega^l \qquad (13.27)$$

$$\sum_{j=1}^{NB} \beta_{ji} = 1 \quad \forall \ ij \in \Omega^l, i \in \Omega^b \qquad (13.37)$$

$$\beta_{ij} = 0 \quad \forall \ j \in \Omega^s, \ ij \in \Omega^l \qquad (13.38)$$

$$\beta_{ij} \ge 0 \quad \forall \ ij \in \Omega^l \qquad (13.39)$$

These equations can meet radiality constraint for distribution networks with different substation and transfer nodes. In order to increase the convergence speed of solution method and easier implementation, these constraints can be written as (13.27), (13.40–13.43).

$$\beta_{ij} + \beta_{ji} = y_{ij} \quad \forall \ ij \in \Omega^l \qquad (13.27)$$

$$\sum_{j=1}^{NB} \beta_{ij} = 1 \quad \forall \ ij \| ji \in \Omega^l, i \in \Omega^b \qquad (13.40)$$

$$\beta_{ij} = 0 \quad \forall \ i \in \Omega^s, \ ij \in \Omega^l \qquad (13.41)$$

$$\beta_{ji} = 0 \forall \ j \in \Omega^s, \ ij \in \Omega^l \qquad (13.42)$$

$$\beta_{ij} \in \{0, 1\} \quad \forall \ ij \in \Omega^l \qquad (13.43)$$

• *Operational constraint*
According to (13.10), there is an equality relationship between apparent power of each branch and its real and imaginary components. In order to build a convex optimization model that can be solved by commercial optimization solvers like CPLEX, this constraint need to be revised as (13.44) [70].

$$|S_{ij}|^2 \ge P_{ij}^2 + Q_{ij}^2 \quad \forall \ ij \in \Omega^l \qquad (13.44)$$

Regarding (13.9) have:

$$|S_{ij}|^2 = |V_j|^2 |I_{ij}|^2 \quad \forall \ ij \in \Omega^l, \ j = 1, 2, \dots, NB \qquad (13.45)$$

- *Active power flow limit*
  This constraint declared that real power flow of each branch has not to violate its operational limits [69].

$$|P_{ij}| \leq P_{ij}^{\max} y_{ij} \quad \forall\ ij \in \Omega^l \tag{13.46}$$

where,

$$P_{ij}^{\max} = V_{\max} I_{ij}^{\max} \tag{13.47}$$

In (13.46), $P_{ij}^{\max}$ is real power flow capacity of branch $i$-$j$ and $|P_{ij}|$ is absolute value of $P_{ij}$. This constraint is not necessary because equations (13.15) and (13.19) can meet this condition. Nevertheless, considering constraint (13.46) besides (13.15) and (13.19) decreases computing time of solution method significantly for large distribution systems.

- *Reactive power flow limit*
  Similar to active power limit, the reactive one can be exhibited as follows [69]:

$$|Q_{ij}| \leq Q_{ij}^{\max} y_{ij} \quad \forall\ ij \in \Omega^l \tag{13.48}$$

where,

$$Q_{ij}^{\max} = V_{\max} I_{ij}^{\max} \tag{13.49}$$

In above equations, $Q_{ij}^{\max}$ is reactive power flow capacity of branch $i$-$j$ and $|Q_{ij}|$ is absolute value of $Q_{ij}$. This constraint is not essential because of the same reasons which were mentioned for active power flow limit, but embedding it in the problem constraints reduces the CPU time considerably for large-scale distribution networks.

## 13.3 OPTIMIZATION TECHNIQUES

In this section, the advantages and disadvantages of several methods were employed toward the DSR problem, and their classifications are presented.

### 13.3.1 MATHEMATICAL METHODS

Mathematical optimization methods are known to be effective to solve simple and linear optimization problems with a relatively small search space, guaranteeing convergence toward the best solution [53]. However, in combinatorial explosion problems with large search space, these methods tend to demand higher and sometimes unaffordable computational efforts. Over time, numerous works regarding DSR have foreground these methods, evidencing their efficiency in this field. In 1975, Merlin and Back [71] were the first researchers who solved the conventional DSR problem based on the minimum energy loss. They formulated the network reconfiguration as a mixed integer non-linear optimization problem and solved it by a discrete branch-and-bound technique (B&B). Then, a knowledge-based software package was developed for load restoration by online network reconfiguration in [72]. For this purpose, database of the SYSRAP (A software for power system operation, analysis and control) was developed to include human knowledge modifications and load flow computation algorithms for feasible reconfiguration strategies. Although implementation of the computation algorithms in the proposed package was easy, their execution was very slow for large distribution networks.

In 1995, a 0-1 integer programing method was proposed to solve the DSR problem aiming loss reduction in ref [73]. In this method, the distribution feeders were classified to several elements (circuits) including some switches. The power loss of each element was presented as quadratic, mutual and linear terms of circuit currents, where situation of each node was defined by integer numbers 0 and 1. The "0" shows that the respective node is not connected to a related circuit while "1" indicates that there is a connection between this node and its respective circuit. However, presenting the losses in this form is not easy for medium and large distribution networks. Afterward, in ref [74], a new approach based on Newton power flow method with second derivatives was used to solve the conventional DSR problem. DSR was formulated as a mixed-integer nonlinear optimization problem, in which the status of switches and the currents flowing through the branches represent the integer and continuous variables, respectively. In order to increase the convergence speed of the proposed algorithm, a loss change estimation method was used. In this method, increase in power loss of a branch was formulated as linear and quadratic terms of its current using the corresponding elements of gradient vector and the Hessian matrix. Although, the approximations used in loss increase estimation caused the method to be very fast, it may prevent the algorithm to find high quality solutions for large-scale distribution networks.

In 2009, Khodr et al. [75] formulated the distribution network reconfiguration as an optimal power flow (OPF) problem by Benders decomposition (BD) in the general algebraic modeling system (GAMS) [76] using CPLEX [77]. The OPF was modeled as loss minimization (in terms of branch complex powers) subjecting to the nodal complex power balance, maximum and minimum complex, active, and reactive power outputs of substation, complex power and capacity limits of the lines, real and reactive power-flow equations, bus voltage magnitude and angle limits, and transformer tap limitations. In BD approach, the DSR was divided into two subproblems (master and slave). In master level, a mixed-integer nonlinear programming (MINLP) problem with a quadratic objective function is solved to determine the radial topology of the distribution network considering constraints related to line power flows. Whereas in slave one, a nonlinear programing problem (NLP) is solved by OPF in order to investigate the feasibility of the master problem solution. Although the computational results demonstrated the effectiveness and robustness of the proposed methodology for practical distribution networks, the efficiency of BD will degrade if nonlinear terms increase.

In 2012, new constraints were introduced to guarantee the radiality constraint for DSR problem using B&B method in ref. [65]. New constraints imposed the radiality condition on reconfiguration of distribution systems with several substation (generation) nodes. The main contribution was to prohibit placement of transfer nodes (nodes without generation and demand) in the end buses of proposed radial topologies.

In simple terms, such nodes had to be connected to at least two branches (lines) or must be isolated form the network. However, these constrains just increase computational burden without guarantee the global optimum and radiality constraint of large-scale DSR problems. Then, Jabr et al. [67] reformulated the conventional DSR problem using mixed-integer conic programming (MICP) and rewriting the nonlinear power flow equations in terms of rotated conic quadratic constraints. Also, they proposed a mixed-integer linear programming (MILP) model for loss reduction by replacing the conic quadratic constraints with their linear approximations in MICP. The network losses were represented in quadratic and mutual terms of nodal voltage magnitudes and sinusoidal terms of nodal voltage angles. The Numerical results obtained by CPLEX indicated that the both proposed methods (MICP and MILP) solved the problem in shorter time than the approach presented in ref. [78].

Later, Taylor, and Hover [69] formulated conventional DSR problem using mixed-integer quadratic programming (mixed-integer QP) [79], quadratically constrained programming (QCP) [80], and second-order cone programming (SOCP) [81] in AMPL [82] by CPLEX. In mixed-integer QP (MIQP) based model, two continues variables were allocated to a line according to its power flow direction (directional variables), in which, sum of them must be equal to unity and status of switch (0 or 1) (i.e., sum of these two variables was zero for open lines and could not be equal to 1 for closed switches). This new constraints forced the algorithm find only radial configurations. In QCP based framework, an inequality constraint describing the relationship of maximum complex power with its quadratic terms (active and reactive powers) was added to problem constraints. In addition, AC power-flow equations presented in ref [83] were estimated by an SOCP approximation. The accuracy of the proposed convex model was verified by comparing its results to those of BD [75] and approaches reported in refs [84–87]. However, allocation of binary values instead of continues amounts to directional variables can reduce the running time of proposed model in AMPL.

Moreover, in ref. [88], the voltage dependency of load was considered for solving the DSR problem, highlighting importance of load type in network reconfiguration. For this, the loads were formulated as quadratic functions of their related nodal voltages and network losses was represented in terms of bus voltages, and active and reactive powers. The analysis of simulation results for different types of loads (industrial, commercial, domestic, and mixed) demonstrated the importance of load modeling on DSR. In 2012, MILP model was presented in ref. [89] to reduce the distribution network losses by reconfiguration. In the proposed MILP model, the estimated power losses (objective function of ref. [88] without voltage levels) and the new inequality constraint introduced in ref. [69] (total square of active and reactive power on a branch is less or equal to its maximum complex power) were approximated by a piecewise linear function of real and reactive powers. In addition, power flow equations were replaced by a set of linearized equations. The voltage drop constraint (the second Kirchhoff' law) was also represented by approximate equation described in ref. [83]. Although, the proposed linear model can be easily solved by commercial optimization software like GAMS and AMPL using CPLEX, the approximations used may degrade the performance of this model to solve highly nonlinear combinatory DSR problems.

One year later, in ref. [90], the proposed model of [89] was developed for solving conventional DSR problem in presence of DG by representing all equations in terms of real and reactive currents. Also, in the proposed model of [90], directional continuous variables introduced in [69] were represented as binary variables. In addition, the quadratic equations proposed in [88] about voltage dependency of loads were approximated by a piecewise linear function. The proposed model could introduce a linear formulation for simultaneous network reconfiguration and DG allocation. Nevertheless, many approximations and uncertain parameters in the proposed model may degrade its efficiency for reconfiguration of real distribution systems. In addition, the radiality constraint considered in [90] may not lead to radial topologies in large-scale networks when the model is solving by traditional optimization software (AMPL and GAMS).

In 2015, static and dynamic reconfiguration problems were solved to increase the DG hosting capacity (penetration) by a multi-period OPF (MPOPF) in [91]. For this purpose, the maximization of power generated by DG sources was included in the objective function of DSR problem. The goal was to reconfigure the distribution network by remote controlled switches in order to increase the presence of DGs in distribution system. The results showed that the static or dynamic reconfiguration is more effective method than network reinforcement to accommodate larger amounts of DG in distribution systems. Furthermore, in [66], a new formulation for considering radiality constraint (spanning tree condition) in DSR problem with objective of loss minimization was proposed using graph theory. In this method, the minimum spanning tree problem [92] was solved by optimizing weighted sum of the edge connecting vertexes of original graph to those of its dual graph subjecting a simple set of equality constraints. Also, the network losses and load demand were represented as quadratic functions of nodal voltages. Although, this method can decrease the CPU time compare to widely used radiality constrain of NB−1 (number of nodes minus 1), its performance for large-scale distribution systems is difficult and may not lead to optimal solutions.

Also, in [93], simple linear current flow (LCF) equations were introduced to formulate the conventional DSR as a MILP problem. The quadratic terms of real and reactive current flow of objective function proposed in [90] were approximated by linear terms. Performing the proposed method in AIMMS [94] platform signified high efficiency of the MILP model compared to other approaches reported in [67, 69] and [95–97] from viewpoints of computation time and accuracy. For improving the models presented in [67] and [69], a binary convex model was developed to optimize separately the power loss, load balancing, and power system restoration in [70]. In the proposed framework, binary variables related to status of switches (on-off) were included completely in the formulation of [69]. The simulation results confirmed that the proposed method is faster than SOCP [69], MICP [67], and QP [69], respectively for loss minimization of large-scale distribution systems. In addition, the performance of proposed model was better than SOCP and QCP [69] for load balancing. However, the running time of binary model was longer than QP approximation for loss reduction of medium size distribution networks.

In 2016, an epsilon-constraint method was used to solve a multi-objective DSR problem, aiming loss reduction and reliability improvement in [19]. The reliability enhancement was formulated as minimization of three objective functions including SAIFI (sustained average interruption frequency index), SAIDI (sustained average interruption duration index), and average energy not supplied (average ENS (AENS)) index. In the first stage, ε-constraint method [98] was applied to solve the proposed multi-objective problem disregarding the complete load flow constraints using lexicographic optimization method. This technique guarantees that the obtained solutions of the individual optimization are Pareto optimal solutions. In the next stage, radial topologies which could not meet the power flow constraints (unfeasible solutions) were identified by forward-backward load flow and eliminated from solution list. In the last stage, TOPSIS [99] that is a multi-attribute decision-making (MADM) method was performed by genetic algorithm (GA) to choose better options among different optimal values of objective functions (make a best decision). This methodology can reduce wide range of optimal solutions and help the decision maker to decide easier. Then, in [33], a mixed-integer SOCP (MISOCP) model was presented to solve a DSR problem by incorporating costs of power loss and reliability criteria such as SAIFI, SAIDI, and ENS in one objective function. At the same time, SAIFI, SAIDI, and ENS indices were included in the problem constrains and limited by their maximum values. The results obtained by CPLEX showed that the MISOCP model can propose a reliable and economic radial topology for reconfiguration of real distribution systems.

In [56], a new robust MILP model was proposed for distribution network reconfiguration with the aim of loss minimization under load and renewable generation uncertainties. The piecewise linear approximated model was solved by a master-slave decomposition algorithm in AMPL. The proposed model mitigated the adverse impacts of improperly located DG units in the distribution system. However, more optimal solutions can be resulted if AC power flow is used to solve this robust model.

One year later, a disjunctive integer linear formulation was proposed for DSR problem aiming maximum DG integration in distribution systems using BD and estimated AC power flow equations in [31]. DG units with combined remote controlled switches were employed to simultaneously minimize DG curtailment, lines congestion, and voltage rise due to DG integration. The results evaluation revealed that solving the problem using the proposed MILP model is much faster than MINLP formulation.

In 2018, the optimal switching time was determined in order to minimize annual power loss during predefined specific periods using BD in [100]. Applying the proposed approach to a real distribution system showed that the determination of optimal switching time in online applications can leads to more efficient and realistic configurations. Also, dynamic DSR problem was solved for power restoration considering load priority and DG outputs using multi agent system (MAS) in [101]. In MAS, the optimal solution is found by interaction between generator agent including substations and DGs, load agent (load points) and bus agent (nodes) based on their communication. The proposed approach can maximize the network serviceability when small load changes occur during continuous monitoring of demand condition.

In [102], a comprehensive MILP model was presented for reconfiguration of radial distribution systems in presence of DG. In this model, network losses have been embedded inside load flow equations so that the deviations between the modeled and exact losses notably are reduced. The simulation results show that network losses can be calculated more precisely if distribution line losses are included in the load flow equations instead of considering it as power injections in each node. In [103], the extended fast decoupled Newton–Raphson method was used to improve voltage profile in reconfiguration problem. Although the fast decoupled approach can decrease considerably the simulation time compared to basic power flow equations, the efficiency of proposed method will be degraded in distribution networks with high R/X ratio lines.

In order to enhance the network reliability and reduce power losses in presence of demand response (DR), a multi-objective model was formulated for network reconfiguration using GAMS in [104]. According to the obtained results, after reconfiguring the distribution system under the proposed model, loss of power in the system is reduced and reliability of a system associated with ENS index is improved. Moreover, a multi-objective stochastic model for optimal DSR and DG allocation was developed in [105] using GAMS. The aim was maximizing the DG owner's profit and minimizing the DISCO's costs (distribution companies' costs). Also, the uncertainties of wind power generation, electricity price, and demand are handled via scenario based approach. It was shown that the contract price of wind energy considerably influences both DG owner and DISCO schedules.

## 13.3.2 HEURISTIC METHODS

As shown, solving DSR through mathematical optimization typically presents processing limitations. These methods tend to be time-consuming, due to the non-convexity of the search spaces, and this issue increases when integer decision variables are considered. Therefore, heuristic methods can be used as a solution to computational limitations. These methods analyze possible options and logically select good-quality solutions, using simple step-by-step search processes. Although heuristic methods can find feasible solutions with low computational effort, they can guarantee neither good quality nor optimality of these solutions.

In 1988, Civanlar et al. [95] formulated the conventional DSR problem using DC load flow and loss change estimation for online applications. In 1989, Shirmohammadi and Hong [84] introduced a heuristic algorithm to minimize the resistive losses by distribution network reconfiguration using Lagrange multipliers and AC load-flow equations.

At the same year, Baran and Wu [83] minimized the power losses and load balancing separately by a simple heuristic search method (branch exchanges) using exact and estimated forward-backward power flows. The total ohmic loss was formulated in terms of active and reactive powers by replacing current of each branch with its end node's complex power over voltage. The load balancing also was modeled as total ratios of end node's complex power of each branch over its maximum capacity (kVA). In branch exchanges method, a loop is allocated to every open branch (switch) by considering as if the branch is closed. Then, a new tree is created by closing an open branch and by opening a closed branch in the loop. In this way, relevant feasible trees are created starting from base spanning trees (parents) and continue by creating new ones (children). In each stage, the best one (the one with the lowest objective function without any constraint violations) among all the possible trees (children) is chosen. However, the proposed method cannot be applied to large scale distribution systems because of point to point searching process. Also, this method cannot guarantee the convergence to the global optimum in small and medium size networks.

Also, in [106], the conventional DSR problem was formulated by the loss change estimation method. In this approach, the loss change is estimated regarding the results of load flow before the switching operation. After switching, if the loss change is negative the corresponding switches are selected but if it is positive the proposal rejected and other options are investigated by the algorithm. This process continues until the best configuration is found. The proposed solution method was very robust, efficient and simple, because only the voltage drops at the extremes of the tie-line and the amount of transferred load (not its topology) were necessary for first evaluation of loss change. However, this method cannot be applied to medium or large test systems because the all switching options are listed and if one of them includes positive loss change, the calculations have to be done again. This fact makes the method time-consuming with considerable computational efforts.

In 1994, Lee et al. [107] formulated the conventional DSR as a linear mixed-integer programming problem using linearized load flow. In the proposed method, a performance index is defined for each section using dividing the active loss of each branch by its rating current. These indices were calculated and ranked for possible configurations by the linearized load flow. Then a heuristic search algorithm used to find the lowest index in order to determine the best configuration. Nevertheless, the performance of the algorithm was tested on small distribution networks.

In order to overcome the size restrictions of previously described reconfiguration techniques, [108] presented network partitioning theory to solve the traditional DSR problem. In this method, the distribution network was divided into groups of buses and the power losses between these groups were minimized. Unlike the previously described reconfiguration algorithms, in the best case scenario of the proposed method, only two load flow solutions are required. Although repetitive load flows takes a short time even for large distribution networks, it makes real-time implementation unfeasible. However, the performance of proposed technique was tested on a small size distribution network.

In 1999, McDermott et al. [109] introduced discrete ascent optimal programming (DAOP) to reduce power loss in distribution network reconfiguration. In DAOP, the optimal configurations with the smallest discrete increase in total losses are selected by picking up and adding the loads as incremental steps. For this, the objective function was defined as incremental losses divided by incremental load served. Although this heuristic algorithm models the reconfiguration problem more accurately, it takes more computational time than other reconfiguration methods.

One year later, Kashem et al. [110] presented a systematic feeder reconfiguration technique in order to maximize loss reduction in network reconfiguration problem. In the proposed strategy, system loops with positive loss reduction are chosen after running the load flow program. Then loss reduction values of their branches are calculated and the appropriate switching-options are arranged in a descending order. After checking constraints, top list switching options are selected for branch exchange and load flow program is run again. This process repeats until there is no loop in the network with positive losses. The proposed technique can find the best configuration with minimum computational effort by eliminating additional load flow calculations. However, its implementation for larger distribution networks is time-consuming.

In order to improve both the computational time and precision of reconfiguration algorithms for loss reduction, a heuristic method was proposed in [85]. In the proposed solution method, the switches are opened successively to eliminate the loops in the meshed distribution system (first, all switches were considered to be closed). Determining minimum total power loss increase by a power flow program is the criterion for opening the switches. The branch exchange (BE) technique was used to find the best configuration during opening operation. Although, reconfiguration of large-scale distribution networks is possible by this heuristic algorithm, its implementation is hard for highly combinational DSR problems. In addition, the running time has increased in comparison with some other reconfiguration algorithms.

In order to reduce the number of power flows and subsequent computational time, ref. [86] embedded the OPF problem and its constraints in the proposed heuristic algorithm of [85]. The OPF was used to determine the sensitivity of the switches that have positions close to open. By comparing the obtained results with the results of [85], it can be seen that the computation time of the heuristic algorithm with OPF has been reduced considerably. This feature may provide possibility of the algorithm implementation over real distribution systems.

In 2008, Raju and Bijwe [111] suggested a two-stage heuristic algorithm to reduce the power loss in distribution systems by reconfiguration. In the first stage, all switches are considered to be closed and the candidate ones with minimum loss increase are opened one by one according to their current magnitude. In simple words, the switch with minimum current magnitude is opened because it causes minimum deviations in other branch currents, which leads to minimum increase in loss. In the second stage, the best switching proposals are selected by branch exchange method. In spite of accuracy and simplicity of the proposed technique for reconfiguration of both balanced and unbalanced distribution networks, repetitive load flows and checking all possible proposals makes it a time-consuming method for reconfiguration of large-scale networks. In order to remove these drawbacks, they [78] presented active power loss as linear terms of loss sensitivity to the branch impedances. In this way, first the candidate branches are ranked based on their sensitivity values and then two-stage heuristic algorithm is applied. Although, the accuracy and efficiency of the proposed method were acceptable, it may not guarantee a global optimum solution because of its heuristics framework.

In 2009, Arun and Aravindhababu [112] introduced a simple heuristic algorithm for the voltage stability improvement. A voltage stability index (VSI) was adopted from [113] in order to measure the sensitivity of a bus to voltage collapse of radial distribution system. The proposed search algorithm chooses better switching combination by considering a tie-switch and its neighboring sectionalized switches, one at a time, and continues the search process until there is no further improvement in VSI.

In addition, [114] presented a new OPF-based MINLP formulation for simultaneous network reconfiguration and capacitor allocation problem under different load levels. In the proposed methodology, a new sensitivity index was introduced to determine the status of switches for loss reduction considering different levels of the daily load curve using Lagrange multipliers. Value of this index for a switch was proportional to its position and network load level. The objective function of OPF was included power loss cost and a sigmoid function of switch positions. In a sequential process, the capacitor allocation problem was solved using a heuristic constructive algorithm (HCA) [115] after and before solving DSR problem by proposed reconfiguration technique based on minimum loss cost and capacitor investment (total cost). Then the best switching strategy was selected according to minimum total costs obtained after and before reconfiguration. The authors mentioned that the proposed approach has low computational effort for simulations reconfiguration and capacitor allocation in large scale distribution systems with multiple load levels.

Moreover, a two-stage heuristic algorithm for reconfiguration of radial distribution systems with aim of minimum loss was proposed in [116]. In the first stage, feasible switch configuration is determined by successively opening the closed switches of meshed network based on minimum branch current. In the next stage, each open switch of the proposed configuration with its best neighborhood switches are updating by neighbour-chain updating process (NCUP). The results evaluation declared that the proposed heuristic method introduces better solutions than heuristic approaches reported in [84] and [111], but there is no comparison about its computing time.

In 2018, a simple heuristic method without using load flow was proposed in [117] to solve simultaneous reconfiguration and DG sizing problem, aiming loss reduction and voltage stability improvement. The voltage stability enhancement was modeled by minimization of fast voltage stability index (FVSI). The results showed that single-objective model including normalized network losses and FVSI will be solved faster than multi-objective one. In addition, it was demonstrated that DG sizing along with reconfiguration can increase loss reduction more than DG sizing after reconfiguration. It should be noted that the proposed method can reduce the losses and FVSI more than approaches reported in [118–120]. However, there is not comparison between computing time of proposed approach with other reconfiguration methods.

In order to compromise between solution speed and quality, an efficient reconfiguration algorithm based on vector shift operation (VSO) for minimizing network losses considering distributed generators was proposed in [121]. In the proposed algorithm, instead of the time-consuming load flow calculations, variation of power loss after branch exchange is calculated based on the power and resistance vectors. The final optimal solution is obtained when the open-branch set remains unchanged after BE. The results show that the VSO can solve the reconfiguration problem much faster than GA, but its implementation is more complicated than GA for reconfiguration of large-scale distribution networks.

A new heuristic methodology based on the Lagrange relaxation approach for minimizing network losses through the dynamic reconfiguration of the distribution network was presented in [122]. However, this method is relatively fast, transformation of nonlinear DSR problem to a linear form can degrade the solutions in large-sized distribution networks.

### 13.3.3 METAHEURISTIC METHODS

Metaheuristics involve applying heuristic techniques iteratively in order to find good quality solutions, using smart-criteria in the optimization process. Although they represent a higher computational burden, they can lead to better solutions when compared to heuristics. On the other hand, metaheuristics tends to find high-quality solutions with a lower computational burden in comparison with mathematical optimization methods, even though they cannot guarantee the global optimum [123].

In 1990, the problem of [83] was formulated as a two-stage multi-objective optimization problem using simulated annealing (SA) in [124] and [125]. In simple words, the objective functions of losses and maximum load balancing were minimized in the first and second stages of the proposed approach, respectively. SA mimics the physical process of annealing in solids (i.e., heating up a solid, and cooling it down until it crystallizes). It is a point-to-point search method with a strong theoretical base; it grants high skills to reach global optimum solutions, making it a robust optimization algorithm. Although, this model presented an optimal switching strategy for both minimum losses and load balancing, repeatedly run of load-flow program during the annealing process has made this approach very time-consuming method.

In 1992, Nara et al. [126] applied the standard (binary) GA [127]–[131] to optimize the distribution network losses in reconfiguration problem. They formulated DSR as 0-1 integer programing problem by representing power loss as quadratic terms of load and branch currents of each network section and inserting an integer decision variable (0 for open and 1 for closed sections) in the objective function. GA is a popular metaheuristic method that is particularly well suited for solving combinatorial optimization problems like DSR. However, the standard GA cannot applied to large-scale distribution systems because many non-feasible (non-radial) solutions (configurations) are created during recombination process that makes it a time-consuming method.

In order to reach the faster and more exact reconfiguration strategy, [132] proposed artificial neural networks (ANN) to solve the traditional DSR problem considering a variable load pattern. In this method, the ANN estimates the load level of each node according to load data and then chooses the best reconfiguration plan. However, the load characteristics of distribution system have been divided into three types (residential, commercial, and industrial loads) to minimize the set of training data. This fact can reduce the quality of the solutions especially for larger distribution networks with more switches and load details.

In 1994, Chang and Kuo [133] replaced the load-flow equations of [124] by a simplified set of equations in order to minimize the power losses. Although these modifications reduced the computing time of SA [125], in hard combinational DSR problems, the quality of the solutions may decrease because of appearing many local optimums in the search space of SA.

In 2001, Su and Lee [134] solved the DSR and capacitor setting problem for loss reduction considering nodal voltage limits using SA. The results showed that the power loss is reduced more by taking into account capacitor settings in reconfiguration problem. However, the proposed optimization method cannot guarantee the optimal solution in large distribution networks. In order to improve the performance of the SA, an efficient simulated annealing was proposed in [135] for network reconfiguration, aiming to minimize the losses. Although, the proposed method by causing to escape the algorithm from local minima gives better solutions than SA, its implementation over large-scale distribution networks is hard and time-consuming. In order to apply the GA for reconfiguration of larger distribution systems, the refined GA (RGA) was presented in [136] to reduce power losses. In this method, unlike the GA presented in [126], only tie-line (not all switches) positions were embedded in the chromosomes (strings) to reduce the string lengths and therefore computational time. In addition, an approximate fitness function was defined by combining the objective function (power loss) and the penalty function (branch current and nodal voltage constraint violations) to reduce computational burden and guide the chromosomes in search space. Also, adaptive mutation process with variable probability was used instead of fixed mutation operator to prevent premature convergence [126]. The simulation results verified that performance of RGA is better than GA (the standard GA is better than SA).

Later, artificial immune system (AIS) was proposed in [137] to solve a multi-objective DSR problem considering network losses and load balancing. The goal was minimization of power loss and loading imbalance. The loading imbalance of distribution system was defined as the deviation of each feeder loading level from the average feeder loading. AIS is a random search method based on an initial population of antibodies (antibody pool) that each antibody includes several antigens, representing problem variables (position of open tie-switches). This algorithm conducts the antibodies toward the best affinities (objectives) using selection, crossover and mutation operators. The best solutions can be obtained by interaction between multi-objective decision maker (DM) and immune algorithm (IA). Although the switching operations obtained from proposed method are almost the same as those determined by the binary integer programing (binary IP) [138], much less computing time is required by the AIS to solve the multi-objective problem. However, the performance of proposed method was not evaluated for reconfiguration of large-scale distribution systems.

Thus, [54] proposed a mixed-integer hybrid differential evolution algorithm (MIHDE) to minimize ohmic losses in distribution network reconfiguration. MIHDE is combination of hybrid differential evolution (HDE) [139] and integer programing that requires relatively less computation burden than the SA method in large network reconfiguration applications.

In addition, the multi-objective DSR problem in order to optimize the network losses and voltage profile using combination of fuzzy theory and evolution programming (EP) was solved in [140]. The power loss was modeled as difference between active power injected by substation nod and total demand of the network. Also, voltage deviation index (VDI) was defined to measure the deviation between load bus voltages after and before reconfiguration. EP [141] is a random search method that is a useful tool for solving the non-linear optimization problems. It conducts the chromosomes toward the best fitness using selection, reproduction, crossover, and mutation. Multi-objective problems can be solved using several-stage approaches (e.g. two-stage SA [125]), inherent multi-objective algorithms (AIS) [138], and integration of all objectives into one function by "weighted sum" or "fuzzy rule-based" methods. Regarding to these facts that two first techniques (several-stages and multi-objective approaches) already presented by expertized literature and identifying the weighting value of each objective is not a trivial task (in weighted sum method), here, a fuzzy rule-based method was used to solve the optimization problem. Although, the fuzzy EP (FEP) is suitable for solving multi-objective optimization problems with discontinuous solution space, accurate definition of the membership functions in fuzzy rule-based methods is a problem. Later [59] solved the problem of [140] by replacing the loss minimization with loadability maximization. Maximum loadability index (MLI) that was stated in terms of active and reactive power of lines (existing loads), indicates the maximum additional load that can be supplied by a feeder. This index helps the radial system operator to know whether the network is close to the maximum loadability limit. However, by comparing the results of [59] with [140], it can be said that maximization of MLI leads to increase network losses and degrade the voltage profile.

In 2005, Delbem et al. [142] proposed an evolutionary algorithm (EA) to minimize the power loss in distribution network reconfiguration. EA is a stochastic search algorithm based on principles of natural selection and recombination that can be used to solve relatively complex optimization problems like DSR. This algorithm evaluates a set (called population) of candidate solutions (called individuals) according to their fitness. Better solutions have a higher probability of being selected to reproduce a new population (generations). After several generations, very fit individuals are selected to find the best fitness function. However, it cannot be directly used for solving DSR problems because of creation of nonradial configurations [143]. Therefore, in order to apply the EA to the DSR problem, a new coding involving graph (a set of trees) modifications was proposed in [142]. Also, in order to increase the efficiency of the EA for reconfiguration of large-scale distribution systems, two genetic operators were developed. The simulation results show that EA can be an efficient alternative for solving the DSR problem. However, the performance of EA is drastically reduced by inadequate tree representation.

Later, [144] employed a fuzzy GA (FGA) [145] based method to solve the problem of [140]. In the proposed FGA, the fuzzy logic was used to control the mutation operator of standard GA in order to improve the convergence characteristic of GA. However, there is no comparison between the presented results with those of [141] to show performance of FGA method. Then, tabu search (TS) algorithm was employed in [146] to solve the conventional DSR problem in presence of DG resources. TS is an iterative search procedure based on movements and memory. A movement is an operation to jump from one solution to another, while memory guides the search to avoid cycles. The simulation results showed that the performance of this metaheuristic algorithm is better than GA and SA from viewpoints of both computational speed and solution accuracy. However, TS may be captured in local minima and cannot find the optimal solution because of high dependency of global search ability on tabu lengths (candidate neighborhoods).

Later, ant colony optimization (ACO) algorithm was suggested to solve the traditional DSR problem in [55]. ACO is a powerful intelligence and population-based method that has been inspired from natural behavior of the ant colonies in finding the food source and bring them back to their nest using the unique trail formation. Comparing the results obtained by this algorithm with those obtained by GA and SA, confirms the better performance of ACO.

In [147], possibility of applying the standard GA to reconfigure the large distribution systems, aiming loss reduction, was provided by restricting the search space of GA using modification of the genetic operators. In this method, fundamental loops were defined for meshed network (when all switches are closed) to generate feasible individuals (radial solutions). However, this approach only searches for the isolation of exterior nodes and does not take into account the isolation of interior nodes of the distribution networks. Therefore this strategy produces many infeasible individuals.

In order to improve the performance of ANN for solving the traditional DSR problem [132], a clustering technique was used in [96] to determine the best training set. The objective function of power loss was presented in terms of line complex powers instead of branch currents. The proposed clustering technique by putting the loads with similar demand in the same groups and identifying unknown features of each group could enhance the performance of ANN. The low processing time of proposed method makes it suitable for online applications. However, clustering the loads based on their values without considering their locations can decrease the precision of the proposed algorithm.

Furthermore, [148] used the FGA to reconfigure the distribution system in order to enhance the voltage stability. An stability index (SI) was defined to measure the voltage deviation of each bus. SI of each node was presented in terms of voltage magnitude of neighboring buses and the power flow (active and reactive powers) between them and related node. The simulation results illustrate the efficiency and effectiveness of the proposed algorithm in comparison with GA for voltage stability enhancement. Nevertheless, the FGA performance for voltage stability improvement of large distribution networks was not studied.

Das [149, 150] formulated the multi-objective DSR problem, aiming load balancing, active power loss minimization, and minimum node voltage deviation and branch current constraint violation using a fuzzy theory. Heuristic rules are also incorporated in the algorithm for minimizing the number of tie-switch operation. The load balancing of each branch was defined as difference of its current with average of all branch currents when a specified switch in a loop is open. Maximization of load balancing increases the loading margin of heavily loaded feeders by transferring part of their loads to lightly loaded feeders. The current loading index of a branch was represented as its current magnitude when a specific branch is open in a loop. If amount of this index exceeds unity it means that the branch current constraint violation has happened. The simulation results show that the currents of heavily loaded feeders are reduced by increasing currents of lightly loaded branches (i.e. more load balancing). Besides, the network losses has decreased significantly after reconfiguration. This model is useful for system operators who want to make a balance among reduction of network losses, increase in branch loadability and voltage stability enhancement. The results demonstrated that the heuristic-based fuzzy method is very faster than FGA and FEP.

In order to increase the precision of TS algorithm for loss reduction in distribution networks, improved TS (ITS) was proposed in [97] to solve the DSR problem. In ITS, mutation operator of GA was used to weaken the dependence of global search ability on tabu length. The mutation operator used in ITS can help TS to scape local minima and causes the less distribution losses. Also, [151] was presented a hybrid algorithm based on AIS and ACO for solving the problem of [149] without minimizing the branch current constraint violation. The authors claimed that performance of hybrid approach (AIS-ACO) is better than HDE [152].

Also, Carreno et al. [153] presented an efficient GA to minimize the network losses in DSR problem. The proposed GA was adopted from [154] with some modifications in codification and recombination (crossover without mutation) operations. In the proposed codification and reconfiguration, only radial topologies could be produced in initial population and next generations, respectively. The authors claimed that none of genetic algorithms such as standard GA, RGA, FGA, and FEP can generate new radial topologies efficiently like the proposed GA. However, they could not show firmly whether the proposed GA is better than other genetic algorithms.

Then, Enacheanu et al. [155] solved the conventional DSR problem by a GA based on Matroid theory (GAMT). In this method, the crossover and mutation operators of standard GA were formulated using Matroid and graph theories to maintain radiality of proposed topologies after genetic operations. Also, the chromosome codification was similar to RGA, i.e. individuals include just tie-switches. The simulation results proved that the proposed GA is much faster than standard GA. However, the performance of the proposed method has not been compared to those of RGA and other GA based methods.

In 2008, discrete particle swarm optimization (DPSO) algorithm [156] was used in [157] for loss reduction and load balancing separately in DSR problem. Particle swarm optimization (PSO) [158, 159] is a useful tool for the optimization of engineering problems, which utilizes swarm intelligence (the cooperation and competition between the particles in a swarm). Original PSO is based on real numbers while the DSR is an integer optimization problem. Therefore, the integer variables have to be codded as discrete or binary. Although, the results obtained by DPSO are almost the same as results of binary PSO (BPSO) [160, 161] for large networks, in ref. [157], discrete PSO was used because of its simple codification and implementation. However, this method is often not efficient because the extremely large number of unfeasible non-radial solutions appearing at each generation will lead to a long computing time before reaching an optimal solution.

Afterward, Chang [57] solved the conventional DSR problem considering capacitor placement using ACO. He concluded that simultaneous feeder reconfiguration and capacitor placement leads to less power loss than when DSR problem is solved without capacitor placement. In addition, it was shown that the average loss reduction ratio and average CPU times obtained by the ACO are less than those of the SA and GA in small distribution networks. However, in real distribution systems, the proposed technique was not faster than SA and GA. In addition, hyper cube

ACO (HC-AC) was proposed in [74] to minimize distribution losses by network reconfiguration. Two heuristic rules (local and global) were used to improve ACO performance. The aim of local heuristic rule is preparing the candidate configurations for successive random selection. Whereas the aim of global one is maintaining some already found successful configurations. Simple implementation and parameter setting are two important features that caused better performance of the proposed method with respect to SA from viewpoints of accuracy and computational time.

Later, [162] proposed a fuzzy logic based method to solve the problem of [112]. The results evaluation revealed that solution methods presented in [112] and [162] can decrease the loss and increase the voltage stability more than FGA [148]. Nevertheless, the efficiency and effectiveness of the proposed algorithms were evaluated in small size distribution systems.

Also, [62] presented a modified BPSO (MBPSO) algorithm for loss reduction in DSR. Some parameters of standard PSO such as inertia weight, number of iterations, and population size were modified in the proposed algorithm. This setting allows the PSO to explore a large area at the start of the simulation and continues its searching in a smaller area when it is near to global optimum. This feature makes the algorithm faster than BPSO and DPSO, but increase probability of its capture in local minima. Although the solutions obtained by proposed PSO were the same with solutions of SA and TS, its computational burden is less than these mentioned algorithms (SA and TS). Later, the problem of [62] was solved by the same authors [63] using a modified TS algorithm (MTS). In the proposed algorithm, unlike TS, the size of tabu list varies with the change of system size and a random multiplicative move is used in the search process to diversify the search toward unexplored regions. A tabu list with variable size helps the algorithm to escape from local optimums by preventing the TS to cycle around these minimum points. Although the precision of MTS algorithm is more than TS, SA, and branch exchange [83], it has no superiority with respect to ITS [97].

In [64], a dedicated GA was used to solve the network reconfiguration and capacitor allocation problem under three different load levels (light, average and heavy). The objective was to minimize the capacitor investment (costs of fixed and automatic capacitor banks) and power loss cost. The conventional GA (both binary and decimal) cannot be directly applied for solving the reconfiguration problem, because the mutation and crossover operators could destroy radial structure of proposals and makes GA very time-consuming method. Therefore, in the proposed GA, the initial population was constructed by a heuristic algorithm based on two sensitivity indices (one for DSR and another for capacitor placement). A triple crossover operator was defined for different parts of chromosome, in which first, second and third parts indicate switch numbers, numbers of fixed and automatic capacitor banks, respectively. For the second and third parts of chromosome, a mutation operator replaced the value of the random chosen gene with pre-established values of capacitors banks, while for the first part a mutation algorithm was defined to avoid the creation of islands or loops in the system (non-radial configurations). Although the sensitivity indices used in dedicated GA significantly reduces the search space and also the computational burden, there is no comparison between the results of this paper with other specialized literature.

In order to improve the solution methods of [140] and [59], [163] presented gray correlation analysis (GCRA) instead of fuzzy logic to integrate different objectives (real power losses, maximum voltage drop, and load balancing) into one objective function in EP. GCRA [164] was employed for quantitative measurements of candidate solution set in selection and competition process of EP. The results evaluation showed that the proposed method is more suitable than EP [141] (weighted sum methods) and fuzzy rule-based methods like FEP and FGA. For easier implementation and less computational time of solution method presented in [142], [165] employed the node-depth encoding (NDE) [166] instead of graph chains representation (GCR) to minimize the network losses. Also, the population in EA was subdivided into subpopulation tables related to different objectives in order to enable the proposed strategy for solving multi-objective reconfiguration problems. The proposed multi-objective EA with NDE (MEAN) could reconfigure very large-scale distribution networks (with thousands of buses and switches) so faster than MIHDE, ACO [151], restricted GA [147], and plant growth simulation (PGS) [167]. Nevertheless, the proposed technique can not guarantee the global optimum because of stochastic nature of EA.

Furthermore, in [168], a new FGA was proposed to minimize real power losses, node voltage deviation, branch current constraint violation, and number of switching operations in a multi-objective framework. The formulation of network losses, voltage deviation, and current constraint violation are similar to [149] and [150]. Minimizing the number of switching operations reduces the operating cost, switching transients and enhances the life of the switchgears. The proposed FGA was combination of fuzzy system [149, 150] and adoptive GA (AGA) method. AGA is a modified version of GA proposed in [85] that, in addition to fundamental loops, used common branches of each node and prohibited group of switches to avoid the generation of any infeasible individuals (non-radial solutions) during run of GA. Hence, the proposed approach could overcome the shortfall of the method proposed in [147] by producing just radial configurations. It was mentioned that the proposed technique (fuzzy AGA) has better performance than SA, FGA and fuzzy enhanced GA (fuzzy EGA) [169], but it is not clear that whether it can be applied to very large distribution systems like MEAN.

Moreover, Wu et al. [170] solved the problem of [83] by ACO considering DG resources. They concluded that lower system loss and better load balancing can be obtained in presence of DG. Also, Chandramohan et al. [171] formulated a multi-objective DSR problem in a deregulated electricity market using non-dominated sorting genetic algorithm (NSGA). The goal was reconfiguration of distribution system in order to minimize operating costs of DISCOs (active and reactive distribution losses multiplied by real and reactive power pricing, respectively), customer interruption cost (maximum reliability), and voltage deviation (VDI index). NSGA is a combination of GA and pareto techniques that enables to evaluate different objectives without integrating them into one objective function. First time, this method was used in [172] for distribution network design by finding more optimal radial configurations from meshed network (a kind of reconfiguration), in which network design costs and energy not supplied (ENS) were optimized in two different objective functions. Although the proposed model could give the decision makers various options, the accuracy of obtained solutions was not validated.

In 2011, fuzzy ACO (FACO) was proposed in [56] to solve the problem of [124]. Comparing the simulation results to those of [125] clears that FACO can reduce power losses and improve load balancing more than SA. Although it was shown that the performance of FACO is better than RGA, FGA, FEP, and PGS from viewpoint of loss minimization, there is no comparison between proposed method with these mentioned algorithms (RGA, FGA, FEP, and PGS) from viewpoint of power quality enhancement (voltage improvement).

Then, Rao et al. [173] applied the harmony search algorithm (HSA) to solve the conventional DSR problem. The HSA is a new meta-heuristic population search algorithm that was derived from the natural phenomena of musicians' behavior when they collectively play their musical instruments (population members) to come up with a pleasing harmony (global optimal solution). The HSA is simple in concept, less in parameters, and easy in implementation that includes some features of other meta-heuristic algorithms such as TS (history of past vectors saved in harmony memory), SA (harmony memory with variable adaptation rate), and GA (management of several vectors simultaneously). The computation results demonstrated that the HSA converged to optimum solution (minimum loss) quickly with better accuracy compared to GA, TS, and RGA. Creation of a new harmony (vector) form all harmonies in the harmony memory (all existing vectors) instead of generation of a new vector from only two vectors (parents) is the main reason of better performance of HSA with respect to genetic algorithms (GA and RGA). However, in HSA, determination of the penalty coefficients of fitness (cost) function is difficult because it does not follow a specific rule.

In order to accelerate DPSO computations and increase MBPSO accuracy, the enhanced integer coded PSO (enhanced ICPSO) was introduced in [174] to minimize distribution losses by reconfiguration. The similar idea of DPSO was used to formulate the position and velocity vectors of each particle (i.e., the particle position was rounded to the nearest discrete value and the particle velocity was updated according to integer values). The main difference between DPSO and ICPSO is related to their codification. In ICPSO each particle includes just integer numbers 0 and 1 that represent statues of switches (on or off) while in DPSO a particle exhibits positive integer values including bus numbers. For enhancement of ICPSO precision, the local optimal solutions found during the evolution process were included in local optimal list (LOL) to use in creation of new generation. Also, for further improvement of the proposed method, the inertia weigh used in MBPSO was employed in EICPSO. The results showed that EICPSO is very faster than DPSO, GA, and MBPSO, respectively. However, its precision like MBPSO and GA is less than DPSO. This feature makes MBPSO an efficient method for online applications. Nevertheless, it is predicted that the quality of solutions decreases with increase in size of distribution network more than other mentioned methods (DPSO, MBPSO, and GA). Afterward, adaptive ACO (AACO) was presented in [175] to decrease distribution loss through network reconfiguration. The conventional ACO was adapted by the graph theory to always create feasible radial topologies during the whole evolutionary process. The authors showed that reconfiguration of the distribution system by AACO can optimize the network losses more than ACO, AIS-ACO, RGA, GAMT, efficient GA [153], and ITS. Nevertheless, the performance of this algorithm has not been compared with that of AIS-ACO and RGA for large distribution systems.

In order to easier implementation and improvement of ACO for distribution network reconfiguration, the power losses were minimized by a HC-ACO algorithm in [176]. Hyper-cube (HC) framework [177] by limiting the range of pheromone variation to the interval [0 1] (changing the pheromone update rules of ACO) increases the robustness of ACO. Although results evaluation revealed that the solutions obtained by HC-ACO are the same as solutions obtained by SA, efficient SA [135], TS, ITS, and MTS, its implementation is easier than ACO and these mentioned algorithms (SAs and TSs). Then, bacterial foraging optimization algorithm (BFOA) was used in [178] to optimize the objective function of [95]. BFOA is a global optimization algorithm that has been inspired by the social foraging behavior of Escherichia coli [179]. This meta-heuristic algorithm using chemotaxis, reproduction, elimination, and dispersal operators guides the bacterium toward best fitness (health). The simulation results indicated that BFOA could decrease the losses more than approaches reported in [84–87, 109], and other metaheuristic algorithms like RGA, ACO, and AGA. However, the computing time of proposed method has not been compared to other reconfiguration algorithms.

Afterward, a new model for multi-objective DSR problem considering network losses and reliability using BPSO was presented in [180]. In the proposed model, the network reliability was formulated based on minimal cut sets between feeder (substation node) and load points, in which each series cut set has consisted of several parallel components. In this manner, the unavailability (reliability index) of the network that has been represented in terms of unavailability of cut sets had to be minimized. This framework can propose configurations with respectively minimum losses and maximum reliability. The authors mentioned that the proposed model can formulate the network reliability better than Monte Carlo simulation (MCS), because MCS is time-consuming method for both on and offline optimization problems like DSR. However, the results of this paper has not been validated by other reconfiguration algorithms.

In 2013, simultaneous distribution network reconfiguration and DG placement problem was solved with objective of loss minimization using HSA in [118]. In this model, sensitivity analysis was used to identify optimal locations for installation of DG units. The simulation results showed that the HSA decreases the network losses more than GA and RGA, respectively in presence of DG. Also, it was shown that solving the DSR problem and DG placement problem simultaneously reduces the power loss more than only DG placement or reconfiguration.

In [181], distributed wind power generation (WPG) and fuel cells (FC) were incorporated in a multi-objective DSR problem considering uncertainty in the WPG output and load demand. The goal was minimization of power loss, bus voltage deviation, generation cost, and total emissions using fuzzy adaptive PSO (fuzzy APSO). A probabilistic power flow was developed based on the point estimate method (PEM) [182] for considering uncertainties of generation and load in reconfiguration of radial distribution networks. The APSO [183, 184] was based on modifications of some features of PSO such as updating inertia weight factor by an iterative chaotic method (the logistic map) [185] and improvement of the swarm population movement and diversity. The simulation results showed that the proposed APSO method performs better than the standard PSO and GA. In addition, the proposed fuzzy APSO (FAPSO) is an efficient method to handle uncertainties in multi-objective DSR problems. Later, in [186], the switching operation cost was included in the objective function of a probabilistic DSR problem in addition to cost of power losses. The probable seasonal power loss was calculated during one year and weighted based on its probability of occurrence. Also, the combined model of generation and demand was presented using load profile and probability of power produced by wind [187], solar [188] and biomass [189–191] DGs. The results obtained by GA demonstrated that significant reductions in the seasonal, annual energy losses and operational costs. Nevertheless, the simulation results have not been validated by other optimization methods.

Also, in [192], a hybrid ACO and GA algorithm was applied to minimize the power loss in distribution network reconfiguration. In the proposed technique, the crossover operator was used to improve the pheromone process of ACO. It was shown that efficiency of the proposed method is similar to AIS-ACO and efficient GA, and better than MIHDE. However, there was no comparison between the effectiveness of the proposed methodology and HC-ACO, because in the both algorithms, the performance of pheromone has been improved. In order to enhance the performance of GA for solving the conventional DSR, new improvements for genetic operators were considered in [193]. In the proposed GA, after producing the initial population using branch-exchange heuristic algorithm, the integer variables are decoded based on *branch list* instead of nodes-branches incidence matrix. Also, selection operator was defined as an exponential function using ecological niche method instead of Monte Carlo and tournament mechanisms. The authors claimed that the implementation of the proposed GA (SOReco) is enough simple to obtain a fast convergence and is complex enough to obtain a good quality solution in comparison with other GAs. However, there is no comparison between the performance of new approach and previous GA-based methods.

One year later, in [194], a method based on AIS and clonal selection [195] (CLONR) was presented to optimize distribution losses through network reconfiguration. The simulation results showed that the proposed method can improve the loss reduction more than MIHDE and the heuristic technique proposed in [78]. Then, the quantum-inspired binary firefly algorithm (QBFA) was applied to improve the power quality and network reliability via DSR in [196]. The objectives were to minimize separately number of voltage sages and reliability criteria like the ASIFI (average system interruption frequency index), SAIFI, and MAIFI (momentary average interruption frequency index). The results revealed that the performance of QBFA was better than binary firefly algorithm (BFA) [197] and binary gravitational search algorithm (BGSA) [198] for power quality and reliability improvement. However, optimization of mentioned objective function separately without considering losses cannot lead to an optimal solution for DSR problem. Later, in [199], a multi-objective framework, aiming minimum losses, DG cost and greenhouse gas emissions, as well as maximum voltage stability was proposed for simultaneous optimal reconfiguration and DG allocation using fuzzy hybrid big bang-big crunch (fuzzy HBB-BC). The HBB-BC is a combination of big bang-big crunch (BB-BC) [200] and PSO algorithm that converges to optimal solution using *center of mass* and the *best position of each solution* operators of BB-BC and the *best global position* operator of PSO. BB-BC is adapted from two universe evolution theories of Big Bang and Big Crunch that has lower computational burden and higher convergence speed than other population-based methods [201]. The simulation results indicated that simultaneous DG allocation and network reconfiguration reduces the losses, DG costs, and emission and increase the voltage stability index (VSI) more than separate DG allocation and reconfiguration. In addition, it was shown that the fuzzy HBB-BC (FHBB-BC) minimizes losses more than HSA and FACO and maximizes VSI more than GA.

Moreover, in [202], the goal was network reconfiguration and capacitor placement simultaneously in a multi-objective framework to reduce power losses using fuzzy BGSA (FBGSA) and fast forward-backward harmonic power flow. The Weibull-Markov stochastic model was applied for reliability assessment instead of time-consuming MCS method and unrealistic Markov models using SAIFI [203] (ASIFI [196]) and system average interruption unavailability index. The system average interruption unavailability index described the average interruption time at the load points (similar to SAIDI). The interruption cost in load point including customer damage function and its probability was minimized to increase reliability level in addition to optimization of power loss and capacitor installation cost. The simulation results showed that installation of shunt capacitors can reduce the losses and improve the network reliability in an economic way.

Then, in [204], multi-objective DSR problem, aiming loss reduction, load balancing, power quality improvement, and minimum number of switching was optimized using fast NSGA (FNSGA) [205]. The proposed FNSGA can improve the convergence speed of standard NSGA by employing the codification method of AGA and guided mutation operator. The results evaluation revealed that the proposed method can find the optimal solution faster than MIHDE, ACO, GAMT, FGA, fuzzy EGA (FEGA), dedicated GA (DGA), fuzzy AGA (FAGA), and non-revisiting GA (NrGA) [206]. Afterward, a CLONR based method was employed to solve a multi-objective DSR problem considering network losses and reliability in [207]. The system reliability enhancement was formulated by minimization of the power interruption equivalent frequency index (PIEFI). In the proposed CLONR method, the Prim's algorithm was used to create radial configurations in initial population. The proposed model is helpful for reconfiguration of distribution networks in restructured power systems, where the reliability enhancement is as important as loss reduction for DISCOs. However, the performance of the proposed method has not been compared with other reconfiguration algorithms.

Furthermore, Naveen et al. [208] showed that modified BFOA (MBFOA) algorithm can optimize the problem of [178] more efficient and faster than BFOA. It was also concluded that MBFOA can find more optimized solutions compared to MIHDE, HDE, HSA, and the heuristic method presented in [209]. Then, a new formulation was proposed by [210] for annual loss reduction in DSR problem considering variability of active and reactive power demand and DG profiles using GA and fuzzy C-means (FCM) [211] methods. FCM that is a popular clustering algorithm for pattern recognition was used to classify the data (annually repetitive patterns) of load and DG profiles. These points (data) were used by GA to find the optimal configuration for loss reduction. The network losses (objective function of [67]) with penalty function of voltage deviation and radiality constraint were incorporated in the fitness function of GA as weighted sum. The results showed that the different annual losses are obtained from the optimal solutions found in previous literature on the field of DSR. This fact indicated the importance of accurate modeling of load and DG outputs.

In order to solve the simultaneous DSR and DG allocation more robust and faster than the previous proposed algorithms, a simple methodology based on standard GA, branch exchange, sensitivity analysis, and backward–forward sweep power flow was presented in [212]. The results evaluation verified the more robustness and less computing time of proposed method compared to GA, RGA, SA, ACO, HSA, and multi-objective honey bee mating optimization (MHBMO) algorithm [213].

In 2016, Asrari et al. [32] proposed a fuzzy adaptive parallel GA (fuzzy adaptive PGA) for solving a multi-objective DSR problem under load variations in presence of DG. The objectives were power loss reduction and minimization of nodal voltage deviations. In order to reduce load scenarios, FCM was applied to provide essential data of load profile for adaptive PGA (APGA). Also, Dandelion coding method [214] was used to find optimal feasible spanning trees (radial configurations) [215]. This coding scheme is very appropriate for solving large-scale DSR problems because its locality is improved with increase of problem size. PGAs have more efficiency and higher speed than GAs because of additional availability of CPU. In the proposed APGA, several populations exchanged their chromosomes occasionally based on a variable (not fixed) *migration* process. Comparing the results obtained by fuzzy APGA (FAPGA) to those obtained by fuzzy PGA (FPGA), FGA, and fuzzy PSO (FPSO) signified the better performance (shorter computation time as well as less power loss and voltage deviation) of the proposed methodology. Nevertheless, the efficiency of the proposed method has not been investigated for large distribution networks.

Moreover, in [216], a discrete teaching-learning based optimization (discrete TLBO) algorithm was employed for loss reduction and voltage profile improvement, separately, in presence of DG. TLBO [217] is a new meta-heuristic optimization technique which is based on the concept of teaching (teacher phase) and learning (learner phase) process in a class. It was concluded that the discrete TLBO (DTLBO) can decrease the power loss and improve the voltage profile more than PSO. Then, an AIS algorithm [218] was applied to solve the conventional DSR problem under a variable load demand in [49]. The simulation results showed that the proposed AIS is an efficient method for reconfiguration of distribution systems with variable loads.

On year later, in [219], distribution system was reconfigured to minimize the power loss and voltage deviation considering coordination of protective devices using a GA [147]. The normalized network losses and total nodal voltage deviations respect to their basic values (loss and voltage deviation before reconfiguration) were integrated in one objective function. In addition, normal constraints of DSR problem with operational and coordination constraints of protection devices were included in the objective function as normalized penalty terms. The authors mentioned that the optimal topologies obtained from conventional reconfiguration (without considering coordination between protective devices such as overcurrent relays, reclosers, and fuses) cannot be used in real distribution networks, because reconfiguration inevitably affects the protection system. Then, in [220], simultaneous network reconfiguration and capacitor switching problem in presence of DG considering switching cost, cost of buying power from the substation, customer's interruption cost and transformer [221] loss of life cost was solved using ACO. The simulation results revealed that simultaneous feeder and capacitor switching can reduce the total grid cost more than conventional reconfiguration or capacitor switching.

Later, voltage security index (NVSI) and node voltage quality index (NVQI) were developed in [222] to find more secure networks by reconfiguration. The objectives were separately maximization of NVSI, NVQI, and loss reduction considering voltage dependency of loads using HSA and a heuristic algorithm. The model is useful for network reconfiguration in deregulated energy markets because the voltage security is an important issue there. Increase of voltage security could improve the loadability limits and reduce the network losses. Also, it was showed that including NVSI and NVQI indices in DSR formulation can improve the convergence speed of solution algorithms. After that, the DSR problem was solved in [48] to increase the annual investment return considering hourly, weekly and seasonal load profiles using DPSO. The return on investment (ROI) was formulated as difference of cost saving due to loss reduction with investment cost (IC). IC included capital cost, construction cost of communication equipment, and annual maintenance cost of remote control switches. The simulation results confirmed that large amount investment can be saved through reconfiguration of distribution systems by automatic switches.

In 2018, reactive power loss was minimized in order to enhance the distribution system loadability limit by reconfiguration using improved HSA (IHSA) [223] and BE in ref. [224]. In the proposed approach, first, the reactive losses is minimized using branch exchange method and after if the system loadability is near to its critical limit, HSA is used to optimize the problem. The simulation results confirmed that the proposed solution methodology is much faster than IHSA and greedy wolf optimizer (GWO) [225].

Later, in ref. [226], $N-1$ security criterion was considered in reconfiguration of distribution systems with objective of loss minimization using an improved binary GA [227]–[236]. Although, power loss has been increased by considering $N-1$ security criterion, more reliable distribution network was achieved.

Furthermore, in ref. [24], a multi-period DSR problem was solved in order to increase the daily hosting capacity of distribution network for DGs by minimizing the number of switching using FCM and a hybrid PSO (combination of BPSO and DPSO). In this approach, FCM was used to cluster the different scenarios of switching operation and output uncertainties of DGs. Nevertheless, the authors mentioned that the computing time and accuracy of solutions can be improved further by new meta-heuristic algorithms and developing a stochastic model.

Also, in ref. [237], a method based on combination of TLBO and epsilon-constraint method was proposed to maximize the loadability of distribution networks by simultaneous reconfiguration and optimal allocation of distributed energy resources. The results demonstrate that the proposed technique is more effective than artificial bee colony (ABC) algorithm [40, 238] and PSO. Although the method cannot solve the problem very fast, but its computational time is shorter than TS, HSA and GWO.

Recently, in ref. [239], DSR problem [240] was solved in presence of power flow controllers like flexible DC device (FDD), using an improved PSO [241–245] by a sequential AC/DC power flow method. Results show that the FDD have stronger ability to promote the performance of distribution system with power flow optimal adjusting and cooperation with traditional switch.

## 13.4 SUMMARY

A review and classification of the most significant works regarding optimization methods of DSR problem was presented in this chapter. This chapter represents a valuable tool for anyone associated with this research field, as it provides a broad literary framework that can be used as a base for any further investigations related to DSR. Therefore, distribution system operators and researchers can read this chapter in order to improve upon previous solution methods, and they can propose more efficient methodology to better exploit existing infrastructure. The presented classification shows that most researchers have focused on mathematical and metaheuristic approaches as good solution methods for the DSR problem. However, due to large-scale combinational feature of the DSR problem, metaheuristics have been the most commonly used solution method in this matter.

## 13.5 ACKNOWLEDGMENT

The work of Hassan Haes Alhelou was supported in part by the Science Foundation Ireland (SFI) through the SFI Strategic Partnership Programme under Grant SFI/15/SPP/E3125, and in part by the UCD Energy Institute. The opinions, findings and conclusions or recommendations expressed in this material are those of the authors and do not necessarily reflect the views of the Science Foundation Ireland.

## REFERENCES

[1] H. Shayeghi, M. Mahdavi, A. Kazemi, and H. A. Shayanfar, "Studying effect of bundle lines on TNEP considering network losses using decimal codification genetic algorithm," Energy Conversion and Management, vol. 51, no. 12, pp. 2685–2691, 2010.

[2] H. Shayeghi and M. Mahdavi, "Genetic algorithm based studying of bundle lines effect on network losses in transmission network expansion planning," Journal of Electrical Engineering, vol. 60, no. 5, pp. 237–245, 2009.

[3] S. Jalilzadeh, A. Kimiyaghalam, M. Mahdavi, and A. Ashouri, "STNEP considering voltage level and uncertainty in demand using IABPSO," International Review of Electrical Engineering, vol. 7, no. 4, pp. 5186–5195, 2012.

[4] M. Mahdavi, A. Bagheri, and E. Mahdavi, "Comparing efficiency of PSO with GA in transmission expansion planning considering network adequacy," WSEAS Transactions on Power Systems, vol. 7, no. 1, pp. 34–43, 2012.

[5] M. Mahdavi and E. Mahdavi, "Evaluating the effect of load growth on annual network losses in TNEP considering bundle lines using DCGA," International Journal on Technical and Physical Problems of Engineering, vol. 3, no. 4, pp. 1–9, 2011.

[6] A. Kimiyaghalam, M. Mahdavi, S. Jalilzadeh, and A. Ashouri, "Improved binary particle swarm optimization based TNEP considering network losses, voltage level, and uncertainty in demand," Journal of Artificial Intelligence in Electrical Engineering, vol. 1, no. 2, pp. 29–42, 2012.

[7] H. Shayeghi and M. Mahdavi "Studying the effect of losses coefficient on transmission expansion planning using decimal codification based GA," International Journal on Technical and Physical Problems of Engineering, vol. 1, no. 1, pp. 58–64, 2009.

[8] M. Mahdavi and H. Monsef "Review of static transmission expansion planning," Journal of Electrical and Control Engineering, vol. 1, no. 1, pp. 11–18, 2011.

[9] A. Kazemi, S. Jalilzadeh, M. Mahdavi, and H. Haddadian, "Genetic algorithm-based investigation of load growth factor effect on the network loss in TNEP," 3rd IEEE Conference on Industrial Electronics and Applications, Singapore, vol. 1, pp. 764–769, 2008.

[10] S. Jalilzadeh, A. Kazemi, M. Mahdavi, and H. Haddadian, "TNEP considering voltage level, network losses and number of bundle lines using GA," Third International Conference on Electric Utility Deregulation and Restructuring and Power Technologies, China, pp. 1580–1585, 2008.

[11] M. Khodayari, H. Monsef, and M. Mahdavi, "Customer reliability based energy & SR scheduling at hierarchical level II," 27th International Power System Conference, Iran, pp. 1–11, 2012.

[12] M. Khodayari, M. Mahdavi, and H. Monsef, "Simultaneous scheduling of energy & spinning reserve considering customer and supplier choice on reliability," 19th Iranian Conference on Electrical Engineering, Iran, pp. 491–496, 2011.

[13] M. Mahdavi and R. Romero, "Transmission system expansion planning based on LCC criterion considering uncertainty in VOLL using DCGA," 32nd International Power System Conference, Iran, pp. 1–7, 2017.

[14] H. Shayeghi, H. Hosseini, A. Shabani, and M. Mahdavi, "GEP considering purchase prices, profits of IPPs and reliability criteria using hybrid GA and PSO," International Journal of Electrical and Computer Engineering, vol. 2, no. 8, pp. 1619–1625, 2008.

[15] M. Mahdavi, L. H. Macedo, and R. Romero "Transmission and generation expansion planning considering system reliability and line maintenance," 26th Iranian Conference on Electrical Engineering, Iran, pp. 1005–1010, 2018.

[16] M. Mahdavi, L. H. Macedo, and R. Romero, "Considering line maintenance and repair in GTEP," 4th Conference on Modern Technologies in Electrical, Telecommunication and Mechatronics Engineering, Iran, pp. 1–6, 2018.

[17] H. Hosseini, S. Jalilzadeh, V. Nabaei, G. R. Z. Govar, and M. Mahdavi, "Enhancing deregulated distribution network reliability for minimizing penalty cost based on reconfiguration using BPSO," 2nd IEEE International Conference on Power and Energy, Malaysia, pp. 983–987, 2008.

[18] O. Badran, S. Mekhilef, H. Mokhlis, and W. Dahalan, "Optimal reconfiguration of distribution system connected with distributed generations: A review of different methodologies," Renewable and Sustainable Energy Reviews, vol. 73, pp. 854–867, Jun 2017.

[19] N. G. Paterakis, A. Mazza, S. F. Santos, O. Erdinç, G. Chicco, A. G. Bakirtzis, and J. P. S. Catalão, "Multi-Objective reconfiguration of radial distribution systems using reliability indices," IEEE Transactions on Power Systems, vol. 31, no. 2, pp. 1048–1062, March 2016.

[20] M. Mahdavi, C. Sabillón, M. Ajalli, H. Monsef, and R. Romero, "A real test system for power system planning, operation, and reliability," Journal of Control, Automation and Electrical Systems, vol. 29, no. 2, pp. 192–208, 2018.

[21] S. S. Duttagupta, A reliability assessment methodology for distribution systems with distributed generation, M.Sc. Theses, Texas A&M University, Electrical Engineering Department, pp. 1–90, May 2006.

[22] M. Mahdavi, H. Monsef, and R. Romero, "Reliability effects of maintenance on TNEP considering preventive and corrective repairs," IEEE Transactions on Power Systems, vol. 32, no. 5, pp. 3768–3781, 2017.

[23] M. Mahdavi, H. Monsef, and R. Romero, "Reliability and economic effects of maintenance on TNEP considering line loading and repair," IEEE Transactions on Power Systems, vol. 31, no. 5, pp. 3381–3393, 2016.

[24] Y. Y. Fu and H. D. Chiang, "Toward optimal multiperiod network reconfiguration for increasing the hosting capacity of distribution networks," IEEE Transactions on Power Delivery, vol. 33, no. 5, pp. 2294–2304, October 2018.

[25] S. Lei, Y. Hou, F. Qiu, and J. Yan, "Identification of critical switches for integrating renewable distributed generation by dynamic network reconfiguration," IEEE Transactions on Sustainable Energy, vol. 9, no. 1, pp. 420–432, January 2018.

[26] H. Xing, and X. Sun, "Distributed generation locating and sizing in active distribution network considering network reconfiguration," IEEE Access, vol. 5, pp. 14768–14774, August 2017.

[27] A. Azizivahed, H. Narimani, M. Fathi, E. Naderi, H. R. Safarpour, and M. R. Narimani, "Multi-objective dynamic distribution feeder reconfiguration in automated distribution systems," Energy, vol. 147, pp. 896–914, March 2018.

[28] J. Shukla, B. Das, and V. Pant, "Stability constrained optimal distribution system reconfiguration considering uncertainties in correlated loads and distributed generations," International Journal of Electrical Power and Energy Systems, vol. 99, pp. 121–133, 2018.

[29] K. Chen, W. Wu, B. Zhang, S. Djokic, and G. P. Harrison, "A method to evaluate total supply capability of distribution systems considering network reconfiguration and daily load curves," IEEE Transactions on Power Systems, vol. 31, no. 3, pp. 2096–2104, May 2016.

[30] H. Shayeghi and M. Mahdavi, "Application of PSO and GA for transmission network expansion planning," in Analysis, Control and Optimal Operations in Hybrid Power Systems, Springer, London, pp. 187–226, 2013.

[31] N. C. Koutsoukis, D. O. Siagkas, P. S. Georgilakis, and N. D. Hatziargyriou, "Online reconfiguration of active distribution networks for maximum integration of distributed generation," IEEE Transactions on Automation Science and Engineering, vol. 14, no. 2, pp. 437–448, April 2017.

[32] A. Asrari, S. Lotfifard, and M. Ansari, "Reconfiguration of smart distribution systems with time varying loads using parallel computing," IEEE Transactions on Smart Grid, vol. 7, no. 6, pp. 2713–2723, November 2016.

[33] J. C. López, M. Lavorato b, and M. J. Rider, "Optimal reconfiguration of electrical distribution systems considering reliability indices improvement," International Journal of Electrical Power and Energy Systems, vol. 78, pp. 837–845, Jun 2016.

[34] L. H. Macedo, J. F. Franco, M. Mahdavi, and R. Romero "A contribution to the optimization of the reconfiguration problem in radial distribution systems," Journal of Control, Automation and Electrical Systems, vol. 29, no. 6, pp. 756–768, 2018.

[35] M. Mahdavi and E. Mahdavi, "Transmission expansion planning considering network adequacy and investment cost limitation using genetic algorithm," International Journal of Electrical and Computer Engineering, vol. 5, no. 8, pp. 1–5, 2011.

[36] M. Mahdavi, H. Monsef, and A. Bagheri, "Transmission lines loading enhancement using ADPSO approach," International Journal of Electrical and Computer Engineering, vol. 4, no. 3, pp. 556–561, 2010.

[37] H. Shayeghi, M. Mahdavi, and A. Kazemi, "Discrete particle swarm optimization algorithm used for TNEP considering network adequacy restriction," International Journal of Electrical and Computer Engineering, vol. 3, no. 3, pp. 521–528, 2009.

[38] H. Shayeghi, M. Mahdavi, and H. Haddadian, "DCGA based-transmission network expansion planning considering network adequacy," International Journal of Electrical and Computer Engineering, vol. 2, no. 12, pp. 2875–2880, 2008.

[39] M. Mahdavi, A. Nazari, V. Hosseinnezhad, and A. Safari, "A PSO-based static synchronous compensator controller for power system stability enhancement," Journal of Artificial Intelligence in Electrical Engineering, vol. 1, no. 1, pp. 18–25, 2012.

[40] A. Kimiyaghalam, M. Mahdavi, A. Ashouri, and M. Bagherivand, "Optimal placement of PMUs for reliable observability of network under probabilistic events using BABC algorithm," 22nd International Conference on Electricity Distribution, Sweden, pp. 1–4, 2013.

[41] H. Shayeghi and M. Mahdavi, "Optimization of transmission lines loading in TNEP using decimal codification based GA," International Journal of Electrical and Computer Engineering, vol. 2, no. 5, pp. 942–947, 2008.

[42] M. Mahdavi, C. S. Antunez, A. Bagheri, and R. Romero, "Line Maintenance within transmission expansion planning: A multistage framework," IET Generation, Transmission and Distribution, vol. 14, no. 13, pp. 3057–3065, 2019.

[43] M. Mahdavi, A. R. Kheirkhah, L. H. Macedo, and R. Romero, "A genetic algorithm for transmission network expansion planning considering line maintenance," IEEE Congress on Evolutionary Computation (CEC), Glasgow, United Kingdom, pp. 1–6, 2020.

[44] M. Mahdavi, A. R. Kheirkhah, and R. Romero, "Transmission expansion planning considering line failures and maintenance," 33rd International Power System Conference, Iran, pp. 1–7, 2018.

[45] M. Mahdavi, H. Monsef, A. Bagheri, and A. Kimiyaghalam, "Dynamic transmission network expansion planning considering network losses, DG sources and operational costs-part 1: review and problem formulation," WULFENIA, vol. 19, no. 10, pp. 242–257, 2012.

[46] M. Mahdavi, H. Monsef, A. Bagheri, and A. Kimiyaghalam, "Dynamic transmission network expansion planning considering network losses, DG sources and operational costs-part 2: solution method and numerical results," WULFENIA, vol. 19, no. 10, pp. 258–273, 2012.

[47] M. Rahmani-Andebili and M. Fotuhi Firuzabad, "An adaptive approach for PEVs charging management and reconfiguration of electrical distribution system penetrated by renewables," IEEE Transactions on Industrial Informatics, vol. 14, no. 5, pp. 2001–2010, May 2018.

[48] Z. Li, S. Jazebi, and F. de León, "Determination of the optimal switching frequency for distribution system reconfiguration," IEEE Transactions on Power Delivery, vol. 32, no. 4, pp. 2060–2069, August 2017.

[49] S. S. F. Souza, R. Romero, J. Pereira, and J. T. Saraiva, "Artificial immune algorithm applied to distribution system reconfiguration with variable demand," International Journal of Electrical Power and Energy Systems, vol. 82, pp. 561–568, November 2016.

[50] M. A. Muhammad, H. Mokhlis, K. Naidu, J. F. Franco, H. A. Illias, and L. Wang, "Integrated database approach in multiobjective network reconfiguration for distribution system using discrete optimisation techniques," IET Generation, Transmission & Distribution, vol. 12, no. 4, pp. 976–986, 2018.

[51] F. Ding and K. A. Loparo, "Feeder reconfiguration for unbalanced distribution systems with distributed generation: A hierarchical decentralized approach," IEEE Transactions on Power Systems, vol. 31, no. 2, pp. 1633–1642, March 2016.

[52] R. A. Jabr, I. Džafić, and I. Huseinagić, "Real time optimal reconfiguration of multiphase active distribution networks," IEEE Transactions on Smart Grid, vol. 9, no. 6, pp. 6829–6839, November 2018.

[53] M. Mahdavi, C. S. Antunez, M. Ajalli, and R. Romero, "Transmission expansion planning: literature review and classification," IEEE Systems Journal, vol. 3, no. 3, pp. 3129–3140, 2019.

[54] C. T. Su and C. S. Lee, "Network reconfiguration of distribution systems using improved mixed-integer hybrid differential evolution," IEEE Transactions on Power Delivery, vol. 18, no. 3, pp. 1022–1027, July 2003.

[55] C. T. Su, C. F. Chang, and J. P. Chiou, "Distribution network reconfiguration for loss reduction by ant colony search algorithm," Electric Power Systems Research, vol. 75, no. 2-3, pp. 190–199, August 2005.

[56] H. Haghighat and B. Zeng, "Distribution system reconfiguration under uncertain load and renewable generation," IEEE Transactions on Power Systems, vol. 31, no. 4, pp. 2666–2675, July 2016.

[57] E. Carpaneto and G. Chicco, "Distribution system minimum loss reconfiguration in the Hyper-Cube Ant Colony Optimization framework," Electric Power Systems Research, vol. 78, no. 12, pp. 2037–2045, December 2008.

[58] A. Saffar, R. Hooshmand, and A. Khodabakhshian, "A new fuzzy optimal reconfiguration of distribution systems for loss reduction and load balancing using ant colony search-based algorithm," Applied Soft Computing, vol. 11, no. 5, pp. 4021–4028, July 2011.

[59] B. Venkatesh, Rakesh Ranjan, and H. B. Gooi "Optimal reconfiguration of radial distribution systems to maximize loadability," IEEE Transactions on Power Systems, vol. 19, no. 1, pp. 260–266, February 2004.

[60] M. Mahdavi, L. H. Macedo, and R. Romero, "A mathematical formulation for distribution network reconfiguration," 7th Regional Conference on Electricity Distribution, Iran, pp. 1–5, 2019.

[61] M. Mahdavi and R. Romero, "Reconfiguration of radial distribution systems: An efficient mathematical model," IEEE Latin America Transactions, Early Access, vol. 100, no. 1e, 2021.

[62] A. Y. Abdelaziz, F.M. Mohammed, S. F. Mekhamer, M. A. L. Badr, "Distribution systems reconfiguration using a modified particle swarm optimization algorithm," Electric Power Systems Research, vol. 79, no. 11, pp. 1521–1530, November 2009.

[63] A. Y. Abdelaziz, F.M. Mohammed, S. F. Mekhamer, M. A. L. Badr, Distribution system reconfiguration using a modified Tabu Search algorithm, Electric Power Systems Research, vol. 80, no. 8, pp. 943–953, August 2010.

[64] M. A. N. Guimarães, C. A. Castro, and R. Romero, "Distribution systems operation optimisation through reconfiguration and capacitor allocation by a dedicated genetic algorithm," IET Generation, Transmission, and Distribution, vol. 4, no. 11, pp. 1213–1222, April 2010.

[65] M. Lavorato, J. F. Franco, M. J. Rider, and R. Romero, "Imposing radiality constraints in distribution system optimization problems," IEEE Transactions on Power Systems, vol. 27, no. 1, pp. 172–179, February 2012.

[66] H. Ahmadi and J. R. Martí, "Mathematical representation of radiality constraint in distribution system reconfiguration problem," International Journal of Electrical Power and Energy Systems, vol. 64, pp. 293–299, January 2015.

[67] R. A. Jabr, R. Singh, and B. C. Pal, "Minimum loss network reconfiguration using mixed-integer convex programming," IEEE Transactions on Power Systems, vol. 27, no. 2, pp. 1106–1115, May 2012.

[68] Y. Wang, C. Liu, M. Shahidehpour, and C. Guo, "Critical Components for Maintenance Outage Scheduling Considering Weather Conditions and Common Mode Outages in Reconfigurable Distribution Systems," IEEE Transactions on Smart Grid, vol. 7, no. 6, pp. 2807–2816, November 2016.

[69] J. A. Taylor and F. S. Hover, "Convex models of distribution system reconfiguration," IEEE Transactions on Power Systems, vol. 27, no. 3, pp. 1407–1413, August 2012.

[70] H. Hijazi and S. Thiébaux, "Optimal distribution systems reconfiguration for radial and meshed grids," International Journal of Electrical Power and Energy Systems, vol. 72, pp. 136–143, November 2015.

[71] A. Merlin and H. Back, "Search for a minimal-loss operating spanning tree configuration in an urban power distribution system," 5th Power System Computation Conference, UK, September 1975, pp. 1–18.

[72] G. Chang, J. Zrida, and J. D. Birdwell, "Knowledge based distribution system analysis and reconfiguration," IEEE Transactions on Power Systems, vol. 5, no. 3, pp. 744–749, August 1990.

[73] N. D. R. Sarma and k. S. P. Rao, "A new 0–1 integer programming method of feeder reconfiguration for loss minimization in distribution systems," Electric Power Systems Research, vol. 33, no. 2, pp. 125–131, May 1995.

[74] H. P. Schmidt, N. Ida, N. Kagan, and J. C. Guaraldo, "Fast reconfiguration of distribution systems considering loss minimization," IEEE Transactions on Power Systems, vol. 20, no. 3, pp. 1311–1319, August 2005.

[75] H. M. Khodr, J. M. Crespo, M. A. Matos, and J. Pereira, "Distribution systems reconfiguration based on OPF using Benders decomposition," IEEE Transactions on Power Systems, vol. 24, no. 4, pp. 2166–2176, October 2009.

[76] "GAMS-The Solver Manuals, GAMS User Notes," GAMS Develop. Corp., Washington, DC, Jan. 2001.

[77] "GAMS/CPLEX 7.0 User's Notes," GAMS Develop. Corp., Washington DC, 2001.

[78] G. K. V. Raju and P. R. Bijwe, "An efficient algorithm for minimum loss reconfiguration of distribution system based on sensitivity and heuristics," IEEE Transactions on Power Systems, vol. 23, no. 3, pp. 1280–1287, August 2008.

[79] D. Bienstock, "Computational study of a family of mixed-integer quadratic programming problems," Mathematical Programming, vol. 74, no. 2, pp. 121–140, August 1996.

[80] S. Boyd and L. Vandenberghe, Convex Optimization. Cambridge University Press, Cambridge, UK, pp. 1–701, 2004.

[81] M. S. Lobo, L. Vandenberghe, S. Boyd, and H. Lebret, "Applications of second-order cone programming," Linear Algebra and Its Applications, vol. 284, no. 1-3, pp. 193–228, November 1998.

[82] R. Fourer, D. M. Gay, and B. W. Kernighan, "A modeling language for mathematical programming," Management Science, vol. 36, no. 5, pp. 519–554, May 1990.

[83] M. E. Baran and F. F. Wu, "Network reconfiguration in distribution systems for loss reduction and load balancing," IEEE Transactions on Power Delivery, vol. 4, no. 2, pp. 1401–1407, April 1989.

[84] D. Shirmohammadi and W. H. Hong, "Reconfiguration of electric distribution networks for resistive line loss reduction," IEEE Transactions on Power Delivery, vol. 4, no. 2, pp. 1492–1498, April 1989.

[85] F. V. Gomes, S. Carneiro Jr., J. L. R. Pereira, M. P. Vinagre, P. A. N. Garcia, and L. R. Araujo, "A new heuristic reconfiguration algorithm for large distribution systems," IEEE Transactions on Power Systems, vol. 20, no. 3, pp. 1373–1378, August 2005.

[86] F. V. Gomes, S. Carneiro Jr., J. L. R. Pereira, M. P. Vinagre, P. A. N. Garcia, and L. R. Araujo, "A new distribution system reconfiguration approach using optimum power flow and sensitivity analysis for loss reduction," IEEE Transactions on Power Systems, vol. 21, no. 4, pp. 1616–1623, November 2006.

[87] S. Goswami and S. Basu, "A new algorithm for the reconfiguration of distribution feeders for loss minimization," IEEE Transactions on Power Delivery, vol. 7, no. 3, pp. 1484–1491, July 1992.

[88] D. Singh and R. K. Misra, "Load type impact on distribution system reconfiguration," International Journal of Electrical Power and Energy Systems, vol. 42, pp. 582–592, November 2012.

[89] F. Llorens-Iborra, J. Riquelme-Santos, and E. Romero-Ramos, "Mixed-integer linear programming model for solving reconfiguration problems in large-scale distribution systems," Electric Power Systems Research, vol. 88, pp. 137–145, July 2012.

[90] J. F. Franco, M. J. Rider, M. Lavorato, and R. Romero, "A mixed-integer LP model for the reconfiguration of radial electric distribution systems considering distributed generation," Electric Power Systems Research, vol. 97, pp. 51–60, April 2013.

[91] F. Capitanescu, L. F. Ochoa, H. Margossian, and N. D. Hatziargyriou, "Assessing the potential of network reconfiguration to improve distributed generation hosting capacity in active distribution systems," IEEE Transactions on Power Systems, vol. 30, no. 1, pp. 346–356, January 2015.

[92] J. C. Williams, "A linear-size zero-one programming model for the minimum spanning tree problem in planar graphs," Networks, vol. 39, no. 1, pp. 53–60, November 2001.

[93] H. Ahmadi and J. R. Martí, "Linear current flow equations with application to distribution systems reconfiguration," IEEE Transactions on Power Systems, vol. 30, no. 4, pp. 2073–2080, March 2015.

[94] M. Roelofs and J. Bisschop, AIMMS–The language reference. AIMMS Inc., Bellevue, WA, USA, pp. 3–698.

[95] S. Civanlar, J. J. Grainger, H. Yin., and S. S. H. Lee, "Distribution feeder reconfiguration for loss reduction," IEEE Transactions on Power Delivery, vol. 4, no. 3, pp. 1217–1223, July 1988.

[96] H. Salazar, R. Gallego, and R. Romero, "Artificial neural networks and clustering techniques applied in the reconfiguration of distribution systems," IEEE Transactions on Power Delivery, vol. 21, no. 1, pp. 1735–1742, July 2006.

[97] D. Zhang, Z. Fu and L. Zhang, "An improved TS algorithm for loss-minimum reconfiguration in large-scale distribution systems," Electric Power Systems Research, vol. 77, no. 5-6, pp. 685–694, April 2007.

[98] G. Mavrotas, "Effective implementation of the epsilon-constraint method in multi-objective mathematical programming problems," Applied Mathematics and Computation, vol. 213, pp. 455–465, 2009

[99] G. H. Tzeng and J. J. Huang, Multiple Attribute Decision Making: Methods and Applications. Boca Raton, FL, USA: CRC Press, pp. 1–352, Jun 2011.

[100] Y. Takenobu, N. Yasuda, S. Kawano, Y. Hayashi, and S. I. Minato, "Evaluation of annual energy loss reduction based on reconfiguration scheduling," IEEE Transactions on Smart Grid, vol. 9, no. 3, pp. 1986–1996, May 2018.

[101] R. K. Mishr and K. S. Swarup, "Adaptive weight-based self reconfiguration of smart distribution network with intelligent agents," IEEE Transactions on Emerging Topics in Computational Intelligence, vol. 2, no. 6, pp. 464–472, December 2018.

[102] D. F. Teshome and K. L. Lian, "Comprehensive mixed-integer linear programming model for distribution system reconfiguration considering DGs," IET Generation, Transmission and Distribution, vol. 12, no. 20, pp. 4515–4523, 2018.

[103] A. G. Fonseca, O. L. Tortelli, and E. M. Lourenço, "Extended fast decoupled power flow for reconfiguration networks in distribution systems," IET Generation, Transmission and Distribution, vol. 12, no. 12, pp. 6033–6040, 2018.

[104] M. A. Tavakoli Ghazi Jahani, P. Nazarian, A. Safari, and M. R. Haghifam, "Multi-objective optimization model for optimal reconfiguration of distribution networks with demand response services," Sustaiable Cities and Society, vol. 47, pp. 1–11, May 2019.

[105] E. kianmeh, S. Nikkhah, and A. Rabiee, "Multi-objective stochastic model for joint optimal allocation of DG units and network reconfiguration from DG owner's and DisCo's perspectives," Renewable Energy, vol. 132, pp. 471–485, March 2019.

[106] C. A. Castro, J. R. Watanabe, and A. A. Watanabe, "An efficient reconfiguration algorithm for loss reduction of distribution systems," Electric Power Systems Research, vol. 19, no. 2, pp. 137–144, August 1990.

[107] T. E. Lee, M. Y. Cho, and C. S. Chen, "Distribution system reconfiguration to reduce resistive losses," Electric Power Systems Research, vol. 30, no. 1, pp. 25–33, June 1994.

[108] R. J. Sárf, M. M. A. Salama, and A. Y. Chikhan, "Distribution system reconfiguration for loss reduction: An algorithm based on network partitioning theory," IEEE Transactions on Power Systems, vol. 11, no. 1, pp. 504–510, February 1996.

[109] T. E. McDermott, I. Drezga, and R. P. Broadwater, "A heuristic nonlinear constructive method for distribution system reconfiguration," IEEE Transactions on Power Systems, vol. 14, no. 2, pp. 478–483, May 1999.

[110] M. A. Kashem, G. B. Jasmon, and V. Ganapathy, "A new approach of distribution system reconfiguration for loss minimization," International Journal of Electrical Power and Energy Systems, vol. 22, no. 4, pp. 269–276, May 2000.

[111] G. K. V. Raju and P. R. Bijwe, "Efficient reconfiguration of balanced and unbalanced distribution systems for loss minimisation," IET Generation, Transmission and Distribution, vol. 2, no. 1, pp. 7–12, January 2008.

[112] M. Arun and P. Aravindhababu, "A new reconfiguration scheme for voltage stability enhancement of radial distribution systems," Energy Conversion and Management, vol. 50, no. 9, pp. 2148–2151, September 2009.

[113] M. Chakravorty and D. Das, "Voltage stability analysis of radial distribution networks," International Journal of Electrical Power and Energy Systems, vol. 23, no. 2, pp. 129–135, February 2001.

[114] L. W. de Oliveira, S. Carneiro Jr., E. J. de Oliveira, J. L. R. Pereira, I. C. Silva Jr., and J. S. Costa, "Optimal reconfiguration and capacitor allocation in radial distribution systems for energy losses minimization," International Journal of Electrical Power and Energy Systems, vol. 32, no. 8, pp. 840–848, October 2010.

[115] I. C. Silva Jr., S. Carneiro Jr., E. J. Oliveira, J. S. Costa, J. L. R. Pereira, P. A. N. Garcia, "A heuristic constructive algorithm for capacitor placement on distribution systems," IEEE Transactions on Power Systems, vol. 23, no. 4, pp. 1619–1626, November 2008.

[116] A. K. Ferdavani, A. A. M. Zin, A. Khairuddin, and M. M. Naeini, "Reconfiguration of distribution system through two minimum-current neighbour-chain updating methods," IET Generation, Transmission and Distribution, vol. 7, no 12, pp. 1492–1497, December 2013.

[117] K. Jasthi and D. Das, "Simultaneous distribution system reconfiguration and DG sizing algorithm without load flow solution," IET Generation, Transmission and Distribution, vol. 12, no. 6, pp. 1303–1313, March 2018.

[118] R. S. Rao, K. Ravindra, K. Satish, and S. V. L. Narasimham, "Power loss minimization in distribution system using network reconfiguration in the presence of distributed generation," IEEE Transactions on Power Systems, vol. 28, no. 1, pp. 317–325, February 2013.

[119] A. M. Imran, M. Kowsalya, and D. P. Kothari, "A novel integration technique for optimal network reconfiguration and distributed generation placement in power distribution networks," International Journal of Electrical Power and Energy Systems, pp. 461–472, vol. 63, December 2014.

[120] H. B. Tolabi, M. H. Ali, and S. B. M. Ayob, and M. Rizwan, "Novel hybrid fuzzy-Bees algorithm for optimal feeder multi-objective reconfiguration by considering multiple-distributed generation," Energy, vol. 71, pp. 507–515, July 2014.

[121] X. Ji, Q. Liu, Y. Yu, S. Fan, and N. Wu, "Distribution network reconfiguration based on vector shift operation," IET Generation, Transmission and Distribution, vol. 12, no. 13, pp. 3339–3345, 2018.

[122] N. V. Kovacki, P. M. Vidovic, A. T. Saric, "Scalable algorithm for the dynamic reconfiguration of the distribution network using the Lagrange relaxation approach," International Journal of Electrical Power and Energy Systems, vol. 94, pp. 188–202, January 2018.

[123] M. Mahdavi, M. Ajalli, M. Mohammadi, The Role of Artificial Intelligence Techniques in Induction Motors, LAMBERT Academic Publishing, Germany, 2017.

[124] H. D. Chiang and R. Jean-Jameau, "Optimal network reconfigurations in distribution systems: Part 1: A new formulation and a solution methodology," IEEE Transactions on Power Delivery, vol. 5, no. 4, pp. 1902–1909, November 1990.

[125] H. D. Chiang and R. Jean-Jameau, "Optimal network reconfigurations in distribution systems: Part 2: Solution algorithms and numerical results," IEEE Transactions on Power Delivery, vol. 5, no. 3, pp.1568–1574, July 1990.

[126] K. Nara, A. Shiose, M. Kitagawa, and T. Ishihara, "Implementation of genetic algorithm for distribution system loss minimum reconfiguration," IEEE Transactions on Power Systems, vol. 7, no. 3, pp. 1044–1051, August 1992.

[127] M. Mahdavi and H. Haddadian, "Evaluation of GA performance in TNEP considering voltage level, network losses and number of bundle Lines," International Journal of Discrete Mathematic, vol. 3, no. 1, pp. 1–10, 2018.

[128] S. Jalilzadeh, A. Kazemi, M. Mahdavi, and H. Haddadian, "DCGA-based evaluation of inflation rate effect on the network loss in transmission expansion planning," 43rd International Universities Power Engineering Conference, Italy, pp. 1–5, 2008.

[129] A. Kazemi, H. Haddadian, and M. Mahdavi,"Transmission network adequacy optimization using genetic algorithm," 5th International Conference on Electrical and Electronics Engineering, Turkey, pp. 1–5, 2007.

[130] S. Jalilzadeh, A. Kazemi, M. Mahdavi, H. Haddadian, "Transmission expansion planning considering voltage level and network loss using genetic algorithm," 5th International Conference on Electrical and Electronics Engineering, Turkey, pp. 1–5, 2007.

[131] A. Kazemi, S. Jalilzadeh, H. Haddadian, and M. Mahdavi, "Transmission expansion planning considering network adequacy and losses using DCGA," IEEE International Conference on Sustainable Energy Technologies, Singapore, pp. 145–150, 2008.

[132] H. Kim, Y. Ko, and K. H. Jung, "Artificial neural-network based feeder reconfiguration for loss reduction in distribution systems," IEEE Transactions on Power Delivery, vol. 8, no. 3, July 1993.

[133] H. C. Chang and C. C. Kuo, "Network reconfiguration in distribution systems using simulated annealing," Electric Power Systems Research, vol. 29, no. 3, pp. 227–238, May 1994.

[134] C. T. Su and C. S. Lee, "Feeder reconfiguration and capacitor setting for loss reduction of distribution systems," Electric Power Systems Research, vol. 58, no. 2, pp. 97–102, Jun 2001.

[135] Y. J. Jeon, J. C. Kim, J. O. Kim, J. R. Shin, and K. Y. Lee, "An efficient simulated annealing algorithm for network reconfiguration in large-scale distribution systems," IEEE Transactions on Power Delivery, vol. 17, no. 4, pp. 1070–1078, October 2002.

[136] J. Z. Zhu, "Optimal reconfiguration of electrical distribution network using the refined genetic algorithm," Electric Power Systems Research, vol. 62, no. 1, pp. 37–42, May 2002.

[137] C. H. Lin, C.S. Chen, C.J. Wu, and M.S. Kang, "Application of immune algorithm to optimal switching operation for distribution-loss minimisation and loading balance," IEE Proceedings-Generation, Transmission and Distribution, vol. 150, no. 2, pp. 183–189, March 2003.

[138] C. S. Chen and M.Y. Cho, "Energy loss reduction by critical switcher," IEEE Transactions on Power Delivery, vol. 8, no. 3, pp. 1246–1253, July 1993.

[139] J. P. Chiou and F. S. Wang, "A hybrid method of differential evolution with application to optimal control problems of a bioprocess system," IEEE International Conference on Evolutionary Computation Proceedings, USA, pp. 627–632, May 1998.

[140] B. Venkatesh, and R. Ranjan, "Optimal radial distribution system reconfiguration using fuzzy adaptation of evolutionary programming," International Journal of Electrical Power and Energy Systems, vol. 25, no. 10, pp. 775–780, December 2003.

[141] Y. T. Hsiao, "Multiobjective evolution programming method for feeder reconfiguration," IEEE Transactions on Power Systems, vol. 19, no. 1, pp. 594–599, February 2004.

[142] A. C. B. Delbem, A. C. P. D. L. F. de Carvalho, and N. G. Bretas, "Main chain representation for evolutionary algorithms applied to distribution system reconfiguration," IEEE Transactions on Power Systems, vol. 20, no. 1, pp. 425–436, February 2005.

[143] A. E. Eiben and G. Rudolph, "Theory of evolutionary algorithm: A bird's eye view," Theoretical Computer Science, vol. 229, no. 1-2, pp. 3–9, November 1999.

[144] K. Prasad, R. Ranjan, N. C. Sahoo, and A. Chaturvedi, "Optimal reconfiguration of radial distribution systems using a fuzzy mutated genetic algorithm," IEEE Transactions on Power Delivery, vol. 20, no. 2, pp. 1211–1213, April 2005.

[145] Y. Y. Hong and S. Y. Ho, "Determination of network configuration considering multiobjective in distribution systems using genetic algorithms," IEEE Transactions on Power Systems, vol. 2o, no. 2, pp. 1062–1069, May 2005.

[146] Y. Mishima, K. Nara, T. Satoh, T. Ito, and H. Kaneda, "Method for minimum-loss reconfiguration of distribution system by tabu search," Electrical Engineering in Japan, vol. 152, no. 2, pp. 18–25, May 2005.

[147] J. Mendoza, R. Lopez, D. Morales, E. Lopez, P. Dessante, and R. Moaaga: "Minimal loss reconfiguration using genetic algorithms with restricted population and addressed operators: real application," IEEE Transaction on Power Systems, vol. 21, no. 2, pp. 948–954, May 2006.

[148] N. C. Sahoo and K. Prasad, "A fuzzy genetic approach for network reconfiguration to enhance voltage stability in radial distribution systems," Energy Conversion and Management, vol. 47, no. 18–19, pp. 3288–3306, November 2006.

[149] D. Das, "A fuzzy multiobjective approach for network reconfiguration of distribution systems," IEEE Transactions on Power Delivery, vol. 21, no. 1, pp. 202–209, January 2006.

[150] D. Das, "Reconfiguration of distribution system using fuzzy multi-objective approach," International Journal of Electrical Power and Energy Systems, vol. 28, no. 5, pp. 331–338, June 2006.

[151] A. Ahuja, S. Das, and A. Pahwa, "An AIS-ACO hybrid approach for multi-objective distribution system reconfiguration," IEEE Transactions on Power Systems, vol. 22, no. 3, pp. 1101–1111, August 2007.

[152] J. P. Chiou, C. F. Chang, and C. T. Su, "Variable scaling hybrid differential evolution for solving network reconfiguration of distribution systems," IEEE Transactions on Power Systems, vol. 20, no. 2, pp. 668–674, May 2005.

[153] E. M. Carreno, R. Romero, and A. Padilha-Feltrin, "An efficient codification to solve distribution network reconfiguration for loss reduction problem," IEEE Transactions on Power Systems, vol. 23, no. 4, pp. 1542–1551, November 2008.

[154] P. C. Chu and J. E. Beasley, "A genetic algorithm for the generalized assignment problem," Computers and Operations Research, vol. 24, no. 1, pp. 17–23, January 1997.

[155] B. Enacheanu, B. Raison, R. Caire, O. Devaux, W. Bienia, and N. HadjSaid, "Radial network reconfiguration using genetic algorithm based on the Matroid theory," IEEE Transactions on Power Systems, vol. 23, no. 1, pp. 186–195, August 2008.

[156] H. Shayeghi, M. Mahdavi, and A. Bagheri, "An improved DPSO with mutation based on similarity algorithm for optimization of transmission lines loading," Energy Conversion and Management, vol. 51, no. 12, pp. 2715–2723, 2010.

[157] S. Sivanagaraju, J. V. Rao, and P. S. Raju, "Discrete particle swarm optimization to network reconfiguration for loss reduction and load balancing," Electric Power Components and Systems, vol. 36, no. 5, pp. 513–524, May 2008.

[158] S. Jalilzadeh, M. Mahdavi, A. Kimiyaghalam, and A. Bagheri, "Using PSO algorithm for parameter estimation of three-phase double-feed induction motor," 6th International Conference on Electrical Engineering/Electronics, Computer, Telecommunications and Information Technology, Thailand, pp. 266–269, 2009.

[159] S. Jalilzadeh, M. Mahdavi, and A. Kimiyaghalam, "Optimal design of single-phase induction motor with permanent capacitor based on maximum efficiency using PSO algorithm," 6th International Conference on Electrical Engineering/Electronics, Computer, Telecommunications and Information Technology, Thailand, pp. 174–177, 2009.

[160] M. Mahdavi, H. Monsef, and A. Bagheri "Lines loading optimization in transmission expansion planning based on binary PSO algorithm," i-manager's Journal on Information Technology, vol. 1, no. 1, pp. 24–32, 2012.

[161] M. Mahdavi and A. Bagheri, "BPSO applied to TNEP considering adequacy criterion," American Journal of Neural Networks and Applications, vol. 4, no. 1, pp. 1–7, 2018.

[162] M. Arun and P. Aravindhababu, "Fuzzy based reconfiguration algorithm for voltage stability enhancement of distribution systems," Expert Systems with Applications, vol. 37, no. 10, pp. 6974–6978, October 2010.

[163] M. S. Tsai and F. Y. Hsu, "Application of grey correlation analysis in evolutionary programming for distribution system feeder reconfiguration," IEEE Transactions on Power Systems, vol. 25, no. 2, pp. 1126–1133, May 2010.

[164] J. H. Wu, M. L. You, and K. L. Wen, "A modified grey relational grade," Journal of Grey System, vol. 11, no. 3, pp. 283–288, 1999.

[165] A. C. Santos, A. C. B. Delbem, J. B. A. London Jr., and N. G. Bretas, "Node-depth encoding and multiobjective evolutionary algorithm applied to large-scale distribution system reconfiguration," IEEE Transactions on Power Systems, vol. 25, no. 3, pp. 1254–1265, August 2010.

[166] A. C. B. Delbem, A. C. P. L. F. Carvalho, C. A. Policastro, A. K. O. Pinto, K. Honda, and A. C. Garcia, "Node-depth encoding for evolutionary algorithms applied to network design," 6th Genetic and Evolutionary Computation Conference, Springer, 2004, pp. 678–687.

[167] C. Wang and H. Z. Cheng, "Optimization of network configuration in large distribution systems using plant growth simulation algorithm," IEEE Transaction on Power Systems, vol. 23, no. 1, pp. 119–126, February 2008.

[168] N. Gupta, A. Swarnkar, K. R. Niazi, R. C. Bansal, "Multi-objective reconfiguration of distribution systems using adaptive genetic algorithm in fuzzy framework," IET Generation, Transmission and Distribution, vol. 4, no. 12, pp. 1288–1298, December 2010.

[169] Y. C. Huang, "Enhanced genetic algorithm-based fuzzy multi-objective approach to distribution network reconfiguration," IET Generation, Transmission and Distribution, vol. 149, no. 5, pp. 615–620, September 2002.

[170] Y. K. Wu, C. Y. Lee, L. C. Liu, and S. H. Tsai, "Study of reconfiguration for the distribution system with distributed generators," IEEE Transaction on Power Delivery, vol. 5, no. 3, pp. 1678–1685, July 2010.

[171] S. Chandramohan, N. Atturulu, R.P. K. Devi, and B. Venkatesh, "Operating cost minimization of a radial distribution system in a deregulated electricity market through reconfiguration using NSGA method," International Journal of Electrical Power and Energy Systems, vol. 32, no. 2, pp. 126–132, February 2010.

[172] F. Mendoza, J. L. Bernal-Agustin, and J. A. Dominguez-Navarro, "NSGA and SPEA applied to multiobjective design of power distribution systems," IEEE Transactions on Power Systems, vol. 21, no. 4, pp. 1938–1945, November 2006.

[173] R. S. Rao, S. V. L. Narasimham, M. R. Raju, and A. S. Rao, "Optimal network reconfiguration of large-scale distribution system using harmony search algorithm," IEEE Transactions on Power Systems, vol. 26, no. 3, pp. 1080–1088, August 2011.

[174] W. C. Wu and M. S. Tsai, "Application of enhanced integer coded particle swarm optimization for distribution system feeder reconfiguration," IEEE Transactions on Power Systems, vol. 26, no. 3, pp. 1591–1599, August 2011.

[175] A. Swarnkar, N. Gupta, and K. R. Niazi, "Adapted ant colony optimization for efficient reconfiguration of balanced and unbalanced distribution systems for loss minimization," Swarm and Evolutionary Computation, vol. 1, no. 3, pp. 129–137, September 2011.

[176] A. Y. Abdelaziz, R. A. Osama, and S. M. El-Khodary, "Reconfiguration of distribution systems for loss reduction using the hyper-cube ant colony optimisation algorithm," IET Generation, Transmission and Distribution, vol. 12, no. 4, pp. 976–986, February 2012.

[177] C. Blum, and M. Dorigo, "The hyper cube framework for ant colony optimization," IEEE Transactions on Systems, Man, and Cybernetics, Part B, vol. 34, no. 2, pp. 1161–1172, April 2004.

[178] K. S. Kumar, and T. Jayabarathi, "Power system reconfiguration and loss minimization for an distribution systems using bacterial foraging optimization algorithm," International Journal of Electrical Power and Energy Systems, vol. 36, pp. 13–17, March 2012.

[179] D. Sambarta, D. Swagatam, A. Ajith, and B. Arijit, "Adaptive computational chemotaxis in bacterial foraging optimization: an analysis," IEEE Transactions on Evolutionary Computation, vol. 13, no. 4, pp. 919–941, August 2009.

[180] B. Amanulla, S. Chakrabarti, and S. N. Singh, "Reconfiguration of power distribution systems considering reliability and power loss," IEEE Transaction on Power Delivery, vol. 27, no. 2, pp. 918–926, April 2012.

[181] A. R. Malekpour, T. Niknam, A. Pahwa, and A. K. Fard, "Multi-objective stochastic distribution feeder reconfiguration in systems with wind power generators and fuel cells using the point estimate method," IEEE Transactions on Power Systems, vol. 28, no. 2, pp. 1483–1492, May 2013.

[182] E. Rosenblueth, "Point estimates for probability moments," Proceedings of the National Academy of Sciences, vol. 72, no. 10, pp. 3812–3814, October 1975.

[183] M. Mahdavi and H. Monsef, "Advanced particle swarm optimization used for optimal design of single-phase induction motor," International Journal on Technical and Physical Problems of Engineering, vol. 1, no. 1, pp. 55–59, 2010.

[184] M. Mahdavi and S. Jalilzadeh, "Advanced particle swarm optimization for parameter identification of three-phase DFIM," The 2009 IEEE International Conference on Intelligent Computing and Intelligent Systems, China, pp. 580–584, 2009.

[185] R. Caponetto, L. Fortuna, S. Fazzino, and M. G. Xibilia, "Chaotic sequences to improve the performance of evolutionary algorithms," IEEE Transactions on Evolutionary Computation, vol. 7, no. 3, pp. 289–304, Jun 2003.

[186] A. Zidan and E. F. El-Saadany, "Distribution system reconfiguration for energy loss reduction considering the variability of load and local renewable generation," Energy, vol. 59, pp. 698–707, September 2013.

[187] S. Jalilzadeh, I. D. Kolagar, M. Mahdavi, and V. Nabaei, "Considering the torsional frequencies of an actual wind turbine to prevent of SSR phenomena," IEEE International Conference on Sustainable Energy Technologies, Singapore, pp. 222–227, 2008.

[188] M. Mahdavi, "Iran energy network overview: Renewable energies in Iran", Presentation, Advanced Topics in Bioenergy, Bioenergy Research Institute, Brazil, 2020. https://www.researchgate.net/publication/346719635.

[189] M. Mahdavi and L. A. Okamura, "Biofules production from petrochemical industry wastewater," Presentation, Advanced Topics in Biorefinery, Bioenergy Research Institute, Brazil, 2020. https://www.researchgate.net/publication/346719761.

[190] M. Mahdavi, "Temperature effects on the methanogenesis and sulfidogenesis in the wastewater treatment," Presentation, Advanced Topics in Biorefinery, Bioenergy Research Institute, Brazil, 2020. https://www.researchgate.net/publication/346719672.

[191] M. Mahdavi and R. A. V. Ramos, "Optimal allocation of bioenergy distributed generators in electrical energy systems," IV Bioenergy Workshop, Bioenergy Research Institute, Brazil, 2020. https://www.researchgate.net/publication/347294531.

[192] A. Ahuja, A. Pahwa, B. K. Panigrahi, and S. Das, "Pheromone-based crossover operator applied to distribution system reconfiguration," IEEE Transactions on Power Systems, vol. 28, no. 4, pp. 4144–4151, November 2013.

[193] B. Tomoiagă, M. Chindris, A. Sumper, R. Villafafila-Robles, and A. Sudria-Andreu,"Distribution system reconfiguration using genetic algorithm based on connected graphs," Electric Power Systems Research, vol. 104, pp. 216–225, November 2013.

[194] L. W. de Oliveira, E. J. de Oliveira, F. V. Gomes, I. C. Silva Jr., A. L.M. Marcato, and P. V.C. Resende, "Artificial immune systems applied to the reconfiguration of electrical power distribution networks for energy loss minimization," International Journal of Electrical Power and Energy Systems, vol. 56, pp. 64–74, March 2014.

[195] L. N. Castro and F. J. V. Zuben, "Learning and optimization using the clonal selection principle," IEEE Transactions on Evolutionary Computation, vol. 6, no. 3, pp. 239–251, Jun 2002.

[196] H. Shareef, A. A. Ibrahim, N. Salman, A. Mohamed, and W. L. Ai, "Power quality and reliability enhancement in distribution systems via optimum network reconfiguration by using quantum firefly algorithm," International Journal of Electrical Power and Energy Systems, vol. 58, pp. 160–169, Jun 2014.

[197] X. S. Yang, "Firefly algorithms for multimodal optimization", In: Proceedings of the stochastic algorithms: foundations and applications, Springer, pp. 169–178, 2009.

[198] E. Rashedi, H. Nezamabadi-pour, and S. Saryazdi, "BGSA: binary gravitational search algorithm," Natural Computing, vol. 9, no. 1, pp. 727–745, September 2010.

[199] M. Sedighizadeh, M. Esmaili, and M. Esmaeili, "Application of the hybrid Big Bang-Big Crunch algorithm to optimal reconfiguration and distributed generation power allocation in distribution systems," Energy, vol. 76, pp. 920–930, November 2014.

[200] O. K. Erol and I. Eksin, "A new optimization method: Big Bang–Big Crunch," Advances in Engineering Software, vol. 37, no. 2, pp. 106–111, February 2006.

[201] A. Kaveh and S. Talatahari, "Size optimization of space trusses using Big Bang–Big Crunch algorithm," Computers and Structures, vol. 87, no. 17-18, pp. 1129–1140, September 2009.

[202] H. R. Esmaeilian and R. Fadaeinedjad, "Distribution system efficiency improvement using network reconfiguration and capacitor allocation," vol. 64, pp. 457–468, January 2015.

[203] L. H. Tsai, "Network reconfiguration to enhance reliability of electric distribution systems," Electric Power Systems Research, vol. 27, no. 2, pp. 135–140, July 1993.

[204] A. M. Eldurssi and R. M. O'Connell, "A fast nondominated sorting guided genetic algorithm for multi-objective power distribution system reconfiguration problem," IEEE Transactions on Power Systems, vol. 30, no. 2, pp. 593–601, March 2015.

[205] K. K. Mishra and S. Harit, "A fast algorithm for finding the non-dominated set in multi-objective optimization," International Journal of Computer Applications, vol. 1, no. 25, pp. 35–39, 2010.

[206] C. Wang and Y. Gao, "Determination of power distribution network configuration using non-revisiting genetic algorithm," IEEE Transactions on Power Systems, vol. 28, no. 4, pp. 3638–3648, November 2013.

[207] F. R. Alonso, D. Q. Oliveira, and A. C. Zambroni de Souza, "Artificial immune systems optimization approach for multiobjective distribution system reconfiguration," IEEE Transactions on Power Systems, vol. 30, no. 2, pp. 840–847, March 2015.

[208] S. Naveen, K. S. Kumar, and K. Rajalakshmi, "Distribution system reconfiguration for loss minimization using modified bacterial foraging optimization algorithm," International Journal of Electrical Power and Energy Systems, vol. 69, pp. 90–97, July 2015.

[209] J. A. Martín and A. J. Gil, "A new heuristic approach for distribution systems loss reduction," Electric Power Systems Research, vol. 78, no. 11, pp. 1953–1958, November 2008.

[210] A. M. Tahboub, V. R. Pandi, and H. H. Zeineldin, "Distribution system reconfiguration for annual energy loss reduction considering variable distributed generation profiles," IEEE Transactions on Power Delivery, vol. 30, no. 4, pp. 1677–1685, August 2015.

[211] J. Bezdek, Pattern Recognition, Handbook of Fuzzy Computation. Boston, MA, USA: IOP, 1998.

[212] H. R. Esmaeilian and R. Fadaeinedjad, "Energy loss minimization in distribution systems utilizing an enhanced reconfiguration method integrating distributed generation," IEEE Systems Journal, vol. 9, no. 4, December 2015.

[213] T. Niknam, "An efficient multi-objective HBMO algorithm for distribution feeder reconfiguration," Expert Systems with Applications, vol. 38, no. 3, pp. 2878–2887, March 2011.

[214] E. Thompson, T. Paulden, and D. K. Smith, "The dandelion code: a new coding of spanning trees for genetic algorithms," IEEE Transactions on Evolutionary Computation, vol. 11, no. 1, pp. 91–100, February 2007.

[215] D. Lim, Y. S. Ong, Y. Jin, B. Sendhoff, and B. S. Lee, "Efficient hierarchical parallel genetic algorithms using grid computing," Future Generation Computer Systems, vol. 23, no. 4, pp. 658–670, May 2007.

[216] A. Lotfipour and H. Afrakhte, "A discrete teaching–learning-based optimization algorithm to solve distribution system reconfiguration in presence of distributed generation," International Journal of Electrical Power and Energy Systems, vol. 82, pp. 264–273, November 2016.

[217] R. V. Rao, V. J. Savsani, and D. P. Vakharia, "Teaching–learning-based optimization: an optimization method for continuous non-linear large scale problems," Information Science, vol. 183, no. 1, pp. 1–15, January 2012.

[218] S. S. F. Souza, R. Romero, and J. F. Franco, "Artificial immune networks Copt-aiNet and Opt-aiNet applied to the reconfiguration problem of radial electrical distribution systems," Electric Power Systems Research, vol. 119, pp. 304–312, February 2015.

[219] B. Khorshid-Ghazani, H. Seyedi, B. Mohammadi-ivatloo, K. Zare, and S. Shargh, "Reconfiguration of distribution networks considering coordination of the protective devices," IET Generation, Transmission & Distribution, vol. 11, no. 1, pp. 82–92, January 2017.

[220] A. Ameli, A. Ahmadifar, M. H. Shariatkhah, M. Vakilian, and M. R. Haghifam, "A dynamic method for feeder reconfiguration and capacitor switching in smart distribution systems," International Journal of Electrical Power and Energy Systems, vol. 85, pp. 200–211, February 2017.

[221] S. Jalilzadeh, M. Jabbari, A. Kimiyaghalam, and M. Mahdavi, "Inrush current identification in power transformers using weight functions," The 2009 IEEE International Conference on Intelligent Computing and Intelligent Systems, Shanghai, China, pp. 795–797, 2009.

[222] Pawan Kumar, I. Ali, M. S. Thomas, and S. Singh, "Imposing voltage security and network radiality for reconfiguration of distribution systems using efficient heuristic and metaheuristic approach," IET Generation, Transmission & Distribution, vol. 11, no. 10, pp. 2457–2467, July 2017.

[223] V. R. Pandi, B. K. Panigrahi, M. K. Mallick, A. Abraham, and S. Das "Improved harmony search for economic power dispatch," Ninth International Conference on Hybrid Intelligent Systems, pp. 403–408, August 2009.

[224] A. Tyagi1, A. Verma1, and P. R. Bijwe, "Reconfiguration for loadability limit enhancement of distribution systems," IET Generation, Transmission and Distribution, vol. 12, no. 1, pp. 88–93, February 2018.

[225] S. Mirjalili, S. M. Mirjalili, and A. Lewis, "Grey wolf optimizer," Advances in Engineering Software, vol. 69, pp. 46–61, March 2014.

[226] G. Zu, J. Xiao, and K. Sun, "Distribution network reconfiguration comprehensively considering N–1 security and network loss," IET Generation, Transmission and Distribution, vol. 12, no. 8, pp. 1721–1728, April 2018.

[227] Z. Ming, B. Pengxiang, L. Jian, and A. Wenyuan, "An improved genetic algorithm for distribution system reconfiguration," in proc. International Conference on Power System Technology, vol. 3, pp. 1734–1738, October 2002.

[228] S. Jalilzadeh, A. Kazemi, M. Mahdavi, and H. Haddadian, "Technical and economic evaluation of voltage level effect on TNEP using genetic algorithm," The 8th International Power Engineering Conference, Singapore, pp. 1034–1039, 2007.

[229] A. Kazemi, S. Jalilzadeh, H. Haddadian, and M. Mahdavi, "Transmission network expansion planning considering network adequacy using genetic algorithm," The 8th International Power Engineering Conference, Singapore, pp. 1040–1044, 2007.

[230] S. Jalilzadeh, A. Kazemi, M. Mahdavi, and H. Haddadian, "Studying of voltage level role in TNEP considering the losses for years after expansion using GA with different trails," 43rd International Universities Power Engineering Conference, Padova, Italy, pp. 1–6, 2008.

[231] A. Kazemi, S. Jalilzadeh, M. Mahdavi, and H. Haddadian, "Evaluation of network losses role in TNEP considering voltage level and number of bundle lines using GA," 3rd IEEE Conference on Industrial Electronics and Applications, Singapore, pp. 1–6, 2008.

[232] S. Jalilzadeh, H. Shayeghi, M. Mahdavi, and H. Haddadian, "A GA based transmission network expansion planning considering voltage level, network losses and number of bundle lines," American Journal of Applied Sciences, vol. 6, no. 5, pp. 987–994, 2009.

[233] H. Shayeghi, S. Jalilzadeh, M. Mahdavi, and H. Haddadian, "Role of voltage level and network losses in TNEP solution," Journal of Electrical Systems, vol. 5, no. 3, pp. 1–18, 2009.

[234] M. Mahdavi, H. Shayeghi, and A. Kazemi, "DCGA based evaluating role of bundle lines in TNEP considering expansion of substations from voltage level point of view," Energy Conversion and Management, vol. 50, no. 8, pp. 2067–2073, 2009.

[235] H. Shayeghi, S. Jalilzadeh, M. Mahdavi, and H. Haddadian, "Studying influence of two effective parameters on network losses in transmission expansion planning using DCGA," Energy Conversion and Management, vol. 49, no. 11, pp. 3017–3024, 2008.

[236] S. Jalilzadeh, A. Kazemi, H. Shayeghi, and M. Mahdavi, "Technical and economic evaluation of voltage level in transmission network expansion planning using GA," Energy Conversion and Management, vol. 49, no. 5, pp. 1119–1125, 2008.

[237] I. A. Quadri, S. Bhowmick, and D. Joshi, "Multi-objective approach to maximise loadability of distribution networks by simultaneous reconfiguration and allocation of distributed energy resources," IET Generation, Transmission and Distribution, vol. 12, no. 21, pp. 5700–5712, 2018.

[238] A. Kimiyaghalam, M. Mahdavi, A. Ashouri, and H. Soheil, "Transmission expansion planning considering uncertainty in fuel price using DABC," The 9th International Energy Conference, Tehran, Iran, pp. 1–10, 2013.

[239] Y. Yang, S. Zhang, W. Pei, J. Sun, and Y. Lu, "Network reconfiguration and operation optimisation of distribution system with flexible DC device," The Journal of Engineering, vol. 2019, no. 16, pp. 2401–2404, 2019.

[240] M. Mahdavi, "New models and optimization techniques applied to the problem of optimal reconfiguration of radial distribution systems," Report, São Paulo Research Foundation (FAPESP), no. 16/12190-7, Brazil, 2019.

[241] A. Bagheri, M. Mahdavi, and H. Monsef, "Suggestion of an economical plan for electrical power transmission," The 9th International Energy Conference, Tehran, Iran, pp. 1–10, 2013.

[242] M. Mahdavi and R. Romero, "A new model for dynamic transmission expansion planning," 4th National Conference on Technology in Electrical and Computer Engineering, Tehran, Iran, pp. 1–7, 2018.

[243] H. Shayeghi, M. Mahdavi, and A. Bagheri, "Discrete PSO algorithm based optimization of transmission lines loading in TNEP problem," Energy Conversion and Management, vol. 51, no. 1, pp. 112–121, 2010.

[244] A. Jalilvand, A. Kimiyaghalam, A. Ashouri, and M. Mahdavi "Advanced particle swarm optimization-based PID controller parameters tuning," 2008 IEEE International Multitopic Conference, Karachi, Pakistan, pp. 429–435, 2008.

[245] H. Shayeghi, H. A. Shayanfar, M. Mahdavi, and A. Bagheri, "Application of binary particle swarm optimization for transmission expansion planning considering lines loadinga," The 2009 International Conference on Artificial Intelligence, Las Vegas, Nevada, USA, pp. 653–659, 2009.

# 14 Application of A Mathematical Programming Language (AMPL) in Distribution Network Reconfiguration Problem

*Meisam Mahdavi[1], Pierluigi Siano[2], and Hassan Haes Alhelou[3]*
[1]São Paulo State University, São Paulo, Brazil
[2]University of Salerno, Fisciano (SA), Italy.
[3]University College Dublin, Dublin 4, Ireland

**CONTENTS**

## 14.1 INTRODUCTION

Distribution network is an important part of the power system infrastructure that links high-voltage transmission system [1]–[16] to end-users of electric energy [17]–[19] in order to deliver electricity produced by the generation system (or generation companies (GENCOs) [20]–[22] in deregulated environments) to individual customers [23, 24]. The purpose of a power system is to provide reliable and safe electricity [25]–[27] to its consumers in an economical way [28, 29].

Power distribution systems in urban areas are constructed as a meshed structure, and usually operated in a radial topology by opening sectional switches (normally closed) and closing tie-lines (normally open) because many operation and control activities (e.g., voltage control and protection) are based on radial structure. In simple terms, operation of distribution networks in radial configuration is cheaper because of easier coordination of protective instruments (simple protection and coordination

347

schemes), lower fault and short circuit currents, and better voltage and power flow control. Tie-lines interconnect ends of radial feeders and/or provide connections to alternative supply points. Some of tie and sectionalizing switches are controlled manually (manual switches), and some others may be controlled automatically (remote controlled switches) [30]–[35]. In this case, the flexibility of distribution network increases by switching operation.

Power is lost in the form of heat during distribution process. Power losses of distribution system (it makes up 70% of the total losses in power system) are more than transmission losses [36]–[42] due to its radial structure and higher ratio of current to voltage. Power loss directly affects the operational cost and the voltage profile, especially in heavily loaded power systems [43]. Therefore, active power loss minimization is important to enhance the distribution system efficiency. Three common methods to reduce the power loss of distribution systems are capacitor placement, distributed generation (DG) [44]–[50] allocation, and system reconfiguration [51]–[53].

Although one the most important tasks of distribution system reconfiguration (DSR) is power loss minimization, nowadays, it includes other objectives such as voltage profile improvement, network adequacy increase [54]–[61], reliability enhancement [62, 63], congestion removal (lines loading optimization) [64]–[66], stability improvement [67], lines maintenance cost mitigation [68, 69], load balancing, DG penetration extension, service restoration, etc. [70]–[72]. During reconfiguration, it is very important that the configuration of system remains radial [73, 74].

Although in radial topology, each consumer has a single source of supply, distribution networks are usually the main cause of outages in the power system (80% of customer interruptions happen in the distribution system) [75]. According to the North American Electric Reliability Corporation [76], almost one-third of these power outages are related to technical causes such as equipment failures, operational errors, supply shortage, and system overloading.

Regarding 37% increase in global energy demand by 2040 and the annual growth rate of 0.5% and 0.8% for electricity consumption [77]–[80] in residential and commercial sectors, respectively (from 2013 to 2040), overloading of the distribution feeders plays a more important role nowadays than a few years ago in power deficiency [75].

It is predicted that power interruptions may appear more and with greater severity in future. For example, in January 2016, Mangla and Tarbela power stations in Pakistan tripped due to overloading. This resulted in a blackout across most regions of the country. Northern, eastern and north-eastern Indian regional grids collapsed for two days in July 2012. The reason behind this was overloading on the Bina-Gwalior-Agra link. Also, electricity breakdowns cost united states economy $80 billion per year [75]. Therefore, finding more optimal and efficient solutions for network reconfiguration problem is necessary.

Since the DSR problem was proposed, classical methods such as mathematical programing [81] has been presented as an important tool for solving the problem. Today, one of the important tools for solving this large-scale combinational optimization problem is A Mathematical Programming Language (AMPL) [82, 83].

In this chapter, the aim is to solve the network reconfiguration problem using AMPL. For this, first, DSR problem is modeled in AMPL, and then the nonlinear problem is converted to a linear optimization problem. Afterward, one of the most important linear solvers of AMPL, i.e., C programming language simplex method (CPLEX) is applied to the proposed model for finding the best switching scenarios. It should be noted that AMPL is applied to several distribution test systems and real distribution networks in order to show its efficiency for solving the problem of distribution network reconfiguration. AMPL is one of the most powerful classical tools for optimization of the DSR problem that are broadly used by researchers and distribution system operators to reconfigure any kind of distribution networks with any size.

## 14.2 MATHEMATICAL MODELING OF DISTRIBUTION SYSTEM RECONFIGURATION

DSR is a complex and combinatorial optimization problem, including integer and continuous variables with nonlinear and linear constraints. An efficient solution for this problem requires the selection of the most appropriate topology among all feasible configurations [84–86]. As mentioned in Chapter 13 about DSR modeling with objective of loss ($P_{\text{Loss}}$) minimization, in order to formulate the DSR problem by convex optimization software such as AMPL, following mixed-integer second-order cone programming (MISOCP) model is presented [35, 87]:

$$\text{Min } P_{\text{Loss}} = \sum_{ij\in\Omega^l} R_{ij}\left|I_{ij}\right|^2 \tag{14.1}$$

subject to:

$$P_i^S + \sum_{k=1}^{NB} P_{ki} = \sum_{j=1}^{NB} P_{ij} + \sum_{j=1}^{NB} R_{ij}\left|I_{ij}\right|^2 + P_i^D \ \ \forall\ ij,\ ki\in\Omega^l,\ i=1,\,2,\,...,\,NB \tag{14.2}$$

$$Q_i^S + \sum_{k=1}^{NB} Q_{ki} = \sum_{j=1}^{NB} Q_{ij} + \sum_{j=1}^{NB} X_{ij}\left|I_{ij}\right|^2 + Q_i^D \ \ \forall\ ij,\ ki\in\Omega^l,\ i=1,\,2,\,...,\,NB \tag{14.3}$$

$$\left|V_i\right|^2 = \left|V_j\right|^2 - 2\left(R_{ij}P_{ij} + X_{ij}Q_{ij}\right) + \left|Z_{ij}\right|^2\left|I_{ij}\right|^2 + b_{ij} \ \ \forall\ ij\in\Omega^l,\ i=1,\,...,NB \tag{14.4}$$

$$\left|b_{ij}\right| \le \left(V_{\max}^2 - V_{\min}^2\right)\left(1 - y_{ij}\right) \ \forall\ ij\in\Omega^l \tag{14.5}$$

$$\beta_{ij} + \beta_{ji} = y_{ij} \ \forall\ ij\in\Omega^l \tag{14.6}$$

$$\sum_{j=1}^{NB} \beta_{ij} = 1 \ \forall\ ij\,\|\,ji\in\Omega^l,\ i\in\Omega^b \tag{14.7}$$

$$\beta_{ij} = 0 \ \forall\ i\in\Omega^s,\ ij\in\Omega^l \tag{14.8}$$

$$\beta_{ji} = 0 \ \forall\ j\in\Omega^s,\ ij\in\Omega^l \tag{14.9}$$

$$\beta_{ij} \in \{0, 1\} \ \forall \ ij \in \Omega^l \tag{14.10}$$

$$y_{ij} \in \{0, 1\} \ \forall \ ij \in \Omega^l \tag{14.11}$$

$$V_{\min}^2 \le |V_j|^2 \le V_{\max}^2 \ \forall \ j = 1, 2, ..., NB \tag{14.12}$$

$$0 \le |I_{ij}|^2 \le \left(I_{ij}^{\max}\right)^2 y_{ij} \ \forall \ ij \in \Omega^l \tag{14.13}$$

$$|V_j|^2 |I_{ij}|^2 \ge P_{ij}^2 + Q_{ij}^2 \ \forall \ ij \in \Omega^l, \ j = 1, 2, ..., NB \tag{14.14}$$

$$|P_{ij}| \le V_{\max} I_{ij}^{\max} y_{ij} \ \forall \ ij \in \Omega^l \tag{14.15}$$

$$|Q_{ij}| \le V_{\max} I_{ij}^{\max} y_{ij} \ \forall \ ij \in \Omega^l \tag{14.16}$$

The proposed DSR model including binary variables ($\beta_{ij}$ and $y_{ij}$) and real variables ($|I_{ij}|, |V_j|, P_{ij}, Q_{ij}, b_{ij}, P_i^S, Q_i^S$) cannot be solved by CPLEX because of non-linear terms of $|I_{ij}|^2$ and $|V_j|^2$. Thus, we have to linearize the model using various linear programing methods or use the variable change technique described in refs. [88] and [89]. The variable change method is easier and more accurate than the linearization because many assumptions and approximations need to be considered in linear optimization methods that can decrease the quality of solutions for large-scale distribution systems. Here, the variable change method was used to resolve the problem by replacing square variables $|I_{ij}|^2$, $|V_j|^2$ and $|V_i|^2$ with $I_{ij}^{sqr}$, $V_j^{sqr}$ and $V_i^{sqr}$, respectively, as follows [82, 83].

$$\text{Min} \ \sum_{ij \in \Omega^l} R_{ij} I_{ij}^{sqr} \tag{14.17}$$

subject to:

$$\beta_{ij} + \beta_{ji} = y_{ij} \ \forall \ ij \in \Omega^l \tag{14.18}$$

$$\sum_{j=1}^{NB} \beta_{ij} = 1 \ \forall \ ij \parallel ji \in \Omega^l, \ i \in \Omega^b \tag{14.19}$$

$$\beta_{ij} = 0 \ \forall \ i \in \Omega^s, \ ij \in \Omega^l \tag{14.20}$$

$$\beta_{ji} = 0 \ \forall \ j \in \Omega^s, \ ij \in \Omega^l \tag{14.21}$$

$$\beta_{ij} \in \{0, 1\} \ \forall \ ij \in \Omega^l \tag{14.22}$$

$$y_{ij} \in \{0, 1\} \ \forall \ ij \in \Omega^l \tag{14.23}$$

$$P_i^S + \sum_{k=1}^{NB} P_{ki} - \sum_{j=1}^{NB} P_{ij} - \sum_{j=1}^{NB} R_{ij} I_{ij}^{sqr} = P_i^D \ \forall ij, \ ki \in \Omega^l, \ i = 1, 2, ..., NB \tag{14.24}$$

$$Q_i^S + \sum_{k=1}^{NB} Q_{ki} - \sum_{j=1}^{NB} Q_{ij} - \sum_{j=1}^{NB} X_{ij} I_{ij}^{sqr} = Q_i^D \ \forall ij, \ ki \in \Omega^l, \ i = 1, 2, ..., NB \tag{14.25}$$

$$V_i^{\text{sqr}} - V_j^{\text{sqr}} = \left(R_{ij}^2 + X_{ij}^2\right)I_{ij}^{\text{sqr}} - 2\left(R_{ij}P_{ij} + X_{ij}Q_{ij}\right) + b_{ij} \quad \forall \ ij \in \Omega^l, \ i = 1, \ldots, NB \quad (14.26)$$

$$-\left(V_{\max}^2 - V_{\min}^2\right)\left(1 - y_{ij}\right) \leq b_{ij} \leq \left(V_{\max}^2 - V_{\min}^2\right)\left(1 - y_{ij}\right) \quad \forall \ ij \in \Omega^l \quad (14.27)$$

$$V_{\min}^2 \leq V_j^{\text{sqr}} \leq V_{\max}^2 \quad \forall \ j = 1, 2, \ldots, NB \quad (14.28)$$

$$0 \leq I_{ij}^{\text{sqr}} \leq \left(I_{ij}^{\max}\right)^2 y_{ij} \quad \forall \ ij \in \Omega^l \quad (14.29)$$

$$V_j^{\text{sqr}} I_{ij}^{\text{sqr}} \geq P_{ij}^2 + Q_{ij}^2 \quad \forall \ ij \in \Omega^l, \ j = 1, \ldots, NB \quad (14.30)$$

$$-V_{\max}I_{ij}^{\max} y_{ij} \leq P_{ij} \leq V_{\max}I_{ij}^{\max} y_{ij} \quad \forall \ ij \in \Omega^l \quad (14.31)$$

$$-V_{\max}I_{ij}^{\max} y_{ij} \leq Q_{ij} \leq V_{\max}I_{ij}^{\max} y_{ij} \quad \forall \ ij \in \Omega^l \quad (14.32)$$

In the above equations, we have:

$\Omega^l$: Set of all branches (including normal and tie lines).

$NB$: Total number of buses.

$R_{ij}, X_{ij}$: Resistance and reactance of branch $ij$.

$|I_{ij}|$: Amplitude of current flow from bus $i$ to $j$.

$P_i^S, Q_i^S$: Active and reactive power provided by substation at bus $i$.

$P_i^D, Q_i^D$: Active and reactive power demand at bus $i$.

$P_{ij}, Q_{ij}$: Active and reactive power flow from bus $i$ to $j$ (branch $ij$).

$|V_i|$: Voltage magnitude of bus $i$.

$V_{\min}, V_{\max}$: Lower and upper bound of bus voltage limits.

$|S_{ij}|, |Z_{ij}|$: Amplitudes of complex power and impedance of branch $ij$.

$b_{ij}$: Slack variable that needs to be zero when branch $ij$ is closed (i.e., Equation (14.4) is satisfied) and a real number when the corresponding branch is open.

$\beta_{ij}$: Orientation variable of branch $ij$.

$I_{ij}^{\max}$: Current rating limit of branch $ij$.

$y_{ij}$: A binary variable that represents the status of each switch (0 and 1 for open and closed switches, respectively).

$P_{ij}^{\max}$: Real power flow capacity of branch $ij$.

$|P_{ij}|$: Absolute value of $P_{ij}$.

$Q_{ij}^{\max}$: Reactive power flow capacity of branch $ij$.

$|Q_{ij}|$: Absolute value of $Q_{ij}$.

## 14.3   EXAMPLES

The proposed DSR model in AMPL (set of Equations (14.17)–(14.32)) is applied to several test systems using CPLEX and the results are presented. It should be mentioned that $V_{\min}$ and $V_{\max}$ are considered to be 0.93 and 1 per unit (p.u.), respectively, in which voltage magnitude of substation nodes has been fixed on 1 p.u.

- **Example 1: 7-Bus Test System**
  Figure 14.1 shows this test system with six branches (closed switches) and one loop (tie-line). All data related to this system can be found in ref. [90].

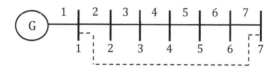

**FIGURE 14.1**  Schematic of 7-bus test system.

The base power is 1 MVA and base voltage is 12.66 kV. Also, the maximum current capacity ($I_{ij}^{max}$) of each branch is considered to be 500 A. Therefore, the proposed model has been applied to the understudied test system, and the results are listed in Tables 14.1 to 14.3. It should be noted that the best voltage magnitudes before and after reconfiguration is 1 p.u. at substation bus 1. Buses 5 and 7 have the worst voltage magnitudes in original and reconfigured networks, respectively.

Furthermore, the solution methods presented for solving DSR problem are ranked based on the power loss decrement and execution time in Tables 14.2 and 14.3, respectively.

**TABLE 14.1**
**Numerical Results Compared to Other Methods for Example 1**

| Items | Open | Power Loss (kW) | | Worst Voltage (p.u.) | | Run Time (s) |
|---|---|---|---|---|---|---|
| Methods | Switches | Before | After | Before | After | |
| [90] | 4 | 1.44 | 1.2 | 0.995 | 0.99 | – |
| CPLEX | 5 | 1.44 | 0.82 | 0.995 | 0.99 | 0.08 |

**TABLE 14.2**
**Ranking of Methods Based on Loss Reduction for Example 1**

| Methods | Power Loss (kW) |
|---|---|
| [90] | 1.2 |
| CPLEX | 0.82 |

**TABLE 14.3**
**Ranking of Methods Based on Execution Time for Example 1**

| Methods | Run Time (s) |
|---|---|
| [90] | - |
| CPLEX | 0.08 |

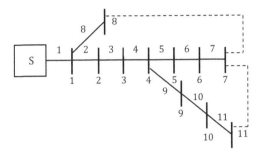

**FIGURE 14.2** AL-Mansoor No.11 network.

The results show that CPLEX could find more optimal solution compared to method reported in ref. [90].

• **Example 2: 12-Bus Test System**
This actual system is a part of Baghdad city network in Iraq (AL-Mansoor No.11), and its single line diagram is shown in Figure 14.2. The feeder nominal (base) voltage and complex power are 11.1 kV and 2250 kVA that is connected to AL-Mansoor substation. The system relevant data are given in ref. [91]. Also, the maximum current capacity ($I_{ij}^{max}$) of each branch is assumed to be 500 A. Tie-lines in this network are branches 7–8 (12) and 7–11 (13).

The proposed method has been implemented on the test system, and the results are provided in Tables 14.4 to 14.7. It should be noted that the best voltage magnitudes before and after reconfiguration is 1 p.u. at bus 1. Bus 10 has the worst voltage magnitude before and after network reconfiguration. Tables 14.5 to 14.7 rank the solution methods based on power loss decrement, voltage profile improvement, and computational time decrease, respectively.

**TABLE 14.4**
**Numerical Results Compared to Other Methods for Example 2**

| Items | | Power Loss (kW) | | Worst Voltage (p.u.) | | |
|---|---|---|---|---|---|---|
| Methods | Open Switches | Before | After | Before | After | Run Time (s) |
| [91] | 12, 13 | 3.55 | 3.55 | 0.9958 | 0.996 | 0.4 |
| CPLEX | 3, 9 | 3.55 | 1.31 | 0.9970 | 0.998 | 0.05 |

**TABLE 14.5**
**Ranking of Methods Based on Loss Reduction for Example 2**

| Methods | Power Loss (kW) |
|---|---|
| [91] | 3.55 |
| CPLEX | 1.31 |

**TABLE 14.6**

**Ranking of Methods Regarding Voltage Improvement for Example 2**

| Methods | Worst Voltage (p.u.) |
|---------|----------------------|
| [91]    | 0.996                |
| CPLEX   | 0.998                |

**TABLE 14.7**

**Ranking of Methods Based on Execution Time for Example 2**

| Methods | Run Time (s) |
|---------|--------------|
| [91]    | 0.4          |
| CPLEX   | 0.05         |

From Tables 14.5 and 14.6, it can be concluded that CPLEX is more efficient method than approach used in [91] to find optimal solution. According to Table 14-7, CPLEX can find the optimal solution more quickly than [91].

- **Example 3: 16-Bus Test System**

  The three-feeder distribution system, including 13 sectional switches (branches) and three tie-lines, has been shown in Figure 14.3. All data of this 23-kV network such as resistance and reactance of branches as well as nodal active and reactive demand have been presented in ref. [92]. The base power and voltage are considered to be 100 MVA and 23 kV. Also, the current-carrying capacity ($I_{ij}^{max}$) of branches 1–4 (11), 2–8 (16), and 18 was considered to be 500 A, 500 A, and 300 A, respectively, while that of all other branches was 250 A [93]. The proposed approaches have been applied to the 16-bus distribution network, and the results are provided in Table 14.8.

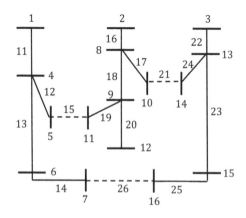

**FIGURE 14.3**  Initial distribution network.

**TABLE 14.8**

**Numerical Results of Proposed Model Compared to Other Methods for Example 3**

| Items Methods | Open Switches | Power Loss (kW) Before | Power Loss (kW) After | Worst Voltage (p.u.) Before | Worst Voltage (p.u.) After | Run Time (s) |
|---|---|---|---|---|---|---|
| Heuristic [94] | 15, 17, 26 | 514 | 488.46 | 0.969 | - | - |
| Heuristic [95] | 15, 17, 26 | 514 | 485.9 | 0.968 | - | - |
| SA[1] [96] | 17, 19, 26 | 511.4 | 466.1 | 0.9693 | 0.9716 | 6 |
| Heuristic [97] | 15, 17, 26 | 514 | 483.88 | 0.969 | - | - |
| SA [98] | 17, 19, 26 | 511.1 | 466 | 0.9693 | 0.9716 | 7.5 |
| RGA[2] [99] | 17, 19, 26 | 511.4 | 466.1 | 0.9693 | - | - |
| MIHDE[3] [100] | 17, 19, 26 | 511.5 | 466.1 | 0.9693 | 0.9716 | 4.5 |
| SA+TS[4] [101] | 17, 19, 26 | 511.43 | 466.1 | 0.9693 | 0.9694 | - |
| TS [102] | 17, 19, 26 | 511.43 | 466.1 | 0.9693 | 0.9694 | - |
| Heuristic [103] | 17, 19, 26 | 511.4 | 466.13 | - | - | - |
| Variable scaling HDE (VSHDE) [92] | 17, 19, 26 | 511.4 | 466.1 | 0.9693 | 0.9716 | > [96] |
| SA [104], [105] | 17, 19, 26 | 511.4 | 466.1 | 0.9693 | 0.9716 | 2.07 |
| GA [104], [105] | 17, 19, 26 | 511.4 | 466.1 | 0.9693 | 0.9716 | 2.32 |
| ACO[5] [104], [105] | 17, 19, 26 | 511.4 | 466.1 | 0.9693 | 0.9716 | 1.81 |
| Efficient GA [106] | 17, 19, 26 | 511.44 | 466.12 | 0.9693 | 0.97158 | 2.0269 |
| Heuristic [107] | 17, 19, 26 | 511.4 | 466.13 | - | - | < [103] |
| Heuristic [108] | 17, 19, 26 | 511.4 | 466.13 | - | - | < [107] |
| GAMT[6] [109] | 17, 19, 26 | 511.5 | 466.1 | 0.9693 | 0.9716 | 1.35 |
| DPSO[7] [110] | 17, 19, 26 | 511.44 | 466.12 | 0.9693 | 0.97158 | 0.156 |
| ABC[8] [111] | 17, 19, 26 | 511.5 | 466.1 | 0.9693 | 0.9716 | 1.7 |
| MTS[9] [112] | 17, 19, 26 | 511.43 | 466.1 | 0.9693 | 0.9694 | - |
| SOReco [113] | 17, 19, 26 | 511 | 465.8 | - | - | 0.212 |
| NrGA[10] [114] | 17, 19, 26 | 511.44 | 466.13 | 0.969 | 0.972 | 0.45 |
| NCUP[11] [115] | 17, 19, 26 | 511.4 | 466.14 | - | - | - |
| AIMMS [116] | 17, 19, 26 | 514 | 468.3 | - | - | 0.16 |
| FNSGA[12] [93] | 17, 19, 26 | 511.44 | 466.13 | 0.969 | 0.972 | 0.27 |
| MIQCP[13] [117] | 17, 19, 26 | 514 | 468.3 | 0.969 | - | - |
| MILP[14] [117] | 17, 19, 26 | 514 | 468.3 | 0.969 | - | - |
| CPLEX | 17, 19, 26 | 511.44 | 465.5 | 0.9693 | 0.9716 | 0.12 |

[1] Simulated annealing
[2] Refined genetic algorithm (Refined GA)
[3] Mixed-integer hybrid differential evolution (Mixed-integer HDE)
[4] Tabu search
[5] Ant colony optimization
[6] GA based on Matroid theory
[7] Discrete particle swarm optimization (Discrete PSO)
[8] Artificial bee colony
[9] Modified TS
[10] Non-revisiting GA
[11] Neighbour-chain updating process
[12] Fast non-dominated sorting GA (Fast NSGA)
[13] Mixed-integer quadratic conic programming
[14] Mixed-integer linear programing

**TABLE 14.9**

**Ranking of Methods Based on Loss Reduction for Example 3**

| Methods | Power Loss (kW) |
| --- | --- |
| [94] | 488.46 |
| [95] | 485.9 |
| [97] | 483.88 |
| [116], [117] | 468.3 |
| [115] | 466.14 |
| [103], [107], [108], [93], [114] | 466.13 |
| [110], [106] | 466.12 |
| [96], [99], [100], [102], [104], [92], [109], [105], [112], [101], [111] | 466.1 |
| [98] | 466 |
| [113] | 465.8 |
| CPLEX | 465.5 |

In Table 14.9, methods have been arranged according to their publication dates. It should be noted that the best (maximum) voltage magnitudes before and after reconfiguration are 1 p.u. at substation buses 1, 2, and 3. From voltage profile aspect, the worst (minimum) voltage magnitude is related to bus 12 in original and reconfigured networks. Furthermore, the solution methods presented for solving DSR problem are ranked based on power loss reduction, worst voltage increase, and computing time in Tables 14.9, 14.10, and 14.11, respectively.

Tables 14.9 and 14.11 show that the CPLEX is more accurate and faster than other reconfiguration methods, while it improves the minimum voltage like most of other methods.

- **Example 4: 28-Bus Test System**
This electrical grid is a part of the electrical power distribution system in the city of Koprivnica, Croatia. It consists of 28 nodes, one transformer

**TABLE 14.10**

**Ranking of Methods Regarding Voltage Improvement for Example 3**

| Methods | Worst Voltage (p.u.) |
| --- | --- |
| [102], [112], [101] | 0.9694 |
| [110], [106] | 0.97158 |
| [92], [96], [98]–[100], [104], [105], [109], [111], CPLEX | 0.9716 |
| [93], [114] | 0.972 |

**TABLE 14.11**

**Ranking of Methods Based on**
**Execution Time for Example 3**

| Methods | Run Time (s) |
| --- | --- |
| [92], [94], [95], [97], [103] | Long |
| [107] | <[103] |
| [108] | <[107] |
| [98] | 7.5 |
| [96] | 6 |
| [100] | 4.5 |
| GA [104], [105] | 2.32 |
| SA [104], [105] | 2.07 |
| [106] | 2.027 |
| ACO [104], [105] | 1.81 |
| [111] | 1.7 |
| [109] | 1.35 |
| [114] | 0.45 |
| [93] | 0.27 |
| [113] | 0.212 |
| [116] | 0.16 |
| [110] | 0.156 |
| CPLEX | 0.12 |

station 110/35 kV, two transformer stations 35/10 kV, 22 loads, and 28 transmission lines. Graph representation of the Koprivnica distribution grid is shown in Figure 14.4, and its data are available in ref. [118]. Lines and transformers [119] that connect nodes are represented as edges of the graph. Full

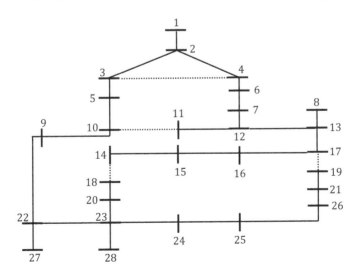

**FIGURE 14.4**    Distribution network of Koprivnica.

lines represent transmission lines that are switched on, while dotted lines represent transmission lines that are switched off in the current topology. Base values for this network are nominal voltages and power. Results for this network are illustrated in Tables 14.12 to 14.15.

It can be seen that CPLEX method can reach the optimal solution quickly.

**TABLE 14.12**

**Numerical Results Compared to Other Methods for Example 4**

| Items | Open | Power Loss (kW) | | Worst Voltage (p.u.) | | |
|---|---|---|---|---|---|---|
| Methods | Switches | Before | After | Before | After | Run Time (s) |
| [118] | 11, 20, 21, 28 | 47 | 40 | 0.9 | 0.92 | - |
| CPLEX | 11, 20, 21, 28 | 47.05 | 39.7 | 0.9151 | 0.936 | 1.11 |

**TABLE 14.13**

**Ranking of Methods Based on Loss Reduction for Example 4**

| Methods | Power Loss (kW) |
|---|---|
| [118] | 40 |
| CPLEX | 39.7 |

**TABLE 14.14**

**Ranking of Methods Regarding Voltage Improvement for Example 4**

| Methods | Worst Voltage (p.u.) |
|---|---|
| [118] | 0.92 |
| CPLEX | 0.936 |

**TABLE 14.15**

**Ranking of Methods Based on Execution Time for Example 4**

| Methods | Run Time (s) |
|---|---|
| [118] | - |
| CPLEX | 1.11 |

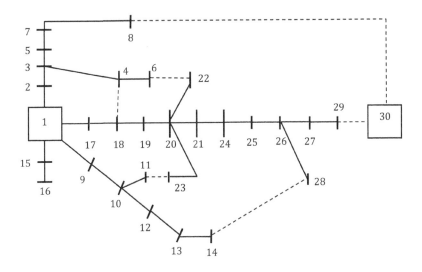

**FIGURE 14.5**   30-bus distribution system.

- **Example 5: 30-Bus Test System**

  Configuration of this test system with six tie-lines and 28 normal branches is shown in Figure 14.5, and its data are available in ref. [120]. The base values are 1 MVA and 18.6 kV. The proposed techniques have been applied to the 29-bus test system, and the results are provided in Tables 14.16 to 14.19. The best voltage magnitudes before and after reconfiguration is 1 p.u. at substation bus 1, while the worst voltage before and after reconfiguration belongs to buses 29 and 14, respectively.

**TABLE 14.16**

**Numerical Results Compared to Other Methods for Example 5**

| Items | | Power Loss (kW) | | Worst Voltage (p.u.) | | Run |
|---|---|---|---|---|---|---|
| Methods | Open Switches | Before | After | Before | After | Time (s) |
| [120] | 3, 7, 10, 25, 32, 33 | 1240 | 368 | 0.8251 | 0.92 | – |
| CPLEX | 3, 7, 10, 25, 32, 33 | 1239 | 367.3 | 0.8251 | 0.93 | 2.02 |

**TABLE 14.17**

**Ranking of Methods Based on Loss Reduction for Example 5**

| Methods | Power Loss (kW) |
|---|---|
| [120] | 368 |
| CPLEX | 367.3 |

**TABLE 14.18**

**Ranking of Methods Regarding Voltage Improvement for Example 5**

| Methods | Worst Voltage (p.u.) |
|---------|----------------------|
| [120]   | 0.92                 |
| CPLEX   | 0.93                 |

**TABLE 14.19**

**Ranking of Methods Based on Execution Time for Example 5**

| Methods | Run Time (s) |
|---------|--------------|
| [120]   | -            |
| CPLEX   | 2.02         |

From Tables 14.17 and 14.18, it can be seen that CPLEX finds more optimal solution compared to the method proposed by [120].

- **Example 6: 33-Bus Test System**

  The system includes two feeder substations with three 12.66 kV-laterals, five tie switches, and 32 normal branches (Figure 14.6). The data of this test system are available in ref. [121]. The MVA and kV base are 1 and 12.66, respectively [122, 123].

  The proposed formulation has been applied to 33-bus test system, and the results are listed in Table 14.20 chronologically. Also, the voltage magnitude of substation bus (node 0) is adjusted in 1 p.u.

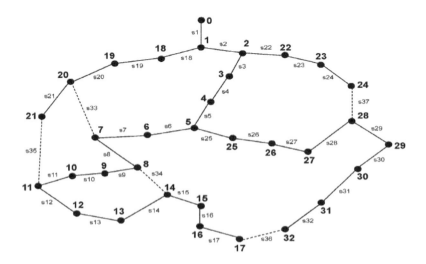

**FIGURE 14.6**   Single diagram of 33-bus distribution system.

**TABLE 14.20**

**Numerical Results of Proposed Model Compared to Other Methods for Example 6**

| Items | | Power Loss (kW) | | Worst Voltage (p.u.) | | Run |
|---|---|---|---|---|---|---|
| Methods | Open Switches | Before | After | Before | After | Time (s) |
| Branch exchanges [121] | 11, 28, 31, 33, 34 | 205.81 | 148.858 | 0.88 | 0.92 | Long |
| Heuristic [124] | 7, 10, 14, 32, 37 | 205.81 | 141.541 | 0.913 | 0.94 | 0.14 |
| Heuristic [95] | 7, 9, 14, 32, 37 | 202.67 | 139.532 | 0.9131 | - | <48 |
| Heuristic [122] | 7, 9, 14, 32, 37 | 205.81 | 140.815 | 0.91253 | 0.9472 | 0.87 |
| Standard GA [125], [126] | 7, 9, 14, 32, 37 | 202.71 | 141.6 | 0.9131 | 0.929 | 19.1 |
| SA [96] | 7, 9, 14, 32, 37 | 203.2 | 139.7 | 0.913 | 0.9378 | 48 |
| DAOP[1] [127] | 7, 9, 14, 32, 37 | 202.67 | 139.55 | 0.913 | 0.94749 | 1.99 |
| Heuristic [128] | 6, 9, 14, 32, 37 | 204.14 | 142.83 | 0.91 | 0.9378 | - |
| Heuristic [129] | 7, 9, 14, 32, 37 | 202.67 | 139.55 | 0.913 | 0.94749 | 647.03 |
| RGA [99], [126], [130] | 7, 9, 14, 32, 37 | 202.7 | 139.55 | 0.9131 | 0.9378 | 7.41 |
| FEP[2] [84] | 7, 9, 14, 32, 37 | 211.22 | 139.83 | 0.9038 | 0.9378 | 55.04 |
| SA+TS [101] | 7, 9, 14, 32, 37 | 203 | 139.8 | 0.9129 | 0.9378 | 6.852 |
| FGA[3] [131] | 9, 28, 33, 34, 36 | 205.81 | 140.6 | 0.842 | 0.95 | - |
| Heuristic [103] | 7, 9, 14, 32, 37 | 202.67 | 139.55 | 0.913 | 0.94749 | 1.66 |
| Restricted GA [132] | 7, 9, 14, 32, 37 | 202.68 | 139.5 | - | - | 6.3 |
| ITS[4] [79], [126] | 7, 9, 14, 36, 37 | 202.71 | 139.28 | 0.9131 | 0.921 | 8.1 |
| Heuristic [108] | 7, 9, 14, 32, 37 | 202.68 | 139.55 | - | - | - |
| PGS[5] [133] | 7, 9, 14, 32, 37 | 202.68 | 139.55 | - | - | <1.66 |
| Heuristic [134] | 7, 9, 14, 32, 37 | 202.68 | 139.55 | 0.9131 | 0.9378 | 0.41 |
| Binary GA (BGA) [109] | 7, 10, 14, 36, 37 | 202.68 | 142.68 | - | - | 160 |
| GAMT [109], [130] | 7, 9, 14, 32, 37 | 202.68 | 139.55 | 0.9131 | 0.9378 | 7.2 |
| Heuristic [135] | 7, 9, 14, 32, 37 | 202.7 | 139.5 | - | - | 0.5 |
| HBMO[6] [136], [137], [138] | 7, 9, 14, 32, 37 | 202.68 | 139.55 | 0.9131 | 0.9378 | 6 |
| DPSO [139], [140] | 7, 9, 14, 32, 37 | 202.67 | 139.53 | 0.913 | 0.9378 | 8 |
| DPSO + HBMO [140] | 7, 9, 14, 32, 37 | 202.67 | 139.53 | 0.913 | 0.9378 | 8 |
| BD[7] [141], [142] | 7, 9, 14, 32, 37 | 202.68 | 139.55 | 0.913 | 0.94749 | 0.11 |
| MBPSO[8] [143], [144], [145] | 7, 9, 14, 32, 37 | 202.6 | 139.5 | 0.9131 | 0.9378 | 5.693 |
| MTS [145] | 7, 9, 14, 32, 37 | 202.6 | 139.5 | 0.9131 | 0.9378 | > 5.693 |
| MHBMO[9] [130], [137] | 7, 9, 14, 32, 37 | 202.68 | 139.55 | 0.9131 | 0.9378 | 8 |
| FACO[10] [146] | 7, 10, 14, 32, 37 | 202.71 | 140.26 | - | - | - |
| HSA[11] [126] | 7, 10, 14, 36, 37 | 202.71 | 142.68 | 0.9131 | 0.9342 | 7.2 |
| EICPSO[12] [139] | 7, 9, 14, 28, 32 | 203.33 | 139.98 | 0.9131 | 0.9378 | 6.343 |
| AACO[13] [130] | 7, 9, 14, 32, 37 | 202.68 | 139.55 | 0.9131 | 0.9378 | 0.3 |
| B&B[14] [147] | 7, 9, 14, 32, 37 | 202.68 | 139.55 | 0.9131 | 0.9378 | 19 |
| HC-ACO[15] [148] | 7, 9, 14, 32, 37 | 203 | 139.8 | 0.9129 | 0.9378 | - |
| MILP [149] | 7, 9, 14, 32, 37 | 202.68 | 139.55 | 0.9131 | 0.9378 | - |
| QP[16] [150] | 7, 9, 14, 32, 37 | 202.7 | 139.5 | 0.913 | 0.9412 | 0.21 |
| QCP[17] [150] | 7, 9, 14, 32, 37 | 202.7 | 139.5 | 0.913 | 0.9412 | 1.43 |
| SOCP[18] [150] | 7, 9, 14, 32, 37 | 202.7 | 139.5 | 0.913 | 0.9412 | 12.8 |
| SOReco [103] | 7, 9, 14, 32, 37 | 202.68 | 139.55 | - | - | 5.704 |

*(Continued)*

**TABLE 14.20 (*Continued*)**

**Numerical Results of Proposed Model Compared to Other Methods for Example 6**

| Items<br>Methods | Open Switches | Power Loss (kW) | | Worst Voltage (p.u.) | | Run Time (s) |
|---|---|---|---|---|---|---|
| | | Before | After | Before | After | |
| NCUP [115] | 7, 9, 14, 32, 37 | 202.68 | 139.55 | - | - | 5.28 |
| CLONR [151] | 7, 9, 14, 32, 37 | 202.68 | 139.55 | - | - | - |
| FHBB-BC[19] [152] | 7, 9, 14, 32, 37 | 202.67 | 139.53 | 0.9131 | 0.9412 | - |
| MIQCP [117] | 7, 9, 14, 32, 37 | 202.68 | 139.5 | 0.9131 | 0.9378 | 35.5 |
| MILP [117] | 7, 9, 14, 32, 37 | 202.68 | 139.5 | 0.9131 | 0.9378 | 0.15 |
| AIMMS [116] | 7, 9, 14, 32, 37 | 202.7 | 139.6 | - | - | 3 |
| MIQP[20] [153] | 7, 9, 14, 32, 37 | 202.68 | 139.5 | 0.9131 | 0.9378 | 3.2 |
| [154] | 7, 9, 14, 32, 37 | 202.7 | 139.5 | 0.9131 | 0.9378 | - |
| Modified PGS [123] | 7, 9, 14, 32, 37 | 202.67 | 139.5 | 0.9052 | 0.9343 | 0.16 |
| ACO [136] | 7, 9, 14, 32, 37 | 202.68 | 139.551 | 0.9131 | 0.9378 | 6.439 |
| HSA [136] | 7, 10, 14, 36, 37 | 202.68 | 139.551 | 0.9131 | 0.9378 | 3.274 |
| GA+ BE [136] | 7, 10, 14, 36, 37 | 202.68 | 139.551 | 0.9131 | 0.9378 | 10.83 |
| Heuristic [155] | 7, 9, 14, 32, 37 | 202.68 | 139.55 | 0.9131 | 0.9378 | 3.7 |
| CPLEX | 7, 9, 14, 32, 37 | 202.68 | 138.56 | 0.9131 | 0.9384 | 3.14 |

[1] Discrete ascent optimal programming
[2] fuzzy evolution programming (fuzzy EP)
[3] Fuzzy GA
[4] Improved TS
[5] Plant growth simulation
[6] Honey bee mating optimization
[7] Branch exchange
[8] Modified binary PSO (Modified BPSO)
[9] Multi-objective HBMO
[10] Fuzzy ACO
[11] Harmony search algorithm
[12] Enhanced integer coded PSO (Enhanced ICPSO)
[13] Adaptive ACO
[14] Branch-and-bound
[15] Hyper-cube ACO
[16] Quadratic programming
[17] Quadratic constrained programming
[18] Second-order cone programming
[19] Fuzzy hybrid big bang-big crunch (Fuzzy HBB-BC)
[20] Mixed-integer QP

The solution methods are arranged in Tables 14.21, 14.22, and 14.23 from viewpoints of convergence speed, voltage stability, and accuracy, respectively. It can be seen that CPLEX solves the DSR problem more precisely than other methods in an acceptable computing time. It is obvious

**TABLE 14.21**

**Methods Based on Computational Time for Example 6**

| Methods | Run Time (s) |
|---|---|
| [129] | 647.03 |
| Binary GA [109] | 160 |
| [84] | 55.04 |
| [96] | 48 |
| MIQCP [117] | 35.5 |
| [125] | 19.1 |
| [147] | 19 |
| SOCP [150] | 12.8 |
| GA+BE [136] | 10.83 |
| ITS [125] | 8.1 |
| [137], [140] | 8 |
| RGA [130] | 7.41 |
| GAMT [109], HSA [125], GAMT [130] | 7.2 |
| [101] | 6.852 |
| ACO [136] | 6.439 |
| EICPSO [139] | 6.343 |
| [132] | 6.3 |
| [138] | 6 |
| [113], MBPSO [139] | 5.7 |
| [115] | 5.28 |
| [155] | 3.7 |
| HSA [136] | 3.274 |
| [144] | 3.2 |
| CPLEX | 3.14 |
| [116] | 3 |
| [127] | 1.99 |
| [103] | 1.66 |
| QCP [150] | 1.43 |
| [122] | 0.87 |
| [135] | 0.5 |
| [134] | 0.41 |
| AACO [130] | 0.3 |
| QP [150] | 0.21 |
| [123] | 0.16 |
| MILP [117] | 0.15 |
| [124] | 0.14 |
| [142] | 0.11 |

**TABLE 14.22**

**Methods According to Voltage Enhancement for Example 6**

| Methods | Minimum Voltage (p.u.) |
|---------|------------------------|
| [121] | 0.92 |
| [125] | 0.929 |
| [125], [126] | 0.934 |
| [84], [96], [99], [101], [109], [117], [125], [127], [130], [134], [136]–[140], [143]–[145], [147]–[149], [153]–[155] | 0.9378 |
| CPLEX | 0.9384 |
| [124], [150], [152] | 0.94 |
| [122] | 0.9472 |
| [103], [127], [129], [141], [142] | 0.9475 |
| [131] | 0.95 |

**TABLE 14.23**

**Arrangement of Methods According to Loss Reduction for Example 6**

| Methods | Power Loss (kW) |
|---------|-----------------|
| [121] | 148.858 |
| [128] | 142.83 |
| GA [109] | 142.68 |
| [125] | 141.6 |
| [124] | 141.54 |
| [122] | 140.815 |
| [131] | 140.6 |
| [146] | 140.26 |
| EICPSO [139] | 139.98 |
| [84] | 139.83 |
| [148], [101] | 139.8 |
| [143] | 139.7 |
| [116] | 139.6 |
| [99], [103], [108], GAMT [109], [113], [115], [127], [129], [133], [134], [136]–[138], [141], [142], [147], [149], [151], [155] | 139.55 |
| [95], [140], [152] | 139.53 |
| [117], [123], [130], [132], [135], [139], [143]–[145], MBPSO [150], [153], [154] | 139.5 |
| [125] | 139.28 |
| CPLEX | 138.56 |

that more accurate solutions cannot be resulted in very short time because
the algorithm needs to search more precisely.

- **Example 7: 49-Bus Test System**

  Figure 14.7 shows the schematic of real distribution network of Baghdad
  city, Iraq. The test system is an 11 kV-network with 49 buses, main feeder,
  six laterals, and seven capacitor in buses 9, 13, 24, 32, 36, 42, and 48 with
  size of 100, 75, 100, 200, 200, 350, and 200 kVAr, respectively. The system data with power demand information are available in ref. [156]. There

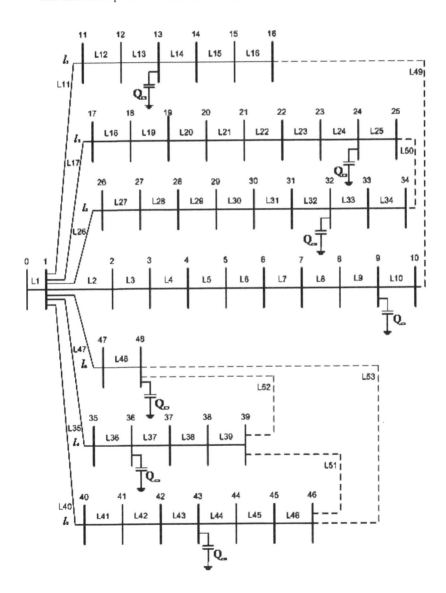

**FIGURE 14.7**   Baghdad distribution network.

**TABLE 14.24**

**Numerical Results Compared to Other Methods for Example 7**

| Items | | Power Loss (kW) | | Worst Voltage (p.u.) | | |
|---|---|---|---|---|---|---|
| Methods | Open Switches | Before | After | Before | After | Run Time (s) |
| [156] | 15, 34, 38, 45, 52 | 10.593 | 8.976 | 0.9957 | 0.9970 | - |
| CPLEX | 34, 39, 45, 49, 51 | 10.48 | 8 | 0.9957 | 0.9970 | 1.94 |

are five looping branches (tie-lines) and 48 sectionalizing switches in this network. The results are obtained according to Tables 14.24 to 14.27 by implementing CPLEX.

Form Tables 14.25 and 14.26, it can be said that CPLEX finds the optimal solution faster than method given in ref. [156].

- **Example 8: 59-Bus Test System**

**TABLE 14.25**

**Ranking of Methods Based on Loss Reduction for Example 7**

| Methods | Power Loss (kW) |
|---|---|
| [156] | 8.976 |
| CPLEX | 8 |

**TABLE 14.26**

**Ranking of Methods Regarding Voltage Improvement for Example 7**

| Methods | Worst Voltage (p.u.) |
|---|---|
| [156] and CPLEX | 0.9970 |

**TABLE 14.27**

**Ranking of Methods Based on Execution Time for Example 7**

| Methods | Run Time (s) |
|---|---|
| [156] | - |
| CPLEX | 1.94 |

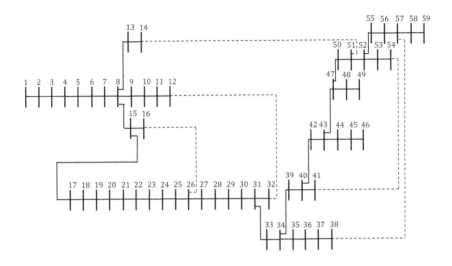

**FIGURE 14.8** Real 59-bus distribution system.

Figure 14.8 shows a 33 kV real distribution network. This system is a region of the distribution network of the city of Ahvaz in the south of Iran. The 59-bus system includes 63 branches, five and 58 tie and sectionalizing switches, respectively. Line and load data for this real distribution network are given in ref. [52].

Tables 14.28 to 14.31 show the simulation results after applying proposed techniques to the above-mentioned case study system.

**TABLE 14.28**
**Numerical Results Compared to Other Methods for Example 8**

| Items | | Power Loss (kW) | | Worst Voltage (p.u.) | | |
|---|---|---|---|---|---|---|
| Methods | Open Switches | Before | After | Before | After | Run Time (s) |
| DTLBO [52] | 25, 31, 39, 40, 62 | 178.66 | 131.69 | 0.9767 | 0.9847 | - |
| CPLEX | 25, 31, 39, 40, 62 | 178.66 | 131.69 | 0.9767 | 0.9847 | 7.8 |

**TABLE 14.29**
**Ranking of Methods Based on Loss Reduction for Example 8**

| Methods | Power Loss (kW) |
|---|---|
| [52] and CPLEX | 131.69 |

**TABLE 14.30**
**Ranking of Methods Regarding Voltage Improvement for Example 8**

| Methods | Worst Voltage (p.u.) |
|---|---|
| [52] and CPLEX | 0.9847 |

**TABLE 14.31**
**Ranking of Methods Based on Execution Time for Example 8**

| Methods | Run Time (s) |
|---|---|
| [52] | - |
| CPLEX | 7.8 |

It can be seen that CPLEX converges to the optimal solution with a good convergence speed.

- **Example 9: 69-Bus Test System**
  This 12.66-kV radial distribution system has 69 nodes and 73 branches, including tie branches, as shown in Figure 14.9. This network has five tie-switches and these tie-switches are open under normal operating condition. Data for this system are given in ref. [76]. Base values for this network are 12.66 kV and 1 MVA. The maximum currents of branches are according to ref. [157].
  The proposed methods have been applied to under study network, and the results are provided in Tables 14.32 to 14.35. Simulation results show that CPLEX can solve the DSR problem more precisely and faster than other reconfiguration methods.

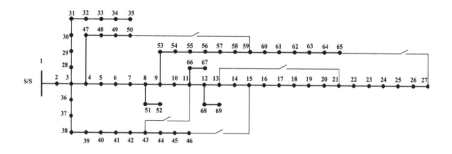

**FIGURE 14.9**   Schematic diagram of 69-bus distribution system.

## TABLE 14.32
## Numerical Results Compared to Other Methods for Example 9

| Items | | Power Loss (kW) | | Worst Voltage (p.u.) | | Run |
|---|---|---|---|---|---|---|
| Methods | Open Switches | Before | After | Before | After | Time (s) |
| SA [96] | 14, 55, 61, 69, 70 | 224.95 | 99.62 | 0.9092 | 0.9428 | 165 |
| [158] | 14, 58, 61, 69, 70 | 225.05 | 99.62 | 0.9092 | 0.9428 | - |
| SA [159] | 14, 57, 61, 69, 70 | 225 | 99.62 | 0.9092 | 0.9428 | 250 |
| Efficient SA [159] | 14, 57, 61, 69, 70 | 225 | 99.62 | 0.9092 | 0.9428 | 20.57 |
| BGA [160] | 12, 20, 58, 64, 69 | 226.13 | 113.8 | 0.9092 | 0.9328 | 25 |
| FGA [160] | 12, 20, 58, 64, 69 | 226.13 | 113.8 | 0.9092 | 0.9328 | 50 |
| [161] | 14, 56, 61, 69, 70 | 224.95 | 99.59 | 0.9092 | 0.9428 | - |
| [108] | 14, 56, 62, 70, 71 | 225 | 99.72 | 0.9092 | 0.9428 | - |
| [134] | 14, 55, 61, 69, 70 | 224.95 | 99.62 | 0.9092 | 0.9428 | - |
| [162] | 14, 58, 61, 69, 70 | 225 | 99.62 | 0.9092 | 0.9310 | - |
| MBPSO [143] | 14, 55, 61, 69, 70 | 224.95 | 99.62 | 0.9092 | 0.9428 | 8 |
| MTS [145] | 14, 55, 61, 69, 70 | 224.95 | 99.62 | 0.9092 | 0.9428 | 150 |
| NSGA [163] | 14, 23, 51, 60, 72 | 226.92 | 136.87 | 0.9092 | 0.9428 | 37 |
| FACO [146] | 12, 57, 61, 69, 70 | 225 | 99.82 | 0.9092 | 0.9428 | - |
| [164] | 15, 56, 62, 70, 71 | 225 | 99.62 | 0.9092 | 0.9428 | - |
| GA [165] | 14, 53, 61, 69, 70 | 225 | 103.29 | 0.9092 | 0.9411 | - |
| RGA [165] | 13, 17, 55, 61, 69 | 225 | 100.28 | 0.9092 | 0.9428 | - |
| NCUP [115] | 14, 59, 62, 70, 71 | 225 | 99.72 | 0.9092 | 0.9428 | - |
| Heuristic [115] | 11, 14, 21, 56, 62 | 225 | 106.67 | 0.9092 | 0.9428 | - |
| GA [166] | 14, 58, 61, 69, 70 | 225.05 | 99.62 | 0.9092 | 0.9428 | 1181 |
| FNSGA [93] | 14, 57, 61, 69, 70 | 224.93 | 99.62 | 0.9092 | 0.9430 | 20.2 |
| MBFOA[1] [167] | 14, 58, 61, 69, 70 | 225.05 | 99.62 | 0.9092 | 0.9428 | 20.69 |
| DTLBO[2] [52] | 14, 57, 61, 69, 70 | 224.9 | 99.62 | 0.9092 | 0.9428 | - |
| Dragonfly [168] | 14, 57, 61, 69, 70 | 225.04 | 99.62 | 0.9092 | 0.9428 | 7 |
| GA [169] | 14, 58, 61, 69, 70 | 225 | 99.62 | 0.9092 | 0.9428 | 12.5 |
| CPLEX | 14, 57, 61, 69, 70 | 225 | 99.62 | 0.9092 | 0.9428 | 6.05 |

[1] Modified bacterial foraging optimization algorithm (Modified BFOA)
[2] Discrete teaching-learning based optimization (Discrete TLBO)

## TABLE 14.33
## Ranking of Methods Based on Loss Reduction for Example 9

| Methods | Power Loss (kW) |
|---|---|
| [163] | 136.87 |
| [160] | 113.8 |
| Heuristic [115] | 106.67 |
| GA [165] | 103.29 |
| RGA [165] | 100.28 |
| [146] | 99.82 |
| NCUP [115] and [108] | 99.72 |
| [52], [93], [96], [134], [143], [145], [158], [159], [164], [166]–[169], CPLEX | 99.62 |

**TABLE 14.34**
**Ranking of Methods Regarding Voltage Improvement for Example 9**

| Methods | Worst Voltage (p.u.) |
|---|---|
| [162] | 0.9310 |
| [160] | 0.9328 |
| GA [165] | 0.9411 |
| [52], [96], [108], [115], [134], [143], [145], [146], [158], [159], [161], [163]–[165], [166]–[169], CPLEX | 0.9428 |
| [93] | 0.9430 |

**TABLE 14.35**
**Ranking of Methods Based on Execution Time for Example 9**

| Methods | Run Time (s) |
|---|---|
| [166] | 1181 |
| SA [159] | 250 |
| [96] | 165 |
| [143] | 150 |
| FGA [160] | 50 |
| [163] | 37 |
| BGA [160] | 25 |
| [167] | 20.69 |
| Efficient SA [159] | 20.57 |
| [93] | 20.2 |
| [169] | 12.5 |
| [143] | 8 |
| [168] | 7 |
| CPLEX | 6.05 |

- **Example 10: 70-Bus Test System**
  The tested system is an 11-kV radial distribution system with two sub-stations, four feeders, 70 nodes, and 76 branches (including tie-lines), as shown in Figure 14.10. Data for this system are given in ref. [170]. Tables 14.36 to 14.39 show simulation results. It can be seen that MILP model presented in ref. [117] can find the best solution much faster than CPLEX and other methods.

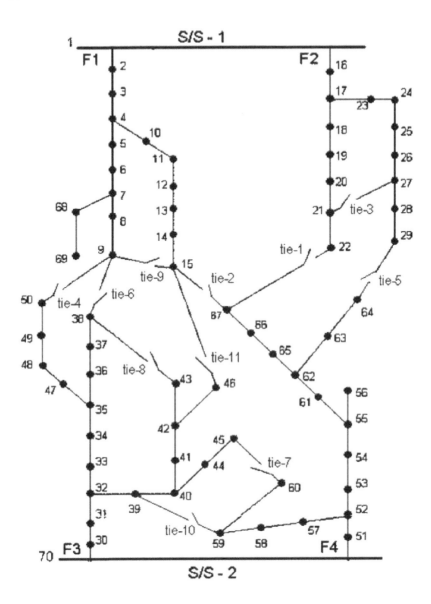

**FIGURE 14.10**   70-bus radial distribution system.

- **Example 11: Distribution System of Taiwan Power Company (TPC)**
  As shown in Figure 14.11, this accrual 11.4-kV network consists of two
  substations, 11 feeders, 83 normal switches, and 13 tie lines, and its
  data are available in ref. [100]. The substation voltage is set at 1.0 p.u.
  and each line current carrying capacity is 410 A [175]. Power base is

**TABLE 14.36**

**Numerical Results Compared to Other Methods for Example 10**

| Items Methods | Open Switches | Power Loss (kW) Before | After | Worst Voltage (p.u.) Before | After | Run Time (s) |
|---|---|---|---|---|---|---|
| [170] | 14, 28, 39, 46, 51, 67, 70, 71, 73, 76, 79 | 227.53 | 205.32 | 0.9052 | 0.9268 | 3 |
| SOReco [113] | 13, 28, 45, 51, 67, 70, 73, 75, 76, 78, 79 | 227.346 | 203.67 | 0.9052 | 0.9268 | 4.64 |
| Brute force [113] | 13, 28, 45, 51, 67, 70, 73, 75, 76, 78, 79 | 227.346 | 203.67 | 0.9052 | 0.9268 | 7633.35 |
| MIQCP [117] | 14, 30, 45, 51, 66, 70, 75–79 | 227.53 | 201.4 | 0.9052 | 0.9268 | 485 |
| MILP [117] | 14, 30, 45, 51, 66, 70, 75–79 | 227.53 | 201.4 | 0.9052 | 0.9268 | 0.91 |
| AACO [130] | 13, 30, 45, 51, 66, 70, 75–79 | 227.53 | 201.39 | 0.9052 | 0.9312 | 91.13 |
| Proposed AACO [130] | 13, 30, 45, 51, 66, 70, 75–79 | 227.53 | 201.39 | 0.9052 | 0.9312 | 19.72 |
| AIMMS [116] | 14, 30, 45, 51, 66, 70, 75–79 | 227.5 | 201.4 | 0.9052 | 0.9268 | 5.7 |
| GAMT [109] | 14, 30, 38, 46, 51, 66, 70, 71, 76, 77, 79 | 227.52 | 203.17 | 0.9052 | 0.9297 | 160 |
| BGA [109] | 13, 30, 45, 51, 66, 70, 75–79 | 227.52 | 201.4 | 0.9052 | 0.9297 | 1900 |
| [171] | 30, 46, 51, 66, 70, 71, 75–79 | 227.53 | 202.18 | 0.9052 | 0.9316 | - |
| [172], [173] | 14, 28, 39, 46, 67, 51, 70, 71, 73, 76, 79 | 227.53 | 205.32 | 0.9052 | 0.9268 | 5.4 |
| [174] | 14, 28, 39, 46, 67, 51, 70, 71, 73, 76, 79 | 227.53 | 205.32 | 0.9052 | 0.9268 | 8 |
| CPLEX | 13, 30, 45, 51, 66, 70, 75–79 | 227.53 | 201.41 | 0.9052 | 0.9311 | 13.73 |

**TABLE 14.37**

**Ranking of Methods Based on Loss Reduction for Example 10**

| Methods | Power Loss (kW) |
|---|---|
| [170], [172]–[174] | 205.32 |
| [113] | 203.67 |
| [109] | 203.17 |
| CPLEX | 201.41 |
| [109], [116], [117] | 201.4 |
| [130] | 201.39 |

**TABLE 14.38**

**Ranking of Methods Regarding Voltage Improvement for Example 10**

| Methods | Worst Voltage (p.u.) |
|---|---|
| [113], [116], [117], [170], [172]–[174] | 0.9268 |
| [109] | 0.9297 |
| CPLEX | 0.9311 |
| [130] | 0.9312 |
| [171] | 0.9316 |

**TABLE 14.39**

**Ranking of Methods Based on Execution Time for Example 10**

| Methods | Run time (s) |
|---|---|
| Brute force [113] | 7633.35 |
| BGA [109] | 1900 |
| MIQCP [117] | 485 |
| GAMT [109] | 160 |
| AACO [130] | 91.13 |
| Proposed AACO [130] | 19.72 |
| CPLEX | 13.73 |
| [174] | 8 |
| [116] | 5.7 |
| [172], [173] | 5.4 |
| SOReco [113] | 4.64 |
| [170] | 3 |
| MILP [117] | 0.91 |

equal to 100 kVA, and voltage base is 11.4 kV. The proposed model has been implemented on under study network, and the results are listed in Tables 14.40 to 14.43.

From Tables 14.41 to 14.43, it can be observed that CPLEX is more efficient than other techniques for reconfiguration of medium size distribution networks because it can converge to an optimal solution faster than other methods.

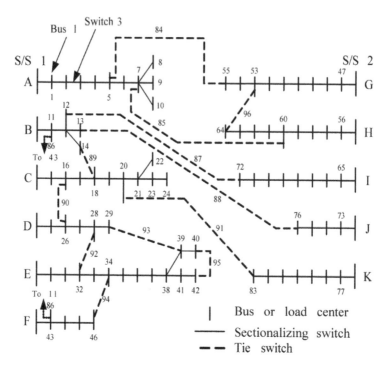

**FIGURE 14.11**   Distribution system of TPC.

**TABLE 14.40**

**Numerical Results of Proposed Model Compared to Other Methods for Example 11**

| Items | | Power Loss (kW) | | Worst Voltage (pu) | | Run |
|---|---|---|---|---|---|---|
| Methods | Open Switches | Before | After | Before | After | Time (s) |
| Heuristic [124] | 7, 13, 34, 39, 41, 55, 62, 72, 83, 86, 89, 90, 92 | 531.99 | 471.44 | 0.9285 | 0.9532 | – |
| SA [100], [175] | 7, 13, 34, 39, 41, 55, 62, 72, 83, 86, 89, 90, 92, | 531.99 | 471.08 | 0.9285 | 0.9532 | 195.21 |
| MIHDE [100] | 7, 13, 34, 39, 41, 55, 62, 72, 83, 86, 89, 90, 92 | 531.99 | 471.08 | 0.9285 | 0.9532 | 36.15 |
| HDE [92] | 7, 13, 34, 39, 41, 55, 62, 72, 83, 86, 89, 90, 92 | 531.99 | 471.08 | 0.9285 | 0.9532 | 318 |
| VSHDE [92], [175] | 7, 13, 34, 39, 41, 55, 62, 72, 83, 86, 89, 90, 92 | 531.99 | 471.08 | 0.9285 | 0.9532 | 585 |
| GA [104], [105], [130] | 7, 13, 34, 39, 41, 55, 62, 72, 83, 86, 89, 90, 92 | 531.99 | 471.08 | 0.9285 | 0.9532 | 303.66 |

*(Continued)*

**TABLE 14.40 (*Continued*)**

**Numerical Results of Proposed Model Compared to Other Methods for Example 11**

| Items | | Power Loss (kW) | | Worst Voltage (pu) | | Run |
|---|---|---|---|---|---|---|
| Methods | Open Switches | Before | After | Before | After | Time (s) |
| ACO [104], [105], [130] | 7, 13, 34, 39, 41, 55, 62, 72, 83, 86, 89, 90, 92 | 531.99 | 471.08 | 0.9285 | 0.9532 | 241.51 |
| Heuristic [103] | 7, 13, 34, 39, 41, 55, 62, 72, 83, 86, 89, 90, 92 | 531.99 | 470.88 | 0.9285 | 0.9532 | - |
| AIS-ACO [176], [130] | 7, 13, 34, 39, 42, 55, 62, 72, 86, 89, 90, 91, 92 | 531.99 | 472.32 | 0.9285 | 0.9479 | 15 |
| Heuristic [108] | 7, 13, 34, 39, 42, 55, 62, 72, 83, 86, 89, 90, 92 | 531.99 | 469.878 | 0.9285 | 0.9532 | 3.92 |
| PGS [133] | 7, 13, 34, 39, 41, 55, 62, 72, 83, 86, 89, 90, 92 | 531.99 | 471.08 | 0.9285 | 0.9532 | 113.25 |
| Heuristic [134] | 7, 13, 34, 39, 42, 55, 62, 72, 83, 86, 89, 90, 92 | 531.99 | 469.878 | 0.9285 | 0.9532 | - |
| Efficient GA [177] | 7, 13, 34, 39, 42, 55, 62, 72, 83, 86, 89, 90, 92 | 531.99 | 469.878 | 0.9285 | 0.9532 | 450 |
| GA [178] | 7, 33, 55, 61, 72, 83, 86, 88, 89, 90, 92, 93, 95 | 531.99 | 471.51 | 0.9285 | 0.9532 | - |
| ACO [178] | 7, 13, 34, 39, 41, 55, 62, 72, 83, 86, 89, 90, 92 | 531.99 | 471.078 | 0.9285 | 0.9517 | - |
| HBMO [137], [130] | 7, 14, 34, 39, 42, 55, 62, 72, 83, 86, 88, 90, 92 | 531.99 | 482.14 | 0.9285 | 0.9529 | 13 |
| AACO [130] | 7, 13, 34, 39, 42, 55, 62, 72, 83, 86, 89, 90, 92 | 531.99 | 469.878 | 0.9285 | 0.9532 | 95.88 |
| MILP [175] | 7, 13, 34, 39, 42, 55, 62, 72, 83, 86, 89, 90, 92 | 531.99 | 469.88 | 0.9285 | 0.9532 | 207.7 |
| MICP [175] | 7, 13, 34, 39, 42, 55, 62, 72, 83, 86, 89, 90, 92 | 531.99 | 469.88 | 0.9285 | 0.9532 | 245.4 |
| MILP [149] | 7, 13, 34, 39, 42, 55, 62, 72, 83, 86, 89, 90, 92 | 531.99 | 469.88 | 0.9285 | 0.9532 | - |
| SOReco [113] | 7, 13, 34, 39, 42, 55, 62, 72, 83, 86, 89, 90, 92 | 531.99 | 469.878 | 0.9285 | 0.9532 | 7 |
| NCUP [115] | 7, 13, 34, 39, 42, 55, 62, 72, 83, 86, 89, 90, 92 | 531.99 | 469.878 | 0.9285 | 0.9532 | 5.28 |
| CLONR [151] | 7, 13, 34, 39, 42, 55, 62, 72, 83, 86, 89, 90, 92 | 531.99 | 469.88 | 0.93 | 0.95 | 160 |
| AIMMS [116] | 7, 13, 34, 39, 42, 62, 72, 83, 84, 86, 89, 90, 92 | 532 | 469.95 | 0.9285 | 0.9532 | 7.8 |
| SA [136] | 7, 13, 34, 39, 42, 62, 72, 83, 84, 86, 89, 90, 92 | 531.99 | 470.513 | 0.9285 | 0.9532 | 22.061 |
| GA [136] | 7, 13, 34, 39, 63, 72, 83, 84, 86, 89, 90, 92, 95 | 531.99 | 470.214 | 0.9285 | 0.9517 | 30.526 |

*(Continued)*

**TABLE 14.40** (*Continued*)
**Numerical Results of Proposed Model Compared to Other Methods for Example 11**

| Items Methods | Open Switches | Power Loss (kW) Before | After | Worst Voltage (pu) Before | After | Run Time (s) |
|---|---|---|---|---|---|---|
| ACO [136] | 7, 13, 34, 39, 42, 55, 62, 72, 83, 86, 89, 90, 92 | 531.99 | 469.878 | 0.9285 | 0.9532 | 19.736 |
| HBMO [136] | 7, 13, 34, 39, 42, 55, 62, 72, 83, 86, 89, 90, 92 | 531.99 | 469.878 | 0.9285 | 0.9532 | 17.104 |
| RGA [136] | 7, 13, 34, 39, 42, 63, 72, 83, 84, 86, 89, 90, 92 | 531.99 | 470.084 | 0.9285 | 0.9517 | 170 |
| HSA [136] | 7, 13, 34, 39, 42, 55, 62, 72, 83, 86, 89, 90, 92 | 531.99 | 469.88 | 0.9285 | 0.9532 | 9.871 |
| GA+BE [136] | 7, 13, 34, 39, 42, 55, 62, 72, 83, 86, 89, 90, 92 | 531.99 | 469.878 | 0.9285 | 0.9532 | 37.983 |
| [179] | 7,13, 34, 39, 42, 55, 62, 72, 83, 86, 89, 90, 92 | 531.99 | 469.89 | 0.9285 | 0.9532 | 7.75 |
| Heuristic [155] | 7, 13, 34, 39, 42, 63, 72, 83, 84, 86, 89, 90, 92 | 531.99 | 470.01 | 0.9285 | 0.9518 | 69 |
| CPLEX | 7, 13, 34, 39, 42, 55, 62, 72, 83, 86, 89, 90, 92 | 531.99 | 469.878 | 0.9285 | 0.9532 | 2.6 |

**TABLE 14.41**
**List of Methods Regarding Loss Minimization in Example 11**

| Methods | Losses (kW) |
|---|---|
| [137] | 482.14 |
| [176] | 472.32 |
| GA [178] | 471.51 |
| [224] | 471.44 |
| [92], [100], [104], [105], ACO [130], VSHDE [175], [177], ACO [178] | 471.08 |
| [103] | 470.88 |
| SA [136] | 470.513 |
| GA [136] | 470.214 |
| RGA [136] | 470.084 |
| [155] | 470.01 |
| [116] | 469.95 |
| [179] | 469.89 |
| HSA [136], [149], [151], [175] | 469.88 |
| [108], [113], [115], efficient GA [130], AACO [130], [134], ACO [136], HBMO [136], GA+BE [136], [177], CPLEX | 469.878 |

## TABLE 14.42
### Methods Based on Nodal Voltage Enhancement of Example 11

| Methods | Minimum Voltage (pu) |
|---|---|
| AIS-ACO [130], [176] | 0.9479 |
| [151] | 0.95 |
| GA [136], RGA [136], ACO [178] | 0.9517 |
| [155] | 0.9518 |
| HBMO [130], [137] | 0.9529 |
| [92], [100], [103]–[105], [113], [115], [116], efficient GA [130], ACO [130], AACO [130], GA [130], [134], SA [136], ACO [136], HBMO [136], GA+BE [136], HSA [136], [149], [175], [177], GA [178], [179], CPLEX | 0.9532 |

## TABLE 14.43
### Classification of Solution Methods According to Running Time in Example 11

| Methods | Computing Time (s) |
|---|---|
| VSHDE [92] | 585 |
| [177] | 450 |
| HDE [92] | 318 |
| GA [104] | 303.66 |
| MICP [175] | 245.4 |
| ACO [104] | 241.51 |
| MILP [175] | 207.7 |
| SA [100] | 195.21 |
| RGA [136] | 170 |
| [151] | 160 |
| [177] | 113.25 |
| AACO [130] | 95.88 |
| [155] | 69 |
| GA+BE [136] | 37.983 |
| MIHDE [100] | 36.15 |
| GA [136] | 30.526 |
| SA [136] | 22.061 |
| ACO [136] | 19.736 |
| HBMO [136] | 17.104 |
| AIS-ACO [130], [176] | 15 |
| HBMO [130], [137] | 13 |
| HSA [136] | 9.871 |
| [116] | 7.8 |
| [179] | 7.75 |
| [113] | 7 |
| [115] | 5.28 |
| [108] | 3.92 |
| CPLEX | 2.6 |

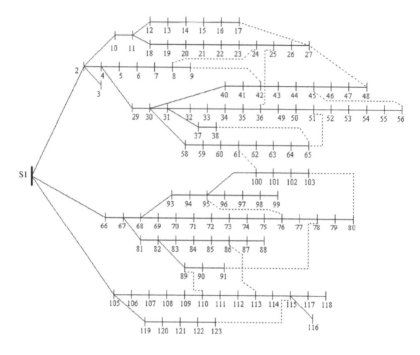

**FIGURE 14.12**   119-bus test system.

- **Example 12: 119-Bus Test System**
  This test system, as shown in Figure 14.12, is an 11 kV distribution network with three feeders, 118 lines, and 12 tie-switches.
  The parameters and related data of this network can be found in ref. [180]. Tables 14.44 to 14.47 show relevant results obtained for 119-bus test system.

## TABLE 14.44
### Numerical Results Compared to Other Methods for Example 12

| Items | | Power Loss (kW) | | Worst Voltage (p.u.) | | |
|---|---|---|---|---|---|---|
| Methods | Open Switches | Before | After | Before | After | Run Time (s) |
| ITS [180] | 24, 27, 35, 40, 43, 52, 59, 72, 75, 96, 98, 110, 123, 130, 131 | 1294.3 | 865.86 | 0.9321 | 0.9323 | 9.38 |
| [134] | 24, 27, 35, 40, 43, 52, 59, 72, 75, 96, 99, 110, 123, 130, 131 | 1298.09 | 870.35 | 0.8783 | 0.9321 | - |
| MTS [145] | 24, 27, 35, 40, 43, 52, 59, 72, 75, 96, 98, 110, 123, 130, 131 | 1294.3 | 865.86 | 0.866 | 0.9323 | 18000 |
| TS [145] | 23, 27, 33, 40, 43, 49, 52, 62, 72, 74, 77, 83, 110, 126, 131 | 1294.3 | 884.163 | 0.866 | 0.9321 | - |
| HAS [125] | 24, 27, 35, 40, 43, 52, 59, 72, 75, 96, 98, 110, 123, 130, 131 | 1298.09 | 865.86 | 0.8783 | 0.9323 | 8.61 |

*(Continued)*

**TABLE 14.44 (*Continued*)**
**Numerical Results Compared to Other Methods for Example 12**

| Items | | Power Loss (kW) | | Worst Voltage (p.u.) | | |
|---|---|---|---|---|---|---|
| Methods | Open Switches | Before | After | Before | After | Run Time (s) |
| GA [125] | 24, 31, 35, 40, 43, 49, 51, 62, 73, 74, 77, 83,110, 120, 126 | 1298.09 | 885.56 | 0.8783 | 0.9321 | 24.45 |
| RGA [125] | 23, 27, 33, 40, 43, 49, 52, 62, 73, 74, 77, 83, 110, 126, 131 | 1298.09 | 883.13 | 0.8783 | 0.9321 | 17.53 |
| AACO [130 ] | 24, 27, 35, 40, 43, 52, 59, 72, 75, 96, 98, 110, 123, 130, 131 | 1298.1 | 865.86 | 0.8688 | 0.9323 | 430.74 |
| CLONR [151] | 24, 26, 35, 40, 43, 51, 62, 72, 74, 77, 83, 110, 122, 126, 131 | 1296.6 | 870.91 | 0.87 | 0.93 | 704.1 |
| AIMMS [116] | 24, 27, 35, 40, 43, 52, 59, 72, 75, 96, 98, 110, 123, 130, 131 | 1298.1 | 869.7 | 0.8783 | 0.9321 | 39.4 |
| MIQCP [117] | 24, 27, 35, 40, 43, 52, 59, 72, 75, 96, 98, 110, 123, 130, 131 | 1298.1 | 869.7 | 0.8783 | 0.9321 | 1009 |
| MILP [117] | 24, 27, 35, 40, 43, 52, 59, 72, 75, 96, 98, 110, 123, 130, 131 | 1298.1 | 869.7 | 0.8783 | 0.9321 | 2.8 |
| GA [181] | 5, 27, 34, 43, 47, 54, 61, 71, 74, 83, 96, 99, 110, 125, 131 | 1298.09 | 980.54 | 0.8687 | 0.9321 | - |
| GA [169] | 24, 27, 35, 40, 43, 52, 59, 72, 75, 96, 98, 110, 123, 130, 131 | 1298.1 | 865.86 | 0.8688 | 0.9323 | 42.13 |
| Heuristic [182] | 24, 27, 35, 40, 43, 52, 59, 72, 75, 96, 98, 110, 123, 130, 131 | 1301.9 | 865.86 | 0.8688 | 0.9323 | 7.64 |
| CPLEX | 24, 27, 35, 40, 43, 52, 59, 72, 75, 96, 98, 110, 123, 130, 131 | 1297.42 | 869.7 | 0.869 | 0.9323 | 70.67 |

**TABLE 14.45**
**Ranking of Methods Based on Execution Time for Example 12**

| Methods | Run Time (s) |
|---|---|
| MTS [145] | 18000 |
| MIQCP [117] | 1009 |
| [151] | 704.1 |
| [130] | 430.74 |
| CPLEX | 70.67 |
| [169] | 42.13 |
| [116] | 39.4 |
| GA [1–25] | 24.45 |
| RGA [125] | 17.53 |
| [180] | 9.38 |
| HAS [125] | 8.61 |
| [182] | 7.64 |
| MILP [117] | 2.8 |

**TABLE 14.46**

**Ranking of Methods Regarding Voltage Improvement for Example 12**

| Methods | Worst Voltage (p.u.) |
|---|---|
| [116], [117], GA [125], RGA [125], [134], TS [145], [181] | 0.9321 |
| HAS [125], [130], MTS [145], [169], [180], [182], CPLEX | 0.9323 |

**TABLE 14.47**

**Ranking of Methods Based on Loss Reduction for Example 12**

| Methods | Power Loss (kW) |
|---|---|
| [181] | 980.54 |
| GA [125] | 885.56 |
| TS [145] | 884.163 |
| RGA [125] | 883.13 |
| [151] | 870.91 |
| [134] | 870.35 |
| [116], [117], CPLEX | 869.7 |
| HAS [125], [130], MTS [145], [169], [180], [182] | 865.86 |

The results reveal that MILP [117] model can find the optimal solution in shorter time than other reconfiguration techniques.

- **Example 13: 136-Bus Test System**
  Data of this actual system with 156 branches and 21 tie-switches, as shown in Figure 14.13, can be found in ref. [183]. It is a part of Tres Lagoas distribution network in Brazil.

  Maximum line current, nominal voltage, and MVA base are 200 A, 13.8 kV, and 1, respectively [175]. The proposed model has been implemented on case study system, and the results are listed in Tables 14.48 to 14.50.

  It should be mentioned that the minimum voltage of all solutions after reconfiguration is almost 0.96 p.u. For this, the arrangement of methods based on the voltage improvement was ignored. According to Tables 14.49 and 14.50, the method proposed in ref. [117] could find the optimal solution for this real large-scale distribution network in a short run time.

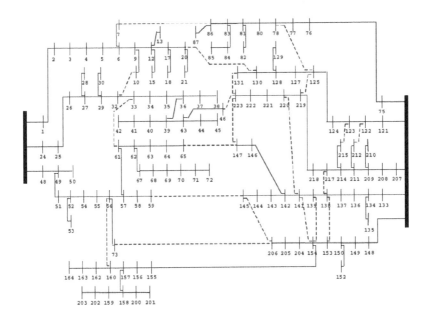

**FIGURE 14.13**  Diagram of 136-bus distribution system.

**TABLE 14.48**
**Numerical Results of Proposed Model Compared to Other Methods for Example 13**

| Items | | Power loss (kW) | | Worst voltage (pu) | | Run |
|---|---|---|---|---|---|---|
| Methods | Open Switches | Before | After | Before | After | time (s) |
| Heuristic [183] | 51, 106, 136–139, 141–152, 154–156 | 320.17 | 285.5 | 0.9307 | 0.9638 | - |
| Heuristic [108] | 7, 9, 38, 51, 55, 90, 92, 95, 104, 106, 120, 126, 128, 135, 138, 141,144–146, 148, 150 | 320.36 | 286 | 0.9307 | 0.963 | 8.96 |
| Decimal GA (DGA) [130] | 7, 51, 53, 84, 90, 96, 106, 118, 126, 128, 137–139, 141, 144, 145, 147, 148,150, 151, 156 | 320.3 | 280.22 | 0.9307 | 0.9605 | 3600 |
| Dedicated GA [184] | 7, 38, 51, 53, 90, 96, 106, 118, 126, 128, 137, 138, 141, 144–148, 150, 151, 156 | 320.3 | 280.4 | 0.9307 | 0.9605 | 227 |
| FAGA [185] | 51, 53, 90, 96, 106, 118, 136–139, 141, 144–148, 150, 151, 154–156 | 320.3 | 281.7 | 0.9307 | 0.9605 | ~ 40 |
| AACO [130] | 7, 35, 51, 90, 95, 106, 118,126, 135, 137, 138, 141, 142,144–148, 150, 151, 155 | 320.3 | 280.19 | 0.9307 | 0.9602 | 894.2 |
| B&B [147] | 7, 35, 51, 90, 96, 106, 118, 126, 135, 137, 138, 141, 142, 144–148, 150, 151, 155 | 320.36 | 280.19 | 0.9307 | 0.9589 | 4473 |
| [175] | 7, 35, 51, 90, 96, 106, 118, 126, 135, 137, 138, 141, 142, 144–148, 150, 151, 155 | 320.36 | 280.19 | 0.9307 | 0.9589 | 1800 |

*(Continued)*

**TABLE 14.48 (*Continued*)**

**Numerical Results of Proposed Model Compared to Other Methods for Example 13**

| Items Methods | Open Switches | Power loss (kW) Before | After | Worst voltage (pu) Before | After | Run time (s) |
|---|---|---|---|---|---|---|
| ACO+GA [186] | 7, 51, 53, 84, 90, 96, 106, 118, 126, 128, 137–139, 141, 144, 145, 147, 148, 150, 151, 156 | 320.36 | 280.22 | 0.9307 | 0.9605 | 5120 |
| [187] | 7, 35, 51, 90, 96, 106, 118, 126, 135, 137, 138, 141, 142, 144–148, 150, 151, 155 | 320.36 | 280.19 | 0.93065 | 0.95892 | - |
| NCUP [115] | 7, 38, 51, 55, 90, 97, 106, 118, 126, 137, 138, 141, 144–148, 150–152, 155 | 320.36 | 282.77 | 0.9307 | 0.963 | 23.93 |
| AIMMS [116] | 7, 38, 51, 54, 84, 90, 96, 106, 118, 126, 128, 135, 137, 138, 141, 144, 145, 147, 148, 150, 151 | 320.4 | 280.38 | 0.9307 | 0.9599 | 132.5 |
| MIQP [153] | 7, 38, 51, 54, 84, 90, 96, 106, 118, 126, 128, 135, 137, 138, 141, 144, 145, 147, 148, 150, 151 | 320.4 | 280.38 | 0.9307 | 0.9599 | 188.4 |
| MIQCP [117] | 7, 35, 51, 90, 96, 106, 118, 126, 135, 137, 138, 141, 142, 144–148, 150, 151, 155 | 320.36 | 280.19 | 0.9307 | 0.9589 | 1785 |
| MILP [117] | 7, 35, 51, 90, 96, 106, 118, 126, 135, 137, 138, 141, 142, 144–148, 150, 151, 155 | 320.36 | 280.19 | 0.9307 | 0.9589 | 5.6 |
| Heuristic [155] | 7, 38, 51, 54, 84, 90, 96, 106, 119, 126, 135, 137, 138, 141, 144, 145, 147, 148, 150, 151, 155 | 320.35 | 281.02 | 0.9307 | 0.9588 | 530 |
| CPLEX | 7, 35, 51, 90, 96, 106, 118, 126, 135, 137, 138, 141, 142, 144–148, 150, 151, 155 | 320.37 | 280.19 | 0.9307 | 0.9589 | 9.89 |

**TABLE 14.49**

**Arrangement of the Results Based on Power Loss Reduction in Example 13**

| Methods | Power Loss (kW) |
|---|---|
| [108] | 286 |
| [183] | 285.5 |
| [115] | 282.77 |
| [185] | 281.7 |
| [155] | 281.02 |
| [184] | 280.4 |
| [116], [153] | 280.38 |
| [186] | 280.22 |
| [117], [130], [147], [175], [187], CPLEX | 280.19 |

**TABLE 14.50**
**Computation Times for Example 13**

| Methods | Time (s) |
|---|---|
| [186] | 5120 |
| [147] | 4473 |
| [175] | 1800 |
| MIQCP [117] | 1785 |
| [130] | 894.2 |
| [155] | 530 |
| [184] | 227 |
| [153] | 188.4 |
| [116] | 132.5 |
| [185] | ~ 40 |
| [115] | 23.93 |
| CPLEX | 9.89 |
| [108] | 8.96 |
| MILP [117] | 5.6 |

- **Example 14: 203-Bus Test System**
  In order to illustrate the proposed methodologies in a large distribution system, as shown in Figure 14.14, CPLEX has been applied to this system, and the results are listed in Tables 14.51 to 14.54. Data of this network are available in refs. [150, 188].

  It can be concluded that CPLEX can find more optimal solution, but not in short time compared to TS technique.
- **Example 15: 415-Bus Test System**
  In order to verify proposed models in very large distribution networks, a real distribution system with 415 buses and nominal voltage of 13.8 kV [190] has been used to demonstrate proposed techniques. The simulation results are given in Table 14.55.

## 14.4  SUMMARY

In this chapter, DSR problem was modeled in AMPL and solved by CPLEX. The proposed reconfiguration model has been applied to several test systems, including small-, medium-, and large-sized distribution networks. Simulation results will show the high effectiveness and robustness of AMPL in finding accurate solutions, especially in small, medium sized, and large distribution networks. It should be noted that AMPL is an efficient optimization tool that can solve distribution network reconfiguration problem in a reasonable and competitive time compared to other mathematical techniques and toolboxes.

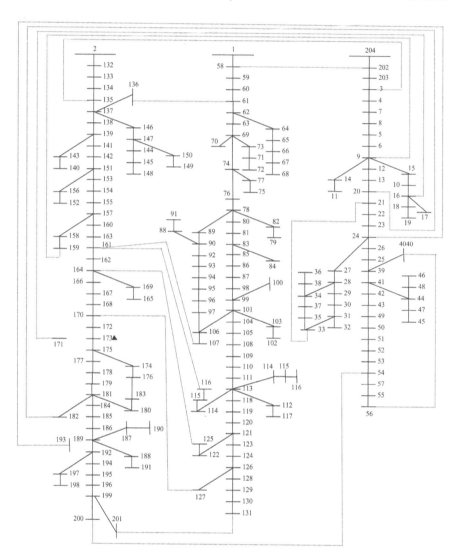

**FIGURE 14.14**   Schematic diagram of 203-bus distribution system.

## TABLE 14.51

## Numerical Results Compared to Other Methods for Example 14

| Items | | Power Loss (kW) | | Worst Voltage (p.u.) | | Run |
|---|---|---|---|---|---|---|
| Methods | Open Switches | Before | After | Before | After | time (s) |
| TS [189] | 29, 66, 74, 83, 111, 118, 125, 131, 135, 136, 140, 177, 199, 202, 208 | 548.89 | 511.5 | 0.9574 | 0.9611 | 49.98 |
| CPLEX | 12, 26, 43, 82, 118, 131, 133, 160, 168, 202, 203, 208, 212, 213, 214 | 548.9 | 511.19 | 0.9574 | 0.9611 | 964.41 |

TABLE 14.52
Ranking of Methods Based on Loss
Reduction for Example 14

| Methods | Power Loss (kW) |
|---------|----------------|
| [189]   | 511.5          |
| CPLEX   | 511.19         |

TABLE 14.53
Ranking of Methods Regarding Voltage
Improvement for Example 14

| Methods | Worst Voltage (p.u.) |
|---------|---------------------|
| [189] and CPLEX | 0.9611      |

TABLE 14.54
Ranking of Methods Based on Execution
Time for Example 14

| Methods | Run Time (s) |
|---------|-------------|
| CPLEX   | 964.41      |
| [189]   | 49.98       |

TABLE 14.55
Numerical Results Compared to Other Methods for Example 15

| Items | | Power Loss (kW) | | Worst Voltage (p.u.+) | | Run |
|-------|--------------|--------|-------|--------|-------|---------|
| Methods | Open Switches | Before | After | Before | After | Time (s) |
| CPLEX | 1, 2, 13, 15, 16, 30, 31, 40, 41, 50, 59, 73, 75, 82, 94, 96, 97, 111, 115, 136, 142, 150, 155, 156, 158, 163, 168, 169, 178, 179, 191, 195, 209, 214, 230, 254, 256, 270, 294, 314, 317, 325, 358, 362, 385, 389, 392, 395, 403, 404, 423, 424, 426, 436, 437, 439, 446, 449, 466 | 708.77 | 582.99 | 0.9453 | 0.9534 | 1721 |

## 14.5  ACKNOWLEDGMENT

The work of Hassan Haes Alhelou was supported in part by the Science Foundation Ireland (SFI) through the SFI Strategic Partnership Programme under Grant SFI/15/SPP/E3125, and in part by the UCD Energy Institute. The opinions, findings and conclusions or recommendations expressed in this material are those of the authors and do not necessarily reflect the views of the Science Foundation Ireland.

## REFERENCES

[1] M. Mahdavi, C. S. Antunez, M. Ajalli, and R. Romero, "Transmission expansion planning: literature review and classification," IEEE Systems Journal, vol. 3, no. 3, pp. 3129–3140, 2019.

[2] M. Mahdavi, C. Sabillón, M. Ajalli, H. Monsef, and R. Romero, "A real test system for power system planning, operation, and reliability," Journal of Control, Automation and Electrical Systems, vol. 29, no. 2, pp. 192–208, 2018.

[3] M. Mahdavi and H. Monsef, "Review of static transmission expansion planning," Journal of Electrical and Control Engineering, vol. 1, no. 1, pp. 11–18, 2011.

[4] M. Mahdavi and R. Romero, "A new model for dynamic transmission expansion planning," 4th National Conference on Technology in Electrical and Computer Engineering, Iran, pp. 1–7, 2018.

[5] A. Kimiyaghalam, M. Mahdavi, A. Ashouri, and H. Soheil, "Transmission expansion planning considering uncertainty in fuel price using DABC," The 9th International Energy Conference, Tehran, Iran, pp. 1–10, 2013.

[6] S. Jalilzadeh, A. Kazemi, M. Mahdavi, and H. Haddadian, "DCGA-based evaluation of inflation rate effect on the network loss in transmission expansion planning," 43rd International Universities Power Engineering Conference, Italy, pp. 1–5, 2008.

[7] S. Jalilzadeh, A. Kazemi, M. Mahdavi, H. Haddadian, "Transmission expansion planning considering voltage level and network loss using genetic algorithm," 5th International Conference on Electrical and Electronics Engineering, Turkey, pp. 1–5, 2007.

[8] S. Jalilzadeh, A. Kazemi, M. Mahdavi, and H. Haddadian, "Technical and economic evaluation of voltage level effect on TNEP using genetic algorithm," The 8th International Power Engineering Conference, Singapore, pp. 1034–1039, 2007.

[9] A. Kazemi, S. Jalilzadeh, H. Haddadian, and M. Mahdavi, "Transmission network expansion planning considering network adequacy using genetic algorithm," The 8th International Power Engineering Conference, Singapore, pp. 1040–1044, 2007.

[10] S. Jalilzadeh, A. Kazemi, M. Mahdavi, and H. Haddadian, "Studying of voltage level role in TNEP considering the losses for years after expansion using GA with different trails," 43rd International Universities Power Engineering Conference, Padova, Italy, pp. 1–6, 2008.

[11] A. Kazemi, S. Jalilzadeh, M. Mahdavi, and H. Haddadian, "Evaluation of network losses role in TNEP considering voltage level and number of bundle lines using GA," 3rd IEEE Conference on Industrial Electronics and Applications, Singapore, pp. 1–6, 2008.

[12] H. Shayeghi, H. A. Shayanfar, M. Mahdavi, and A. Bagheri, "Application of binary particle swarm optimization for transmission expansion planning considering lines loadinga," The 2009 International Conference on Artificial Intelligence, Las Vegas, Nevada, USA, pp. 653–659, 2009.

[13] M. Mahdavi, H. Shayeghi, and A. Kazemi, "DCGA based evaluating role of bundle lines in TNEP considering expansion of substations from voltage level point of view," Energy Conversion and Management, vol. 50, no. 8, pp. 2067–2073, 2009.

[14] H. Shayeghi, S. Jalilzadeh, M. Mahdavi, and H. Haddadian, "Studying influence of two effective parameters on network losses in transmission expansion planning using DCGA," Energy Conversion and Management, vol. 49, no. 11, pp. 3017–3024, 2008.

[15] H. Shayeghi, M. Mahdavi, and A. Bagheri, "Discrete PSO algorithm based optimization of transmission lines loading in TNEP problem," Energy Conversion and Management, vol. 51, no. 1, pp. 112–121, 2010.

[16] S. Jalilzadeh, A. Kazemi, H. Shayeghi, and M. Mahdavi, "Technical and economic evaluation of voltage level in transmission network expansion planning using GA," Energy Conversion and Management, vol. 49, no. 5, pp. 1119–1125, 2008.

[17] M. Mahdavi and R. Romero, "Transmission system expansion planning based on LCC criterion considering uncertainty in VOLL using DCGA," 32nd International Power System Conference, Tehran, Iran, pp. 1–7, 2017.

[18] M. Khodayari, M. Mahdavi, and H. Monsef, "Simultaneous scheduling of energy & spinning reserve considering customer and supplier choice on reliability," 19th Iranian Conference on Electrical Engineering, Tehran, Iran, pp. 491–496, 2011.

[19] M. Khodayari, H. Monsef, and M. Mahdavi, "Customer reliability based energy & SR scheduling at hierarchical level II," 27th International Power System Conference, Teharn, Iran, pp. 1–11, 2012.

[20] H. Shayeghi, H. Hosseini, A. Shabani, and M. Mahdavi, "GEP considering purchase prices, profits of IPPs and reliability criteria using hybrid GA and PSO," International Journal of Electrical and Computer Engineering, vol. 2, no. 8, pp. 1619–1625, 2008.

[21] M. Mahdavi, L. H. Macedo, and R. Romero, "Considering line maintenance and repair in GTEP," 4th Conference on Modern Technologies in Electrical, Telecommunication and Mechatronics Engineering, Iran, pp. 1–6, 2018.

[22] M. Mahdavi, L. H. Macedo, and R. Romero "Transmission and generation expansion planning considering system reliability and line maintenance," 26th Iranian Conference on Electrical Engineering, Mashhad, Iran, pp. 1005–1010, 2018.

[23] O. Badran, S. Mekhilef, H. Mokhlis, and W. Dahalan, "Optimal reconfiguration of distribution system connected with distributed generations: A review of different methodologies," Renewable and Sustainable Energy Reviews, vol. 73, pp. 854–867, June 2017.

[24] N. G. Paterakis, A. Mazza, S. F. Santos, O. Erdinç, G. Chicco, A. G. Bakirtzis, and J. P. S. Catalão, "Multi-objective reconfiguration of radial distribution systems using reliability indices," IEEE Transactions on Power Systems, vol. 31, no. 2, pp. 1048–1062, March 2016

[25] M. Mahdavi, H. Monsef, and R. Romero, "Reliability effects of maintenance on TNEP considering preventive and corrective repairs," IEEE Transactions on Power Systems, vol. 32, no. 5, pp. 3768–3781, 2017.

[26] M. Mahdavi, H. Monsef, and R. Romero, "Reliability and economic effects of maintenance on TNEP considering line loading and repair," IEEE Transactions on Power Systems, vol. 31, no. 5, pp. 3381–3393, 2016.

[27] M. Mahdavi, C. S. Antunez, A. Bagheri, and R. Romero, "Line Maintenance within transmission expansion planning: A multistage framework," IET Generation, Transmission and Distribution, vol. 14, no. 13, pp. 3057–3065, 2019.

[28] S. S. Duttagupta, A reliability asssessment methodology for distribution systems with distributed generation, M.Sc. Theses, Texas A&M University, Electrical Engineering Department, pp. 1–90, May 2006.

[29] A. Bagheri, M. Mahdavi, and H. Monsef, "Suggestion of an economical plan for electrical power transmission," The 9th International Energy Conference, Tehran, Iran, pp. 1–10, 2013.

[30] Y. Y. Fu and H. D. Chiang, "Toward optimal multiperiod network reconfiguration for increasing the hosting capacity of distribution networks," IEEE Transactions on Power Delivery, vol. 33, no. 5, pp. 2294–2304, October 2018.

[31] S. Lei, Y. Hou, F. Qiu, and J. Yan, "Identification of critical switches for integrating renewable distributed generation by dynamic network reconfiguration," IEEE Transactions on Sustainable Energy, vol. 9, no. 1, pp. 420–432, January 2018.

[32] H. Xing and X. Sun, "Distributed generation locating and sizing in active distribution network considering network reconfiguration," IEEE Access, vol. 5, pp. 14768–14774, August 2017.

[33] A. Azizivahed, H. Narimani, M. Fathi, E. Naderi, H. R. Safarpour, M. R. Narimani, "Multi-objective dynamic distribution feeder reconfiguration in automated distribution systems," Energy, vol. 147, pp. 896–914, March 2018.

[34] J. Shukla, B. Das, and V. Pant, "Stability constrained optimal distribution system reconfiguration considering uncertainties in correlated loads and distributed generations," International Journal of Electrical Power and Energy Systems, vol. 99, pp. 121–133, 2018.

[35] K. Chen, W. Wu, B. Zhang, S. Djokic, and G. P. Harrison, "A method to evaluate total supply capability of distribution systems considering network reconfiguration and daily load curves," IEEE Transactions on Power Systems, vol. 31, no. 3, pp. 2096–2104, May 2016.

[36] H. Shayeghi, M. Mahdavi, A. Kazemi, and H. A. Shayanfar, "Studying effect of bundle lines on TNEP considering network losses using decimal codification genetic algorithm," Energy Conversion and Management, vol. 51, no. 12, pp. 2685–2691, 2010.

[37] H. Shayeghi and M. Mahdavi, "Genetic algorithm based studying of bundle lines effect on network losses in transmission network expansion planning," Journal of Electrical Engineering, vol. 60, no. 5, pp. 237–245, 2009.

[38] S. Jalilzadeh, H. Shayeghi, M. Mahdavi, and H. Haddadian, "A GA based transmission network expansion planning considering voltage Level, network Losses and number of bundle lines," American Journal of Applied Sciences, vol. 6, no. 5, pp. 987–994, 2009.

[39] H. Shayeghi, S. Jalilzadeh, M. Mahdavi, and H. Haddadian, "Role of voltage level and network losses in TNEP Solution," Journal of Electrical Systems, vol. 5, no. 3, pp. 1–18, 2009.

[40] M. Mahdavi and H. Haddadian, "Evaluation of GA performance in TNEP considering voltage level, network losses and number of bundle Lines," International Journal of Discrete Mathematic, vol. 3, no. 1, pp. 1–10, 2018.

[41] H. Shayeghi and M. Mahdavi "Studying the effect of losses coefficient on transmission expansion planning using decimal codification based GA," International Journal on Technical and Physical Problems of Engineering, vol. 1, no. 1, pp. 58–64, 2009.

[42] S. Jalilzadeh, A. Kazemi, M. Mahdavi, and H. Haddadian, "TNEP considering voltage level, network losses and number of bundle lines using GA," Third International Conference on Electric Utility Deregulation and Restructuring and Power Technologies, China, pp. 1580–1585, 2008.

[43] M. Rahmani-Andebili and M. Fotuhi Firuzabad, "An adaptive approach for PEVs charging management and reconfiguration of electrical distribution system penetrated by renewables," IEEE Transactions on Industrial Informatics, vol. 14, no. 5, pp. 2001–2010, May 2018.

[44] M. Mahdavi, H. Monsef, A. Bagheri, and A. Kimiyaghalam, "Dynamic transmission network expansion planning considering network losses, DG sources and operational costs-part 1: review and problem formulation," WULFENIA, vol. 19, no. 10, pp. 242–257, 2012.

[45] M. Mahdavi, H. Monsef, A. Bagheri, and A. Kimiyaghalam, "Dynamic transmission network expansion planning considering network losses, DG sources and operational costs-part 2: solution method and numerical results," WULFENIA, vol. 19, no. 10, pp. 258–273, 2012.

[46] M. Mahdavi, "Iran energy network overview: Renewable energies in Iran," Presentation, Advanced Topics in Bioenergy, Bioenergy Research Institute, Brazil, 2020. https://www.researchgate.net/publication/346719635.

[47] M. Mahdavi and R. A. V. Ramos, "Optimal allocation of bioenergy distributed generators in electrical energy systems," IV Bioenergy Workshop, Bioenergy Research Institute, Brazil, 2020. https://www.researchgate.net/publication/347294531.

[48] M. Mahdavi and L. A. Okamura, "Biofules production from petrochemical industry wastewater," Presentation, Advanced Topics in Biorefinery, Bioenergy Research Institute, Brazil, 2020. https://www.researchgate.net/publication/346719761.

[49] M. Mahdavi, "Temperature effects on the methanogenesis and sulfidogenesis in the wastewater treatment," Presentation, Advanced Topics in Biorefinery, Bioenergy Research Institute, Brazil, 2020. https://www.researchgate.net/publication/346719672.

[50] S. Jalilzadeh, I. D. Kolagar, M. Mahdavi, and V. Nabaei, "Considering the torsional frequencies of an actual wind turbine to prevent of SSR phenomena," IEEE International Conference on Sus-tainable Energy Technologies, Singapore, pp. 222–227, 2008.

[51] M. Mahdavi, "New models and optimization techniques applied to the problem of optimal reconfiguration of radial distribution systems," Report, São Paulo Research Foundation (FAPESP), no. 16/12190-7, Brazil, 2019.

[52] A. Lotfipour and H. Afrakhte, "A discrete Teaching–Learning-Based Optimization algorithm to solve distribution system reconfiguration in presence of distributed generation," International Journal of Electrical Power and Energy Systems, vol. 82, pp. 264–273, November 2016.

[53] L. H. Macedo, J. F. Franco, M. Mahdavi, and R. Romero "A contribution to the optimization of the reconfiguration problem in radial distribution systems," Journal of Control, Automation and Electrical Systems, vol. 29, no. 6, pp. 756–768, 2018.

[54] M. Mahdavi and A. Bagheri, "BPSO applied to TNEP considering adequacy criterion," American Journal of Neural Networks and Applications, vol. 4, no. 1, pp. 1–7, 2018.

[55] M. Mahdavi, A. Bagheri, and E. Mahdavi, "Comparing efficiency of PSO with GA in transmission expansion planning considering network adequacy," WSEAS Transactions on Power Systems, vol. 7, no. 1, pp. 34–43, 2012.

[56] M. Mahdavi and E. Mahdavi, "Transmission expansion planning considering network adequacy and investment cost limitation using genetic algorithm," International Journal of Electrical and Computer Engineering, vol. 5, no. 8, pp. 1–5, 2011.

[57] M. Mahdavi, H. Monsef, and A. Bagheri, "Transmission lines loading enhancement using ADPSO approach," International Journal of Electrical and Computer Engineering, vol. 4, no. 3, pp. 556–561, 2010.

[58] H. Shayeghi, M. Mahdavi, and A. Kazemi, "Discrete particle swarm optimization algorithm used for TNEP considering network adequacy restriction," International Journal of Electrical and Computer Engineering, vol. 3, no. 3, pp. 521–528, 2009.

[59] H. Shayeghi, M. Mahdavi, and H. Haddadian, "DCGA based-transmission network expansion planning considering network adequacy," International Journal of Electrical and Computer Engineering, vol. 2, no. 12, pp. 2875–2880, 2008.

[60] A. Kazemi, S. Jalilzadeh, H. Haddadian, and M. Mahdavi, "Transmission expansion planning considering network adequacy and losses using DCGA," IEEE International Conference on Sustainable Energy Technologies, Singapore, pp. 145–150, 2008.

[61] A. Kazemi, H. Haddadian, and M. Mahdavi, "Transmission network adequacy optimization using genetic algorithm," 5th International Conference on Electrical and Electronics Engineering, Turkey, pp. 1–5, 2007.

[62] H. Hosseini, S. Jalilzadeh, V. Nabaei, G. R. Z. Govar, and M. Mahdavi, "Enhancing deregulated distribution network reliability for minimizing penalty cost based on reconfiguration using BPSO," 2nd IEEE International Conference on Power and Energy, Malaysia, pp. 983–987, 2008.

[63] A. Kimiyaghalam, M. Mahdavi, A. Ashouri, and M. Bagherivand, "Optimal place-ment of PMUs for reliable observability of network under probabilistic events using BABC algorithm," 22nd International Conference on Electricity Distribution, Sweden, pp. 1–4, 2013.

[64] H. Shayeghi, M. Mahdavi, and A. Bagheri, "An improved DPSO with mutation based on similarity algorithm for optimization of transmission lines loading," Energy Conversion and Management, vol. 51, no. 12, pp. 2715–2723, 2010.

[65] M. Mahdavi, H. Monsef, and A. Bagheri "Lines loading optimization in transmis-sion expansion planning based on binary PSO algorithm," i-manager's Journal on Information Technology, vol. 1, no. 1, pp. 24–32, 2012.

[66] H. Shayeghi and M. Mahdavi, "Optimization of transmission lines loading in TNEP using decimal codification based GA," International Journal of Electrical and Computer Engineering, vol. 2, no. 5, pp. 942–947, 2008.

[67] M. Mahdavi, A. Nazari, V. Hosseinnezhad, and A. Safari "A PSO-based static synchronous compensator controller for power system stability enhancement," Journal of Artificial Intelligence in Electrical Engineering, vol. 1, no. 1, pp. 18–25, 2012.

[68] M. Mahdavi, A. R. Kheirkhah, L. H. Macedo, and R. Romero, "A genetic algo-rithm for transmission network expansion planning considering line maintenance," IEEE Congress on Evolutionary Computation, Glasgow, United Kingdom, pp. 1–6, 2020.

[69] M. Mahdavi, A. R. Kheirkhah, and R. Romero, "Transmission expansion planning con-sidering line failures and maintenance," 33rd International Power System Conference, Tehran, Iran, pp. 1–7, 2018.

[70] A. Asrari, S. Lotfifard, and M. Ansari, "Reconfiguration of smart distribution systems with time varying loads using parallel computing," IEEE Transactions on Smart Grid, vol. 7, no. 6, pp. 2713–2723, November 2016.

[71] Z. Li, S. Jazebi, and F. de León, "Determination of the optimal switching frequency for distribution system reconfiguration," IEEE Transactions on Power Delivery, vol. 32, no. 4, pp. 2060–2069, August 2017.

[72] S. S. F. Souza, R. Romero, J. Pereira, and J. T. Saraiva, "Artificial immune algo-rithm applied to distribution system reconfiguration with variable demand," International Journal of Electrical Power and Energy Systems, vol. 82, pp. 561–568, November 2016.

[73] M. A. Muhammad, H. Mokhlis, K. Naidu, J. F. Franco, H. A. Illias, and L. Wang, "Integrated database approach in multiobjective network reconfiguration for distribu-tion system using discrete optimisation techniques," IET Generation, Transmission & Distribution, vol. 12, no. 4, pp. 976–986, 2018.

[74] F. Ding and K. A. Loparo, "Feeder reconfiguration for unbalanced distribution systems with distributed generation: A hierarchical decentralized approach," IEEE Transactions on Power Systems, vol. 31, no. 2, pp. 1633–1642, March 2016.

[75] B. Sultana, M. W. Mustafa, U. Sultana, and A. R. Bhatti, "Review on reliability improvement and power loss reduction in distribution system via network reconfigura-tion," Renewable and Sustainable Energy Reviews, vol. 66, pp. 297–310, 2016.

[76] P. Hines, K. Balasubramaniam, and E. C. Sasnchez, "Cascading failures in power grids," IEEE Potentials, pp. 24–30, September/October 2009.

[77] S. Jalilzadeh, A. Kimiyaghalam, M. Mahdavi, and A. Ashouri, "STNEP consider-ing voltage level and uncertainty in demand using IABPSO," International Review of Electrical Engineering, vol. 7, no. 4, pp. 5186–5195, 2012.

[78] M. Mahdavi and E. Mahdavi, "Evaluating the effect of load growth on annual net-work losses in TNEP considering bundle lines using DCGA," International Journal on Technical and Physical Problems of Engineering, vol. 3, no. 4, pp. 1–9, 2011.

[79] A. Kimiyaghalam, M. Mahdavi, S. Jalilzadeh, and A. Ashouri, "Improved binary particle swarm optimization based TNEP considering network losses, voltage level, and uncertainty in demand," Journal of Artificial Intelligence in Electrical Engineering, vol. 1, no. 2, pp. 29–42, 2012.

[80] A. Kazemi, S. Jalilzadeh, and M. Mahdavi, H. Haddadian, "Genetic algorithm-based investigation of load growth factor effect on the network loss in TNEP," 3rd IEEE Conference on Industrial Electronics and Applications, Singapore, vol. 1, pp. 764–769, 2008.

[81] H. Shayeghi and M. Mahdavi, "Application of PSO and GA for transmission network expansion plan-ning," in Analysis, Control and Optimal Operations in Hybrid Power Systems, Book chapter, Springer, London, pp. 187–226, 2013.

[82] M. Mahdavi and R. Romero, "Reconfiguration of radial distribution systems: An efficient mathematical model," IEEE Latin America Transactions, Early Access, vol. 100, no. 1e, 2021.

[83] M. Mahdavi, L. H. Macedo, and R. Romero, "A mathematical formulation for distribution network reconfiguration," 7th Regional Conference on Electricity Distribution, Iran, pp. 1–5, 2019.

[84] B. Venkatesh, and R. Ranjan, "Optimal radial distribution system reconfiguration using fuzzy adaptation of evolutionary programming," International Journal of Electrical Power and Energy Systems, vol. 25, no. 10, pp. 775–780, December 2003.

[85] Y. T. Hsiao, "Multiobjective evolution programming method for feeder reconfiguration," IEEE Transactions on Power Systems, vol. 19, no. 1, pp. 594–599, February 2004.

[86] B. Venkatesh, Rakesh Ranjan, and H. B. Gooi "Optimal reconfiguration of radial distribution systems to maximize loadability," IEEE Transactions on Power Systems, vol. 19, no. 1, pp. 260–266, February 2004.

[87] A. C. B. Delbem, A. C. P. D. L. F. de Carvalho, and N. G. Bretas, "Main chain representation for evolutionary algorithms applied to distribution system reconfiguration," IEEE Transactions on Power Systems, vol. 20, no. 1, pp. 425–436, February 2005.

[88] J. C. López, M. Lavorato b, and M. J. Rider, "Optimal reconfiguration of electrical distribution systems considering reliability indices improvement," International Journal of Electrical Power and Energy Systems, vol. 78, pp. 837–845, June 2016.

[89] J. C. López, M. Lavorato, J. F. Franco, and M. J. Rider, "Robust optimisation applied to the reconfiguration of distribution systems with reliability constraints," IET Generation, Transmission & Distribution, vol. 10, no. 4, pp. 917–927, March 2016.

[90] K. Masteri and B. Venkatesh, "A new optimization method for distribution system reconfiguration," IEEE 27th Canadian Conference on Electrical and Computer Engineering, Canada, May 2014, pp. 1–4.

[91] J. S. R. K. AL-Sujada and O. Y. M. Al-Rawi, "Novel load flow algorithm for multiphase balanced/unbalanced radial distribution systems," International Journal of Advanced Research in Electrical, Electronics and Instrumentation Engineering, vol. 3, no. 7, pp. 10928–10942, July 2014.

[92] J. P. Chiou, C. F. Chang, and C. T. Su, "Variable scaling hybrid differential evolution for solving network reconfiguration of distribution systems," IEEE Transactions on Power Systems, vol. 20, no. 2, pp. 668–674, May 2005.

[93] A. M. Eldurssi and R. M. O'Connell, "A fast nondominated sorting guided genetic algorithm for multi-objective power distribution system reconfiguration problem," IEEE Transactions on Power Systems, vol. 30, no. 2, pp. 593–601, March 2015.

[94] S. Civanlar, J. J. Grainger, H. Yin., and S. S. H. Lee, "Distribution feeder reconfiguration for loss reduction," IEEE Transactions on Power Delivery, vol. 4, no. 3, pp. 1217–1223, July 1988.

[95] C. A. Castro, J. R. Watanabe, and A. A. Watanabe, "An efficient reconfiguration algorithm for loss reduction of distribution systems," Electric Power Systems Research, vol. 19, no. 2, pp. 137–144, August 1990.

[96] H. C. Chang and C. C. Kuo, "Network reconfiguration in distribution systems using simulated annealing," Electric Power Systems Research, vol. 29, no. 3, pp. 227–238, May 1994.

[97] R. J. Sárf, M. M. A. Salama, and A. Y. Chikhan, "Distribution system reconfiguration for loss reduction: An algorithm based on network partitioning theory," IEEE Transactions on Power Systems, vol. 11, no. 1, pp. 504–510, February 1996.

[98] C. T. Su and C. S. Lee, "Feeder reconfiguration and capacitor setting for loss reduction of distribution systems," Electric Power Systems Research, vol. 58, no. 2, pp. 97–102, June 2001.

[99] J. Z. Zhu, "Optimal reconfiguration of electrical distribution network using the refined genetic algorithm," Electric Power Systems Research, vol. 62, no. 1, pp. 37–42, May 2002.

[100] C. T. Su and C. S. Lee, "Network reconfiguration of distribution systems using improved mixed-integer hybrid differential evolution," IEEE Transactions on Power Delivery, vol. 18, no. 3, pp. 1022–1027, July 2003.

[101] Y. J. Jeon and J. C. Kim, "Application of simulated annealing and tabu search for loss minimization in distribution systems," International Journal of Electrical Power and Energy Systems, vol. 26, no. 1, pp. 9–18, January 2004.

[102] Y. Mishima, K. Nara, T. Satoh, T. Ito, and H. Kaneda, "Method for minimum-loss reconfiguration of distribution system by tabu search," Electrical Engineering in Japan, vol. 152, no. 2, pp. 18–25, May 2005.

[103] F. V. Gomes, S. Carneiro Jr., J. L. R. Pereira, M. P. Vinagre, P. A. N. Garcia, and L. R. Araujo, "A new heuristic reconfiguration algorithm for large distribution systems," IEEE Transactions on Power Systems, vol. 20, no. 3, pp. 1373–1378, August 2005.

[104] C. T. Su, C. F. Chang, and J. P. Chiou, "Distribution network reconfiguration for loss reduction by ant colony search algorithm," Electric Power Systems Research, vol. 75, no. 2–3, pp. 190–199, August 2005.

[105] C. F. Chang, "Reconfiguration and capacitor placement for loss reduction of distribution systems by ant colony search algorithm," IEEE Transactions on Power Systems, vol. 23, no. 4, pp. 1747–1755, November 2008.

[106] S. Sivanagaraju, Y. Sreekanth, and E. J. Babu, "An efficient genetic algorithm for loss minimum distribution system reconfiguration," Electric Power Components and Systems, vol. 34, no. 3, pp. 249–258, March 2006.

[107] F. V. Gomes, S. Carneiro Jr., J. L. R. Pereira, M. P. Vinagre, P. A. N. Garcia, and L. R. Araujo, "A new distribution system reconfiguration approach using optimum power flow and sensitivity analysis for loss reduction," IEEE Transactions on Power Systems, vol. 21, no. 4, pp. 1616–1623, November 2006.

[108] G. K. V. Raju and P. R. Bijwe, "Efficient reconfiguration of balanced and unbalanced distribution systems for loss minimisation," IET Generation, Transmission and Distribution, vol. 2, no. 1, pp. 7–12, January 2008.

[109] B. Enacheanu, B. Raison, R. Caire, O. Devaux, W. Bienia, and N. HadjSaid, "Radial network reconfiguration using genetic algorithm based on the Matroid theory," IEEE Transactions on Power Systems, vol. 23, no. 1, pp. 186–195, August 2008.

[110] S. Sivanagaraju, J. V. Rao, and P. S. Raju, "Discrete particle swarm optimization to network reconfiguration for loss reduction and load balancing," Electric Power Components and Systems, vol. 36, no. 5, pp. 513–524, May 2008.

[111] N. T. Linh and N. Q. Anh, "Application of artificial bee colony algorithm (ABC) for reconfiguring distribution network," Second International Conference on Computer Modeling and Simulation, China, January 2010, pp. 102–106.

[112] A.Y. Abdelaziz, F.M. Mohammed, S. F. Mekhamer, M. A. L. Badr, "Distribution system reconfiguration using a modified Tabu Search algorithm," Electric Power Systems Research, vol. 80, no. 8, pp. 943–953, August 2010.

[113] B. Tomoiagă, M. Chindris, A. Sumper, R. Villafafila-Robles, and A. Sudria-Andreu,"Distribution system reconfiguration using genetic algorithm based on connected graphs," Electric Power Systems Research, vol. 104, pp. 216–225, November 2013.

[114] C. Wang and Y. Gao, "Determination of power distribution network configuration using non-revisiting genetic algorithm," IEEE Transactions on Power Systems, vol. 28, no. 4, pp. 3638–3648, November 2013.

[115] A. K. Ferdavani, A. A. M. Zin, A. Khairuddin, and M. M. Naeini, "Reconfiguration of distribution system through two minimum-current neighbour-chain updating methods," IET Generation, Transmission and Distribution, vol. 7, no 12, pp. 1492–1497, December 2013.

[116] H. Ahmadi and J. R. Martí, "Mathematical representation of radiality constraint in distribution system reconfiguration problem," International Journal of Electrical Power and Energy Systems, vol. 64, pp. 293–299, January 2015.

[117] H. Ahmadi and J. R. Martí, "Linear current flow equations with application to distribution systems reconfiguration," IEEE Transactions on Power Systems, vol. 30, no. 4, pp. 2073–2080, March 2015.

[118] B. Novoselnik, M. Bolfek, M. Bošković, and M. Baotić, "Electrical power distribution system reconfiguration: Case Study of a Real-life grid in Croatia," IFAC PapersOnLine, vol. 50, no. 1, pp. 61–66, 2017.

[119] S. Jalilzadeh, M. Jabbari, A. Kimiyaghalam, and M. Mahdavi, "Inrush current identification in power transformers using weight functions," The 2009 IEEE International Conference on Intelli-gent Computing and Intelligent Systems, Shanghai, China, pp. 795–797, 2009.

[120] Nelson Kagan, "Configuração de redes de distribuição através de algoritmos genéticos e tomada de decisão Fuzzy", PhD Thesis, University of São Paulo, São Paulo, Brazil, pp. 1–187, 1999.

[121] M. E. Baran and F. F. Wu, "Network reconfiguration in distribution systems for loss reduction and load balancing," IEEE Transactions on Power Delivery, vol. 4, no. 2, pp. 1401–1407, April 1989.

[122] S. Goswami and S. Basu, "A new algorithm for the reconfiguration of distribution feeders for loss minimization," IEEE Transactions on Power Delivery, vol. 7, no. 3, pp. 1484–1491, July 1992.

[123] R. Rajaram, K. Sathish Kumar, and N. Rajasekar, "Power system reconfiguration in a radial distribution network for reducing losses and to improve voltage profile using modified plant growth simulation algorithm with Distributed Generation (DG)," Energy Reports, vol. 1, pp. 116–122, November 2015.

[124] D. Shirmohammadi and W. H. Hong, "Reconfiguration of electric distribution networks for resistive line loss reduction," IEEE Transactions on Power Delivery, vol. 4, no. 2, pp. 1492–1498, April 1989.

[125] K. Nara, A. Shiose, M. Kitagawa, and T. Ishihara, "Implementation of genetic algorithm for distribution system loss minimum reconfiguration," IEEE Transactions on Power Systems, vol. 7, no. 3, pp. 1044–1051, August 1992.

[126] R. S. Rao, S. V. L. Narasimham, M. R. Raju, and A. S. Rao, "Optimal network reconfiguration of large-scale distribution system using harmony search algorithm," IEEE Transactions on Power Systems, vol. 26, no. 3, pp. 1080–1088, August 2011.

[127] T. E. McDermott, I. Drezga, and R. P. Broadwater, "A heuristic nonlinear constructive method for distribution system reconfiguration," IEEE Transactions on Power Systems, vol. 14, no. 2, pp. 478–483, May 1999.

[128] M. A. Kashem, G. B. Jasmon, and V. Ganapathy, "A new approach of distribution system reconfiguration for loss minimization," International Journal of Electrical Power and Energy Systems, vol. 22, no. 4, pp. 269–276, May 2000.

[129] A. B. Morton and I. M. Y. Mareels, "An efficient brute-force solution to the network reconfiguration problem," IEEE Transactions on Power Delivery, vol. 15, no. 3, pp. 996–1000, July 2000.

[130] A. Swarnkar, N. Gupta, and K. R. Niazi, "Adapted ant colony optimization for efficient reconfiguration of balanced and unbalanced distribution systems for loss minimization," Swarm and Evolutionary Computation, vol. 1, no. 3, pp. 129–137, September 2011.

[131] Y. Y. Hong and S. Y. Ho, "Determination of network configuration considering multiobjective in distribution systems using genetic algorithms," IEEE Transactions on Power Systems, vol. 20, no. 2, pp. 1062–1069, May 2005.

[132] J. Mendoza, R. Lopez, D. Morales, E. Lopez, P. Dessante, and R. Moaaga, "Minimal loss reconfiguration using genetic algorithms with restricted population and addressed operators: real application," IEEE Transaction on Power Systems, vol. 21, no. 2, pp. 948–954, May 2006.

[133] C. Wang and H. Z. Cheng, "Optimization of network configuration in large distribution systems using plant growth simulation algorithm," IEEE Transaction on Power Systems, vol. 23, no. 1, pp. 119–126, February 2008.

[134] G. K. V. Raju and P. R. Bijwe, "An efficient algorithm for minimum loss reconfiguration of distribution system based on sensitivity and heuristics," IEEE Transactions on Power Systems, vol. 23, no. 3, pp. 1280–1287, August 2008.

[135] J. A. Martín and A. J. Gil, "A new heuristic approach for distribution systems loss reduction," Electric Power Systems Research, vol. 78, no. 11, pp. 1953–1958, November 2008.

[136] H. R. Esmaeilian and R. Fadaeinedjad, "Energy loss minimization in distribution systems utilizing an enhanced reconfiguration method integrating distributed generation," IEEE Systems Journal, vol. 9, no. 4, December 2015.

[137] T. Niknam, "An efficient multi-objective HBMO algorithm for distribution feeder reconfiguration," Expert Systems with Applications, vol. 38, no. 3, pp. 2878–2887, March 2011.

[138] T. Niknam, "Application of honey bee mating optimization on distribution state estimation including distributed generators," Journal of Zhejiang University Science A, vol. 9, no. 12, pp. 1753–1764, December 2008.

[139] W. C. Wu and M. S. Tsai, "Application of enhanced integer coded particle swarm optimization for distribution system feeder reconfiguration," IEEE Transactions on Power Systems, vol. 26, no. 3, pp. 1591–1599, August 2011.

[140] T. Niknam, "A new hybrid algorithm for multiobjective distribution feeder reconfiguration," Energy Conversion and Management, vol. 50, no. 8, pp. 2074–2082, August 2009.

[141] H. M. Khodr, J. M. Crespo, M. A. Matos, and J. Pereira, "Distribution systems reconfiguration based on OPF using Benders decomposition," IEEE Transactions on Power Systems, vol. 24, no. 4, pp. 2166–2176, October 2009.

[142] H. M. Khodr and J. J. M. Crespo, "Integral methodology for distribution systems reconfiguration based on optimal power flow using Benders decomposition technique," IET Generation, Transmission & Distribution, vol. 3, no. 1, pp. 521–534, June 2009.

[143] A.Y. Abdelaziz, F. M. Mohammed, S. F. Mekhamer, and M. A. L. Badr, "Distribution systems reconfiguration using a modified particle swarm optimization algorithm," Electric Power Systems Research, vol. 79, no. 11, pp. 1521–1530, November 2009.

[144] S. A. Yin and C. N. Lu, "Distribution feeder scheduling considering variable load profile and outage costs," IEEE Transactions on Power Systems, vol. 24, no. 2, pp. 652–660, May 2009.

[145] A. Y. Abdelaziz, F. M. Mohammed, S. F. Mekhamer, and M. A. L. Badr, "Distribution system reconfiguration using a modified Tabu Search algorithm," Electric Power Systems Research, vol. 80, no. 8, pp. 943–953, August 2010.

[146] A. Saffar, R. Hooshmand, and A. Khodabakhshian, "A new fuzzy optimal reconfiguration of distribution systems for loss reduction and load balancing using ant colony search-based algorithm," Applied Soft Computing, vol. 11, no. 5, pp. 4021–4028, July 2011.

[147] M. Lavorato, J. F. Franco, M. J. Rider, and R. Romero, "Imposing radiality constraints in distribution system optimization problems," IEEE Transactions on Power Systems, vol. 27, no. 1, pp. 172–179, February 2012.

[148] A. Y. Abdelaziz, R. A. Osama, and S. M. El-Khodary, "Reconfiguration of distribution systems for loss reduction using the hyper-cube ant colony optimisation algorithm," IET Generation, Transmission and Distribution, vol. 12, no. 4, pp. 976–986, February 2012.

[149] F. Llorens-Iborra, J. Riquelme-Santos, and E. Romero-Ramos, "Mixed-integer linear programming model for solving reconfiguration problems in large-scale distribution systems," Electric Power Systems Research, vol. 88, pp. 137–145, July 2012.

[150] J. A. Taylor and F. S. Hover, "Convex models of distribution system reconfiguration," IEEE Transactions on Power Systems, vol. 27, no. 3, pp. 1407–1413, August 2012.

[151] L. W. de Oliveira, E. J. de Oliveira, F. V. Gomes, I. C. Silva Jr., A. L.M. Marcato, and P. V.C. Resende, "Artificial Immune Systems applied to the reconfiguration of electrical power distribution networks for energy loss minimization," International Journal of Electrical Power and Energy Systems, vol. 56, pp. 64–74, March 2014.

[152] M. Sedighizadeh, M. Esmaili, and M. Esmaeili, "Application of the hybrid Big Bang-Big Crunch algorithm to optimal reconfiguration and distributed generation power allocation in distribution systems," Energy, vol. 76, pp. 920–930, November 2014.

[153] H. Ahmadi and J. R. Martí, "Distribution system optimization based on a linear power flow formulation," IEEE Transactions on Power Delivery, vol. 30, no. 1, pp. 25–33, February 2015.

[154] I. Ali, M. S. Thomas, and P. Kumar, "Energy efficient reconfiguration for practical load combinations in distribution systems," IET Generation, Transmission & Distribution, vol. 9, no. 1, pp. 1051–1060, August 2015.

[155] K. Jasthi and D. Das, "Simultaneous distribution system reconfiguration and DG sizing algorithm without load flow solution," IET Generation, Transmission and Distribution, vol. 12, no. 6, pp. 1303–1313, March 2018.

[156] Rana Ali Abttan, "Electric distribution system reconfiguration for loss reduction using genetic algorithm," Journal of Baghdad College of Economic Sciences University, vol. 0, no. 43, pp. 457–478, 2015.

[157] R. Parasher, "Load flow analysis of radial distribution network using linear data structure," MSc. Dissertation, Department of Computer Science & Engineering, Yagyavalkya Institute of Technology, Jaipur, pp. 1–38, October 2013.

[158] M.A. Kashem, V. Ganapathy, and G.B. Jasmon, "A geometrical approach for network reconfiguration based loss minimization in distribution systems," International Journal of Electrical Power and Energy Systems, vol. 21, no. 4, pp. 295–304, May 2001.

[159] Y. J. Jeon, J. C. Kim, J. O. Kim, J. R. Shin, and K. Y. Lee, "An efficient simulated annealing algorithm for network reconfiguration in large-scale distribution systems," IEEE Transactions on Power Delivery, vol. 17, no. 4, pp. 1070–1078, October 2002.

[160] N.C. Sahoo and K. Prasad, "A fuzzy genetic approach for network reconfiguration to enhance voltage stability in radial distribution systems," Energy Conversion and Management, vol. 47, no. 18-19, pp. 3288–3306, November 2006.

[161] J. S. Savier and Debapriya Das, "Impact of network reconfiguration on loss allocation of radial distribution systems," IEEE Transactions on Power Delivery, vol. 22, no. 4, pp. 2473–2480, October 2007.

[162] M. Arun and P. Aravindhababu, "A new reconfiguration scheme for voltage stability enhancement of radial distribution systems," Energy Conversion and Management, vol. 50, no. 9, pp. 2148–2151, September 2009.

[163] S. Chandramohan, N. Atturulu, R.P. K. Devi, and B. Venkatesh, "Operating cost minimization of a radial distribution system in a deregulated electricity market through reconfiguration using NSGA method," International Journal of Electrical Power and Energy Systems, vol. 32, no. 2, pp. 126–132, February 2010.

[164] A. A. M. Zin, A. K. Ferdavani, A. B. Khairuddin, M. M. Naeini, "Reconfiguration of radial electrical distribution network through minimum-current circular-updating-mechanism method," IEEE Transactions on Power Systems, vol. 27, no. 2, pp. 968–974, 2012.

[165] R. S. Rao, K. Ravindra, K. Satish, and S. V. L. Narasimham, "Power loss minimization in distribution system using network reconfiguration in the presence of distributed generation," IEEE Transactions on Power Systems, vol. 28, no. 1, pp. 317–325, February 2013.

[166] S. A. Taher and M. H. Karimi, "Optimal reconfiguration and DG allocation in balanced and unbalanced distribution systems," Ain Shams Engineering Journal, vol. 5, no. 3, pp. 735–749, September 2014.

[167] S. Naveen, K. S. Kumar, and K. Rajalakshmi, "Distribution system reconfiguration for loss minimization using modified bacterial foraging optimization algorithm," International Journal of Electrical Power and Energy Systems, vol. 69, pp. 90–97, July 2015.

[168] A. V. S. Reddy and M. D. Reddy, "Optimization of distribution network reconfiguration using dragonfly algorithm," Journal of Electrical Engineering, vol. 16, no. 4, pp. 1–10, 2016.

[169] B. Khorshid-Ghazani, H. Seyedi, B. Mohammadi-ivatloo, K. Zare, S. Shargh, "Reconfiguration of distribution networks considering coordination of the protective devices," IET Generation, Transmission & Distribution, vol. 11, no. 1, pp. 82–92, January 2017.

[170] D. Das, "A fuzzy multiobjective approach for network reconfiguration of distribution systems," IEEE Transactions on Power Delivery, vol. 21, no. 1, pp. 202–209, January 2006.

[171] T. Niknam and E.A. Farsani, "A hybrid self-adaptive particle swarm optimization and modified shuffled frog leaping algorithm for distribution feeder reconfiguration," Engineering Applications of Artificial Intelligence, vol. 23, no. 8, pp. 1340–1349, December 2010.

[172] T. Niknam, "An efficient hybrid evolutionary algorithm based on PSO and ACO for distribution feeder reconfiguration," European Transactions on Electrical Power, vol. 20, no. 5, pp. 575–590, July 2010.

[173] T. Niknam, R. Khorshidi, and B. Bahmani Firouzi, "A hybrid evolutionary algorithm for distribution feeder reconfiguration," Sadhana, vol. 35, no. 2, pp. 139–162, April 2010.

[174] T. Niknam, "An efficient hybrid evolutionary algorithm based on PSO and HBMO algorithms for multi-objective distribution feeder reconfiguration," Energy Conversion and Management, vol. 50, no. 8, pp. 2074–2082, August 2009.

[175] R. A. Jabr, R. Singh, and B. C. Pal, "Minimum loss network reconfiguration using mixed-integer convex programming," IEEE Transactions on Power Systems, vol. 27, no. 2, pp. 1106–1115, May 2012.

[176] A. Ahuja, S. Das, and A. Pahwa, "An AIS-ACO hybrid approach for multi-objective distribution system reconfiguration," IEEE Transactions on Power Systems, vol. 22, no. 3, pp. 1101–1111, August 2007.

[177] E. M. Carreno, R. Romero, and A. Padilha-Feltrin, "An efficient codification to solve distribution network reconfiguration for loss reduction problem," IEEE Transactions on Power Systems, vol. 23, no. 4, pp. 1542–1551, November 2008.

[178] Y. K. Wu, C. Y. Lee, L. C. Liu, and S. H. Tsai, "Study of reconfiguration for the distribution system with distributed generators," IEEE Transaction on Power Delivery, vol. 5, no. 3, pp. 1678–1685, July 2010.

[179] Pawan Kumar, I. Ali, M. S. Thomas, and S. Singh, "Imposing voltage security and network radiality for reconfiguration of distribution systems using efficient heuristic and metaheuristic approach," IET Generation, Transmission & Distribution, vol. 11, no. 10, pp. 2457–2467, July 2017.

[180] D. Zhang, Z. Fu and L. Zhang, "An improved TS algorithm for loss-minimum reconfiguration in large-scale distribution systems," Electric Power Systems Research, vol. 77, no. 5-6, pp. 685–694, April 2007.

[181] R. Tapia-Juárez, E. Espinosa-Juárez, "Reconfiguration of radial distribution networks by applying a multi-objective technique," Int'l Conf. Artificial Intelligence, pp. 131–137, 2015.

[182] S. Ghasemi, "Balanced and unbalanced distribution networks reconfiguration considering reliability indices," Ain Shams Engineering Journal, vol. 9, no. 4, pp. 1567–1579, December 2018.

[183] J. R. S. Mantovani, F. Casari, and R. A. Romero, "Reconfiguração de sistemas de distribuição radiais utilizando o critério de queda de tensão," SBA Controle & Automação, vol. 11, no. 3, pp. 150–159, 2000.

[184] M. A. N. Guimarães, C. A. Castro, and R. Romero, "Distribution systems operation optimisation through reconfiguration and capacitor allocation by a dedicated genetic algorithm," IET Generation, Transmission, and Distribution, vol. 4, no. 11, pp. 1213–1222, April 2010.

[185] N. Gupta, A. Swarnkar, K. R. Niazi, R. C. Bansal, "Multi-objective reconfiguration of distribution systems using adaptive genetic algorithm in fuzzy framework," IET Generation, Transmission and Distribution, vol. 4, no. 12, pp. 1288–1298, December 2010.

[186] A. Ahuja, A. Pahwa, B. K. Panigrahi, and S. Das, "Pheromone-based crossover operator applied to distribution system reconfiguration," IEEE Transactions on Power Systems, vol. 28, no. 4, pp. 4144–4151, November 2013.

[187] J. F. Franco, M. J. Rider, M. Lavorato, and R. Romero, "A mixed-integer LP model for the reconfiguration of radial electric distribution systems considering distributed generation," Electric Power Systems Research, vol. 97, pp. 51–60, April 2013.

[188] L. H. Tsai, "Network reconfiguration to enhance reliability of electric distribution systems," Electric Power Systems Research, vol. 27, no. 2, pp. 135–140, July 1993.

[189] M. N. A. Guimarães and C. A. Castro, "Reconfiguration of distribution systems for loss reduction using Tabu Search," The 15th Power Systems Computation Conference, Belgium, pp. 1–6, August 2005.

[190] I. J. Ramírez-Rosado and J. L. Bernal-Augustín, "Genetic algorithms applied to the design of large power distribution systems," IEEE Transactions on Power Systems, vol. 13, no. 2, pp. 696–703, May 1998.

# Index